D1210176

Pheromones and Animal Behaviour
Communication by Smell and Taste

We are entering one of the most exciting periods in the study of chemical communication since the first pheromones were identified some 40 years ago. The rapid progress that has been made is reflected in this book for advanced undergraduates and researchers, which is the first to cover the whole animal kingdom at this level for 25 years. The importance of chemical communication is illustrated with examples from a diverse range of animals including humans, marine copepods, *Drosophila*, *Caenorhabditis elegans*, moths, snakes, goldfish, elephants and mice. The book is designed to be advanced and up to date, but at the same time accessible to readers whatever their scientific background. For students of ecology, evolution and behaviour, it gives an introduction to the rapid progress in our understanding of olfaction at the molecular and neurological level. In addition, it offers chemists, molecular biologists and neurobiologists an insight into the ecological, evolutionary and behavioural context of olfactory communication.

TRISTRAM WYATT is a senior researcher in the Department of Zoology at the University of Oxford, where he works on pheromones and behaviour. He is also Oxford University's Director of Online and Distance Learning and a Fellow of Kellogg College, Oxford.

Pheromones and Animal Behaviour

Communication by Smell and Taste

Tristram D. Wyatt

University of Oxford

CAMBRIDGE
UNIVERSITY PRESS

PUBLISHED BY THE PRESS SYNDICATE OF THE UNIVERSITY OF CAMBRIDGE
The Pitt Building, Trumpington Street, Cambridge, United Kingdom

CAMBRIDGE UNIVERSITY PRESS
The Edinburgh Building, Cambridge CB2 2RU, UK
40 West 20th Street, New York, NY 10011-4211, USA
477 Williamstown Road, Port Melbourne, VIC 3207, Australia
Ruiz de Alarcón 13, 28014 Madrid, Spain
Dock House, The Waterfront, Cape Town 8001, South Africa

http://www.cambridge.org

© Cambridge University Press 2003

This book is in copyright. Subject to statutory exception
and to the provisions of relevant collective licensing agreements,
no reproduction of any part may take place without
the written permission of Cambridge University Press.

First published 2003

Printed in the United Kingdom at the University Press, Cambridge

Typeface Swift 9/13pt   *System* QuarkXPress® [TB]

A catalogue record for this book is available from the British Library

Library of Congress Cataloguing in Publication data

Wyatt, Tristram D., 1956–
Pheromones and animal behaviour: communication by smell and
taste / Tristram D. Wyatt.
     p.   cm.
Includes bibliographical references (p.).
ISBN 0 521 48068 X – ISBN 0 521 48526 6 (pb.)
1. Animal communication.   2. Pheromones.   3. Chemical senses.   I. Title.
QL776 .W93   2002
591.59 – dc21   2002024628

ISBN 0 521 48068 X hardback
ISBN 0 521 48526 6 paperback

AUSTIN COMMUNITY COLLEGE
LIBRARY SERVICES

The publisher has used its best endeavours to ensure that the URLs for external web sites referred to in this book are correct and active at time of going to press. However, the publisher has no responsibility for the web sites and can make no guarantee that a site will remain live or that the content is or will remain appropriate.

Dedicated to Martin C. Birch, Joan and Vivian, and Robert

I should think that we might fairly gauge the future of biological science, centuries ahead, by estimating the time it will take to reach a complete, comprehensive understanding of odor. It may not seem to be a profound enough problem to dominate all the life sciences, but it contains, piece by piece, all the mysteries.

Thomas L. (1983). *Late night thoughts on Mahler's ninth symphony*. New York: Viking Press.

Until recently, the chemical senses have engendered, relative to the other major senses, comparatively little attention on the part of the scientific and medical communities. This is due to a number of factors, including (1) the lack of a simple physical dimension analogous to wave length that correlates with olfactory or taste quality, (2) the fact that chemosensory dysfunction rarely produces obvious influences on such everyday activities as locomotion and social interaction, and (3) the widespread belief that chemical senses are of little importance to humans.

Doty, R. L. (1995) Introduction and historical perspective. In *Handbook of olfaction and gustation*, ed. R. L. Doty, pp. 1–32. New York: Marcel Dekker.

Contents

Preface   *page* xiii

Acknowledgements   xiv

I Animals in a chemical world   1

  1.1 Introduction   1

  1.2 What are pheromones?   1

  1.3 Evolution of chemical cues into signals   6

  1.4 Secretory organs for pheromones   9

  1.5 Functional signal design: contrasting different signal modalities   12

  1.6 Specificity   16

  1.7 Composite signals: pheromones working in concert with other modalities   17

  1.8 Primer and releaser pheromones   18

  1.9 Cost of signalling   20

  1.10 Pheromones in humans?   22

  1.11 Conclusion   22

  1.12 Further reading   22

2 Discovering pheromones   23

  2.1 Introduction   23

  2.2 How are pheromones identified?   25

  2.3 Bioassays   25

  2.4 Collection of chemical signals   29

  2.5 Separating the chemicals, finding the active components, and identifying them   31

  2.6 Multi-component pheromones and synergy   34

  2.7 New tools in pheromone research   35

  2.8 Conclusion   36

  2.9 Further reading   36

3 Sex pheromones: finding and choosing mates   37

3.1 Introduction   37

3.2 Mate choice and sexual selection   38

3.3 Which sex should call?   40

3.4 External fertilisation and chemical duets   41

3.5 Sexual selection, scramble and contest   43

3.6 Sexual selection, mate quality and courtship   46

3.7 Leks   57

3.8 Conflict between the sexes revealed in signalling   59

3.9 Alternative mating strategies   60

3.10 Sperm competition and mate guarding   61

3.11 Sex pheromones and speciation   64

3.12 Conclusion   72

3.13 Further reading   73

4 Coming together and keeping apart: aggregation and host-marking pheromones   74

4.1 Introduction   74

4.2 Aggregation pheromones and Allee effects, the advantages of group living   74

4.3 Host-marking pheromones   83

4.4 Conclusion   85

4.5 Further reading   86

5 Scent marking and territorial behaviour   87

5.1 Introduction   87

5.2 Why scent mark?   90

5.3 Scent fence hypothesis   91

5.4 Scent matching hypothesis   91

5.5 Border maintenance hypothesis   96

5.6 Economics of scent marking patterns in territories   97

5.7 Dear enemies   99

5.8 Over-marking   99

5.9 Scent marking in non-territorial mammals   100

5.10 Conclusion   101

5.11 Further reading   101

6 Pheromones and social organisation   102

6.1 Introduction   102

6.2 Colony, kin, family and individual recognition   103

6.3 Pheromones and reproduction in social groups: control or signalling?   113

6.4	Conclusion	128
6.5	Further reading	128

7 Pheromones and recruitment communication — 129

7.1	Introduction	129
7.2	Foraging ecology and evolution of recruitment communication	133
7.3	Social insects as self-organising systems	141
7.4	Conclusion	144
7.5	Further reading	145

8 Fight or flight: alarm pheromones — 146

8.1	Introduction	146
8.2	Evolution of alarm signals between related individuals	147
8.3	Evolution of alarm signals in unrelated individuals	157
8.4	Conclusion	162
8.5	Further reading	163

9 Perception and action of pheromones: from receptor molecules to brains and behaviour — 164

9.1	Introduction	164
9.2	Chemical cues: perception and interpretation by the brain	166
9.3	Vertebrate dual olfactory system	178
9.4	Moths and sex pheromones	186
9.5	Factors affecting behavioural and physiological responses to pheromones	188
9.6	Primer pheromones and reproduction	192
9.7	Olfactory cues and recognition learning	198
9.8	Developmental paths or metamorphosis prompted by pheromones	202
9.9	Conclusion	205
9.10	Further reading	205

10 Finding the source: pheromones and orientation behaviour — 206

10.1	Introduction	206
10.2	Investigating orientation behaviour mechanisms	207
10.3	Ranging behaviour: search strategies for finding odour plumes, trails or gradients	209
10.4	Finding the source: orientation to pheromones	210
10.5	Conclusion	227
10.6	Further reading	228

11 Breaking the code: illicit signallers and receivers of semiochemical signals 229

11.1 Introduction 229

11.2 Eavesdropping 230

11.3 Chemical communication in mutualisms 237

11.4 Deception by aggressive mimicry of sex pheromones 240

11.5 Propaganda 241

11.6 Specialist relationships of predators, guests and parasites of social insects 244

11.7 Conclusion 249

11.8 Further reading 249

12 Using pheromones: applications 251

12.1 Introduction 251

12.2 Pheromones used with beneficial and domestic animals 251

12.3 Pheromones in pest management 255

12.4 Pest resistance to pheromones? 267

12.5 Commercialisation – problems and benefits of pheromones 267

12.6 Conclusion 269

12.7 Further reading 269

13 On the scent of human attraction: human pheromones? 270

13.1 Introduction 270

13.2 Cultural and social aspects of odours and humans 273

13.3 Evidence that olfaction is important in human behaviour and biology 274

13.4 Candidate compounds for human pheromone odours 285

13.5 Perception of odours 291

13.6 Putting human odours to use: applications 295

13.7 Conclusion 299

13.8 Further reading 300

Appendix A1   An introduction to pheromones for non-chemists 302

Appendix A2   Isomers and pheromones 304

Appendix A3   Further reading on pheromone chemical structure 309

References 310

List of credits 359

Index 371

Preface

Pheromones offer exceptional opportunities to study fundamental biological problems. Recent progress in the field is rapid. The excitement comes from the convergence of powerful techniques from different areas of science including chemistry and animal behaviour, combined with new techniques in genomics and molecular biology. For perhaps the first time, we can now investigate questions at every level: molecular, neurobiological, hormonal, behavioural, ecological, and evolutionary.

The discoveries from molecular biologists are likely to greatly expand our knowledge of the evolutionary biology of olfactory communication. Equally, molecular biology only makes sense in the context of evolution. Pheromone research almost always brings together biologists of many kinds and a rich diversity of chemists – each is approaching the other parts of the study as a non-specialist. This book is designed to bridge those gaps and to bring together people already working on pheromones and to encourage others to take up the challenge.

Different parts of the book emphasise examples from different taxa. For example, mammals feature more strongly than invertebrates in the sections on individual variation and hormonal effects of pheromones, but invertebrates dominate the chapter on searching behaviour. Because of pressure of space, the literature citations in the text are more to offer a way into the current literature than to give full credit for discoveries.

Acknowledgements

I warmly thank Martin C. Birch for his generous assistance and encouragement at all stages of this project. Tracey Sanderson, Commissioning Editor at CUP, has skilfully guided the writing of the book from its conception. I am very grateful to Everild Haynes, the copy-editor, for her perceptive, helpful and thorough contributions. The librarians at the Zoology Department and the Radcliffe Science Libraries in Oxford helped to obtain a diverse set of literature for me.

I thank the University of Oxford, the British Council and the Royal Society (London) for grants that enabled me to visit pheromone laboratories in the USA, Canada, Australia, New Zealand, and continental Europe while researching the book. Many of the illustrations were done by Carolyn Sharp. I am grateful to Oxford University Department for Continuing Education for a small grant towards these costs.

Numerous friends and colleagues have read part of the text in draft, including Richard Adeney, Manuel Berdoy, Christina Büesching, Martin Birch, John Clarke, Kathy Dorries, Heather Eisthen, Jean-Claude Grégoire, Jane Haigh, Katherine Houpt, David W. Kelly, Heine Kiesby, Albert Minks, Allen Moore, Elsie Owusu, Christian Peeters, Jane Parker, George Preti, Julio Rojas, Ruth Russell, Judy Stamps, Robert Taylor, and Stephen Jones and my other colleagues in Zoology Group 8. I would particularly like to thank Joan and Vivian Wyatt, and Wladimir Alonso for reading the whole manuscript. The remaining errors are mine of course, and I would welcome comments and suggestions (tristram.wyatt@zoo.ox.ac.uk).

I am enormously grateful to all the scientists who advised me and kindly sent reprints and pre-prints of their work. The book would not have been possible without their help and generosity. In keeping the range of animal groups represented as wide as possible, I have had to be selective. Inevitably I have not been able to include many examples that I would have liked to. I apologise to authors whose research I was not able to describe here despite its high quality.

I should like to thank the publishers and societies listed at the end of the book for permission to reproduce figures, tables, photographs and data, either in their original form or as redrawn and modified versions. The list does not include non-copyright material, or individual acknowledgements (which are made in captions where appropriate).

Chapter 1

Animals in a chemical world

1.1 Introduction

Elephants and moths are unlikely mates, so scientists and the general public were surprised when it was discovered recently that one of the world's largest living land animals, the Asian elephant (*Elephas maximus*), shares its female sex pheromone with some 140 species of moth (Rasmussen *et al.* 1996). The compound is a small, volatile molecule (Z)-7-dodecen-1-yl acetate (Fig. 1.1). But before explaining why elephants and moths are not likely to be confused, I should introduce pheromones in general.

1.2 What are pheromones?

Pheromones are the molecules used for communication between animals. A broader term for chemicals involved in animal communication is **semiochemical** (from the Greek *semeion* sign) (Law & Regnier 1971). Strictly speaking, **pheromones** are a subclass of semiochemicals, used for communication *within* the species (intraspecific chemical signals). Pheromones were originally defined as 'substances secreted to the outside by an individual and received by a second individual of the same species in which they release a specific reaction, for instance a definite behaviour [releaser pheromone] or developmental process [primer pheromone]' (Karlson & Lüscher 1959); the division into primer and releaser pheromones is discussed in Section 1.8. The word pheromone comes from the Greek *pherein*, to carry or transfer, and *hormōn*, to excite or stimulate. The action of pheromones *between* individuals is contrasted with the action of hormones as internal signals within an individual organism. Pheromones are often divided by function, for example into sex

1

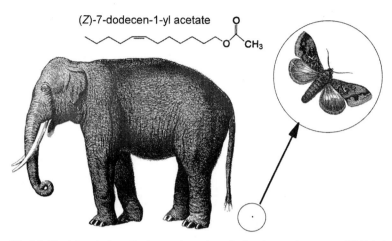

(Z)-7-dodecen-1-yl acetate

Fig. 1.1. The Asian elephant *Elephas maximus*, shares its female sex pheromone, (Z)-7-dodecen-1-yl acetate (top), with some 140 species of moth (Rasmussen *et al.* 1996). Animal figures from Harter (1979).

pheromones and aggregation pheromones. This functional division underlies the structure of the book, which has an emphasis on the ways evolved by different kinds of animals to solve the same communication needs. Communication itself is hard to define; various approaches are outlined in Box 1.1.

Individuals from other species can perceive signals broadcast to the wider world (Chapter 11). Semiochemicals acting between individuals from different species are called **allelochemicals** and are further divided depending on the costs and benefits to signaller and receiver (Nordlund 1981). Pheromone signals can be eavesdropped ('overheard') by unintended recipients: for example, in the way specialist predatory beetles use the pheromones of their bark beetle prey to locate them. The predators are using the bark beetle pheromones as **kairomones**. Animals of one species can emit signals that benefit themselves at the cost of the receiving species. Chemical signals used in such deceit or propaganda are termed **allomones**: for example, bolas spiders synthesise particular moth pheromones to lure male moths of those species into range for capture. Semiochemicals benefiting both signaller and receiver in mutualisms, such as those between sea anemones and anemone fish (clownfish), are termed **synomones**.

Classifications of semiochemicals rapidly become complicated, not least because the same chemical may be used as a pheromone within a species but may be exploited by specialist predators as a kairomone to locate their prey.

Just as communication is hard to define, the term **pheromone** leads to many questions. I have taken a broad and generous approach that includes many important examples of behaviours mediated or influenced by

Box 1.1 | What is communication?

Wilson (1970) defined biological communication as 'action on the part of one organism (or cell) that alters the probability pattern of behaviour in another organism (or cell) in an adaptive fashion. By adaptive I mean that the signalling, or the response, or both have been genetically programmed to some extent by natural selection.' The mention of natural selection acting on signaller or receiver allows inclusion of deception and eavesdropping, the allomones and kairomones of earlier sections. This definition is more useful than ones which limit 'true communication' to situations where the signal is the function of the behaviour of the sender and the response is adaptive for the receiver (e.g. Dusenbery 1992, p. 37). Dusenbery's definition would confine 'true communication' to signals within a species, beneficial to the receiver. However, given the conflicts even between the sexes *within* a species, communication which is equally adaptive for both signaller and receiver is likely to be rare. Most biologists exclude incidental information, such as a twitch of a grasshopper's leg alerting a predator. In the context of behaviour mediated by chemicals, prey waste products such as CO_2 would not count as kairomones if they have no signal function for the prey.

Put more simply, 'communication occurs when one animal's behaviour can be shown to have an effect on the behaviour of another. 'Signals' are the means by which these effects are achieved' (Dawkins 1995, p. 73).

Communication is one of most contentious issues in animal behaviour and there is no commonly agreed definition (see discussion in Dawkins 1995 and Bradbury & Vehrencamp 1998). The complications come in refining the definition and attempting to include all communication in one definition.

Signals may often be ritualised, that is made conspicuous and exaggerated (Dawkins 1995). In the context of pheromones, ritualisation could be the evolution of pre-existing chemicals as a pheromone (for example in the way that goldfish sex pheromones have evolved from hormones leaking out across the gills, see Section 1.1). However, not all signals evolve to be conspicuous. Pheromone signals such as recognition cues in social insects and mammals may be subtle and complex.

Pheromones can be used as honest signals (Zahavi 1975) which provide reliable information because they accurately reflect the signaller's ability or resources (Guilford 1995). For example, female tiger moths (*Utetheisa ornatrix*) choose a male with the most pheromone. His pheromone is derived from the same plant poisons, used to protect the eggs, which he will pass to the female at mating. His pheromone load is correlated with the gift he will give (Chapter 3) (Eisner & Meinwald 1995). The males of another arctiid moth, *Creatonotus gangis*, display inflated coremata, releasing pheromone. The size of the coremata, which can be up to 1.5 times a male's body length, is directly related to the amount of plant poisons the male sequestered as a larva (Fig. 1.5) (Boppré & Schneider 1985). In garter snake females, the levels of skin pheromones reflect evidence of the previous season's fertility. Male garter snakes court larger snakes, which have more pheromone (Chapter 3). In mammals, production of pheromone is directly related to hormone levels (Chapter 3) and so scent marks will tend to be honest. Animals such as mammals and lizards that scent mark their territories (Chapter 5) leave signals that are inherently reliable – only if the owner does own the territory will his marks exclusively cover it. Where pheromones effectively have the role of badges of status as, for example, in cockroaches (Moore *et al.* 1997), queenless ants (Peeters 1997), or mice (Hurst & Rich 1999), the major cost may be that of maintaining the advertised status (see Chapters 3 and 6).

chemical cues that would currently fall outside a rigid definition of pheromone. First, I include the chemical cues used for social recognition in both mammals and social insects (see Chapter 6), which do not fit the original pheromone criterion of a 'defined chemical mixture eliciting particular behaviour or other response'. The cues used for social recognition of kin, clans, colony members and the like are complex, greatly varied mixtures of many compounds (Box 6.1). The differences between the odour mixtures *are* the message. For example, as well as sex pheromones, each elephant will produce its own highly individual odour mixtures and this complex bouquet can be used by other elephants for recognition of kin, clan or social group, and perhaps individuals. Elephants spend much time sniffing each other (incidentally, people are also good at recognising their own family by smell, see Chapter 13).

Second, while we tend to think of pheromones as being detected by 'sniffing' air or water after travelling some distance from the signaller, many chemical cues are detected by contact chemoreception, as in the case of an ant tapping its antennae on a fellow ant to detect the complex mixtures of chemicals on its cuticle that differ between colonies and allow distinction of nestmates from strangers. Pheromones may be transferred directly from signaller to receiver. For example, male Queen butterflies (*Danaus gilippus*) deposit crystals of the pheromone danaidone from their hair pencils directly onto the antennae of the female (see Eisner & Meinwald 1995). The male of the terrestrial salamander (*Plethodon jordani*) directly transfers his high molecular weight glycopeptide pheromone from his chin gland to the nostrils of the female (Rollmann *et al.* 1999). The male of the related salamander, *Desmognathus ochrophaeus*, takes this a stage further by directly 'injecting' his pheromone into her capillary blood supply, using elongated teeth to pierce the female skin, thus bypassing her chemosensory system (Houck & Reagan 1990). In this same continuum I have included molecules passed, together with sperm, to the female during mating in many species: for example, the fruit fly *Drosophila melanogaster*, and garter snakes (Chapter 3).

ELEPHANTS AND MOTHS – CONVERGENT PHEROMONES

The discovery that elephants and some moths share the sex pheromone (Z)-7-dodecen-1-yl acetate is particularly interesting because it illustrates important points emerging about pheromones in mammals and insects, and animals in general.

First, it illustrates the ubiquity of pheromones. Across the animal kingdom, more interactions are mediated by pheromones than by any other kind of signal.

Second, the shared use of a compound as a signal illustrates a relatively common phenomenon of independent evolution of particular molecules as

Table 1.1. *Biochemical convergence of pheromones among ants, bees, moths and termites and other animals including mammals*

In some cases, the same compound is used for similar functions in different species. More commonly, the arbitrary nature of signals is revealed by different uses for same compound. See other chapters for more details of the functions of these pheromones.

Compound	Function	Occurrence	
		Family	Genus
Benzaldehyde	Trail pheromone	Bee, Apidae	*Trigona*
	Defence	Ant, Formicidae	*Veromessor*
	Male sex pheromone	Moth, Amphipyrinae	*Pseudaletia*
2-Tridecanone	Alarm pheromone	Ant, Formicidae	*Acanthomyops*
	Defence	Termite, Rhinotermitidae	*Schedorhinotermes*
Dehydro-exo-brevicomin	Male sex pheromone	Mammal	Mouse, *Mus*
Exo-brevicomin	Aggregation pheromone	Insect	Bark beetle, *Dendroctonus*
(Z)-7-Dodecen-1-yl acetate	Female sex pheromone	{ Mammal	Asian elephant *Elephas maximus*
		{ Insect	~140 species of moth (as one component of a multi-component pheromone)

After Blum (1982), with additional information from Kelly (1996).

signals by species that are not closely related (Table 1.1) (Kelly 1996). Such co-incidences are a consequence of the common origin of life: basic enzyme pathways are common to all multicellular organisms, and most classes of molecule are found throughout the animal kingdom.

However, despite sharing an attraction to (Z)-7-dodecen-1-yl acetate, male moths and elephants are unlikely to be confused. Apart from the mating difficulties should they try, male moths are unlikely to be attracted by the pheromones in female elephant urine because moth pheromones are multi-component (Section 1.6). The (Z)-7-dodecen-1-yl acetate would be only one of perhaps five or six other similar compounds making up a precise blend for each moth species. Male elephants are unlikely to be attracted to a female moth because she releases such small quantities (picograms per hour) that they would not be noticed by a male elephant (but can be tracked by the specialised sensory system of a male moth).

Third, it is an important illustration that, like insects, mammals can use small molecules, singly or in simple mixtures, as pheromones for sexual signalling. It is harder to identify mammalian pheromones than those of insects (Chapter 2) but this does not necessarily mean that their pheromones are more complicated. One difference may be that unlike insects, mammals may increase the activity of their pheromones by interaction with carrier proteins in the urine, as is the case in the elephant and also in mice.

1.3 | Evolution of chemical cues into signals

Chemical senses are the oldest, shared by all organisms including bacteria, so animals are pre-adapted to detect chemical signals in the environment (Wilson 1970). Chemical information is used to locate potential food sources and to detect predators as well as to receive the chemical signals in the social interactions that form the focus of this book.

Signals are derived from movements, body parts or molecules already in use and are subsequently changed in the course of evolution to enhance their signal function. Thus pheromones evolve from compounds originally having other uses or significance, for example from hormones, host plant odours, chemicals released on injury, or waste products. There is selection for functional signal features such as longevity and specificity (Section 1.5). There is also evolution in the senses and response of the receiver. The original functions of the chemicals may or may not be eventually lost.

The ubiquity and extraordinary diversity of pheromones are the evolutionary consequence of the powerful and flexible way the olfactory system is organised (Chapter 9); taste does not have this flexibility. Most animal olfactory systems have a large range of relatively *non*-specific olfactory receptors which means that almost any chemical in the rich chemical world of animals will stimulate some olfactory sensory neurons and can potentially evolve into a pheromone. If detection of a particular chemical cue leads to greater reproductive success or survival, there can be selection for receptors more sensitive to it or expressed in greater numbers. In some cases animals may evolve a finely tuned system, including specialised sensory organs and brain circuits, such as those of male moths used to detect and respond to female pheromones (Chapter 9).

Any pheromone signal that overlaps the receiver's pre-existing sensory sensitivities, for example for food odours, is likely to be selected over others. This is the phenomenon of sensory drive (Fig. 1.2) (see reviews in Endler & Basolo 1998, Ryan 1998). For example, female moths use plant odours to find host plants when egg laying, so their olfactory system is already tuned to these odours – and male pheromones have evolved to exploit the sensory bias of females (Fig. 1.2) (Chapter 3) (Birch *et al.* 1990; Phelan 1997).

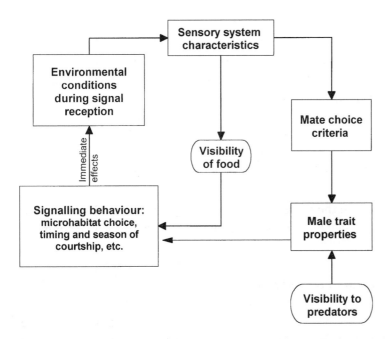

Fig. 1.2. Sensory drive. (Left) Signals that exploit the existing senses of the receiver will be selected for: this is the phenomenon of sensory drive. In the diagram, arrows indicate evolutionary influences (except for the one labelled 'immediate effects'). Predators can have a counter-selection pressure on conspicuous signals. Different authors have emphasised different, partially overlapping, aspects of the phenomenon, terming them 'sensory traps', 'preexisting bias', 'sensory drive', 'sensory exploitation', 'receiver psychology', 'hidden preference', and 'perceptual drive' (Endler & Basolo 1998). Figure after Endler (1992).

(Right) A male oriental fruit moth, *Grapholitha molesta*, displays its hair pencils in courtship to a female. The male's hair pencils are loaded with plant-derived pheromones including ethyl-*trans*-cinnamate, a signal which may have evolved through sensory drive exploiting female sensitivity for odours present in their fruit food (Löfstedt *et al.* 1989). The females prefer males with the most cinnamate. Photograph by Tom Baker.

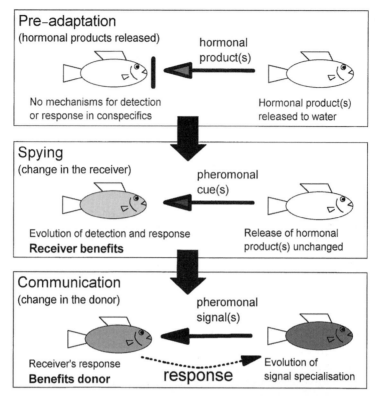

Fig. 1.3. Proposed stages in the evolution of a communication function for hormonal pheromones from pre-existing hormones by intraspecific eavesdropping or spying. In the 'spying phase' only the receiver benefits. The transition to bilateral benefit to both sender and receiver could occur later if there is a selective advantage to the sender. In the 'spying phase' there need not be changes in the signal released by the sender (see Fig. 1.11). Figure redrawn from Sorensen & Stacey (1999).

Pheromones evolved from leaking hormones or from compounds used in defence provide illustrations of the way that evolution can act on available chemical cues.

1.3.1 Pheromones evolved from leaking hormones or other metabolites

Coordinating reproduction is very important, particularly for externally fertilising animals, which must release gametes (sperm and eggs), at the same time as their partner(s). Molecules in body cavity fluids normally released with the sperm or by sexually mature adults may become pheromones. For example, marine polychaete worms release sex pheromones with their gametes, which immediately prompt the other sex to release its gametes (Chapter 3).

Hormones or other molecules associated with reproductive cycles have evolved into pheromones by eavesdropping ('spying') in many animals (Fig. 1.3) (Sorensen & Stacey 1999). In terrestrial animals such as elephants and mice, some pheromones are excreted in the urine. In aquatic animals such as fish and lobsters, pheromones may have evolved from molecules excreted in urine or leaking into the water across permeable membranes such as gills.

Hormone-based sex pheromones in the goldfish *Carassius auratus* provide a good model system (Sorensen & Stacey 1999). A chance observation revealed that male goldfish are extraordinarily sensitive (with picomolar thresholds) to steroid and prostaglandin hormones and their metabolites released into the water by females (see Chapter 2). The released molecules reflect the blood concentrations of the hormones in the female and are a reliable indicator of her biological state. In the evening while the female matures her eggs before release, rising levels of the steroid 4-pregnen-17α-, 20β-diol-3-one (17,20β-P) in her blood leak into the water. The pheromone stimulates physiological responses in the male (Section 1.8). When the female spawns the next morning, males respond to other hormone pheromones released by the female: blood prostaglandin F2α (PGF2α) and 15-keto-PGF2α.

1.3.2 Alarm pheromones and compounds released by fighting or injured animals

Many alarm pheromones, which provoke fight or flight in receivers, appear to have evolved from compounds released by fighting or injured conspecifics (Chapter 8). There will be a selective advantage to potential receivers sensitive to these compounds and responding appropriately. Over evolutionary time, defensive compounds may gain a signal function: for example, most ant species use the same chemicals for defence and alarm, to repel enemies and to alert and recruit nestmates (Hölldobler & Wilson 1990, p. 260). This pattern is shown across the arthropods (Blum 1985).

In other animals, alarm pheromones may derive from compounds evolved to make the flesh unpalatable or toxic to predators (Chapter 8). These compounds would be released by an injured animal, for example anthopleurine in the sea anemone and the bufotoxins and larval skin extracts which elicit an alarm response in toad tadpoles. The alarm pheromone of fish is not an antifeedant but may have evolved with a primary function such as control of skin pathogens.

1.4 | Secretory organs for pheromones

The independent and multiple evolution of pheromones is illustrated not only by the diversity of compounds produced but also by the enormous variety of specialised secretory glands among male mammals and male Lepidoptera (moths and butterflies). The variety is probably largely the

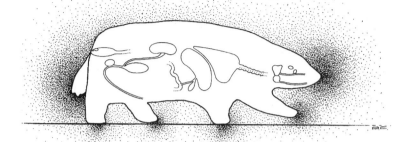

Fig. 1.4. A summary diagram of an imaginary mammal illustrating a variety of possible odour sources. The stippling indicates a potential distribution of odorants in the immediate environment. The following are shown: salivary glands and glands associated with the eye (e.g. preorbital glands); the lungs and trachea; the liver, gall bladder, bile duct, and portion of the small intestine; the kidney, ureter, bladder, urethra, and male accessory gland; the rectum; and an anal sac. The female genital system could be readily substituted for the male (plus specific glands on the feet and legs and on skin in many parts). Figure from Flood (1985).

Fig. 1.5. The expanded coremata of a displaying male of an arctiid lekking moth, *Creatonotus gangis*. Photograph by M. Boppré.

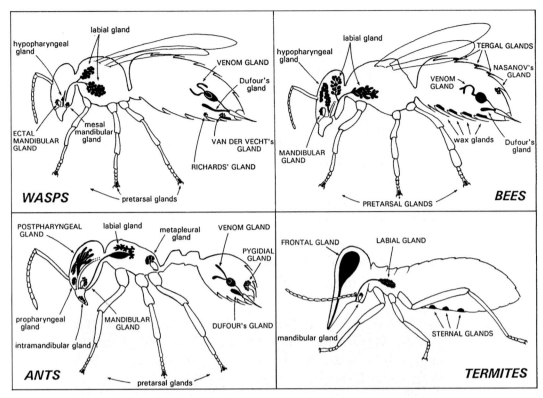

Fig. 1.6. Schematic profile drawings of the commonly found exocrine glands in wasps, bees, ants, and termites. Glands with a pheromonal function are given capital lettering. Pheromones may be identified in future from other glands. Figure from Billen & Morgan (1998).

result of sexual selection (Andersson 1994) (Chapter 3). In mammals there are secretory glands in species-specific positions such as the flanks, around the eye, around the genitals and anus (Fig. 1.4). The equivalents in male Lepidoptera are a profusion of specialised brushes, fans, inflatable balloons (coremata) and other structures on the wings, legs, and abdomen for exposing the pheromones produced in associated glands (Fig. 1.5) (Chapter 3). There is convergence in the special lattice-like hairs to soak up and emit pheromone in some moths and mammals. Among ants, the range of alarm pheromones is matched by the extreme diversity of glands involved (Fig. 1.6) (Hölldobler & Wilson 1990).

While the signaller itself synthesises and secretes most pheromones, some components of pheromones may be collected or gained in other ways. For example, the complete aggregation pheromone for some bark beetle species includes components produced by the host tree as well as components produced by symbiotic bacteria in the beetle gut (Chapter 4). Similarly, many mammal pheromones are produced by the action of bacterial fermentation, for example of fatty acid secretions in the anal glands of foxes (Albone & Perry 1976).

Plants containing pyrrolizidine alkaloids (PA), which are poisonous to most vertebrates and other insects, are sought out and fed on by specialist moth and butterfly species (Lepidoptera), a behaviour known as pharmacophagy (Boppré 1990). In some species only the larvae sequester the alkaloids, in others, such as the milkweed danaine butterflies (Nymphalidae), adults also feed on these PAs. Courtship in PA-sequestering species usually involves display of hair pencils or coremata loaded with derivatives of these alkaloids (Box 1.1). Males without chemical riches are rejected (see Chapter 3). Likewise, to be successful in attracting mates, male euglossine bees in the American tropics must collect perfume oils such as mono- and sesqui-terpenes from orchid flowers (Schemske & Lande 1984).

1.5 | Functional signal design: contrasting different signal modalities

The chemical senses of olfaction and taste are very different from vision, hearing and touch, which detect energy in the form of light, sound or pressure: chemical senses rely on the physical movement of odour molecules from the signaller to the receiver (Table 1.2).

Table 1.2. *Characteristics of different sensory channels of communication*

Feature of channel	Type of signal			
	Chemical	Acoustical	Visual	Tactile
Range	Long	Long	Medium	Very short
Transmission rate	Slow	Fast	Fast	Fast
Flow round barrier?	Yes	Yes	No	No
Locatability of sender	Difficult	Medium	High	High
Energetic cost to sender	Low	High	Low to moderate	Low
Longevity (fade-out)	Variable, potentially high	Instantaneous	Instantaneous	Short
Use in darkness	Yes	Yes	No (unless make own light)	Yes
Specificity	Potentially very high	High	More limited	Limited

After Alcock (1989).

Pheromones are not effectively instantaneous like sound or light, because the molecules of odour have to reach the nose or other sense organ of the receiving animal. For longer range attraction of mates, for example, pheromones offer many advantages over other sensory channels because chemical signals go round barriers, can be carried for long distances in the wind or water currents, and have low production costs (see below) (Thornhill & Alcock 1983). However, locating the pheromone source can be challenging (Chapter 10).

The use of pheromones as the dominant communication channel may be related to habitat and daily activity rhythm. For example, the grysbok (*Raphicerus melanotis*), a small antelope, occupies dense scrub and relies more heavily on olfactory communication than antelope species in more open habitats (Albone 1984). As might be expected, nocturnal animals such as bushbabies (*Galago* spp.) use odour or sound rather than visual signals.

For chemical signals used for marking, the whole function of the signal is the association of a message with a place. Pheromone deposited on a territory marker by an antelope or a lizard remains after the animal has moved on (Chapter 5). Whereas sound and visual signals only act at the time they are made, chemical messages can 'shout' long after the signaller has gone. This highlights one of the unique elements of chemical signals – that in evolutionary time, different lengths of signal life can be selected for: selection can work on the chemical characteristics of pheromones such as volatility and stability, giving signal durations from seconds to years.

1.5.1 Pheromones in terrestrial habitats

The characteristics of pheromones can be related to their signalling function and the signalling environment. Wilson & Bossert (1963) suggested that pheromones used in air should have 5–20 carbon atoms and a molecular weight (MW) of between 80 and 300. Below the lower limit, only a few permutations of molecule can be created; above it, molecular diversity increases very rapidly (Wilson 1970). For example, alarm pheromones have low molecular weights to allow fast diffusion from the source but a quick decline once the danger has past. In ants, most, but not all, alarm pheromones have molecular weights of between 100 and 200 (Fig. 1.7) (Hölldobler & Wilson 1990). Sex pheromones would be expected to be larger molecules to allow more specificity – in insects they typically have molecular weights of 200–300 (Fig. 1.7).

Alberts (1992) tested these ideas on mammalian chemical signals and found that in each of five orders, sex pheromones had lower molecular weights than territorial marking pheromones and would thus be more volatile and shorter lived. The signal life of mammalian pheromones used as territory markers was increased by carrier components in secretions such as sebum, a lipid-rich, oily

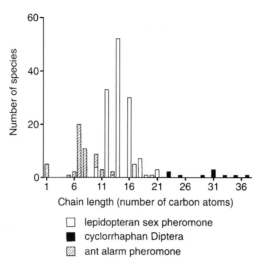

Fig. 1.7. Chain lengths of aliphatic hydrocarbons used as alarm pheromones by ants (short chain, low molecular weight), long range sex attractants by moths, and contact pheromones by cyclorrhaphan Diptera such as *Drosophila* (see Chapter 10 for discussion of molecular weight and diffusion). Figure from Chapman (1998).

substance produced by sebaceous glands (Alberts 1992). Higher temperatures and humidities increase evaporation rates, reducing the persistence of chemical signals. This may explain why the molecular weights of marking pheromones used by tropical forest species were higher than those from temperate forests, with temperate grassland species lowest of all (Fig. 1.8).

Another way, used by mice and other mammals, to increase the longevity of small volatile pheromones is to associate them with urinary protein molecules which release them slowly (Chapter 9).

Signal design must sometimes meet conflicting selection pressures. For example, the hot habitat of the desert iguana *Dipsosaurus dorsalis* presents a challenge to long-lasting but effective marks: volatile pheromones would fade quickly, but longer lasting, less volatile chemical signals would not be detected by intruders. This lizard has evolved high molecular weight scent marks which strongly absorb long-wave ultraviolet light, which lizards can see (Alberts 1990). Lizards orient towards these conspicuous visual 'flags' and, once at the mark, they tongue-flick to pick up the non-volatile pheromone molecules. The advantage of this composite signal is that the high information content from the non-volatile molecules is combined with long-distance visual attraction. In an analogous way, male butterflies use visual cues to find females at long range, and then at short range, in many species, they communicate with pheromones (Chapter 3) (Vane-Wright & Boppré 1993). Some animals, such as dogs, add their marks to conspicuous sites or landmarks.

The effect of habitat on signals is shown by comparison with another iguana species, the green lizard *Iguana iguana*, which lives in humid tropical

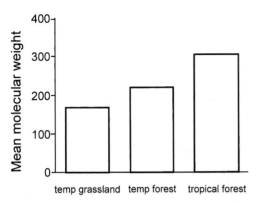

Fig. 1.8. Comparison of mean molecular weight of territorial scent marks from mammals living in temperate grassland, temperate forest, and tropical forest habitats. Data are pooled across seven orders of mammals. Figure redrawn from Alberts (1992).

forest. The green lizard has a higher percentage and more varied volatile lipids in its scent marks than the desert iguana, but does not include UV-absorbing molecules in its marks (Alberts 1993).

1.5.2 Aquatic pheromones

In water, the size rules for potential diversity of molecules still apply, but whereas volatility is a key signal design feature of pheromones in air, solubility of molecules is perhaps the functional equivalent in water. There appear to be two main types of molecules used as pheromones in aquatic species. First, there are soluble molecules similar in size to those used as pheromones on land, such as the steroid-based pheromones used as fish sex pheromones (Sections 1.3.1 and 1.8) and barnacle egg hatching pheromone (Section 4.2.2). Second, large, polar molecules can be used, which despite their size can be highly soluble. For example, anthopleurine, the alarm pheromone of a sea anemone (Chapter 8) is a large cation. Many other marine invertebrates use polypeptides as chemical signals (Chapter 4) (Zimmer & Butman 2000). These peptide pheromones are involved in gregarious settlement by oyster and barnacle larvae, precisely timed larval release by crabs to coincide with tides, and a sex pheromone of the sea-slug mollusc *Aplysia* (Painter *et al.* 1998). The barnacle cues are glycoproteins and water-borne peptides (Clare & Matsumura 2000). The crab 'pumping pheromone' is a short peptide, with activity at concentrations as low as 10^{-21}M (Pettis *et al.* 1993). The activity of synthetic peptides can be predicted by quantitative structure-activity relationships (Browne *et al.* 1998).

The first peptide pheromone to be identified in a vertebrate was the decapeptide, sodefrin, in the Asian red-bellied newt (*Cynops pyrrhogaster*) (Fig. 1.9) (Kikuyama *et al.* 1995).

Fig. 1.9. Two phases of the underwater courtship behaviour of the red-bellied newt, *Cynops pyrrhogaster*. (Top) The male wafts a decapeptide pheromone, sodefrin, from his cloacal glands towards the female's nostrils. (Bottom) The male (left) then moves in front of the female and she follows him with her snout in contact with his tail. Figure from Kikuyama & Toyoda (1999). Photograph by Sakae Kikuyama & Fumiyo Toyoda.

1.6 | Specificity

In some biological signalling systems, selection pressures from sexual selection and speciation (Chapter 3) will lead to species-specific sex pheromones and responses. In other situations, such as alarm pheromones, there is little need for privacy in communication and alarm pheromones are often not species specific (Chapter 8). If a predator has alarmed individuals of another species, and it is of benefit to respond, then cross-species responses will evolve. Within a prey guild with shared predators, all species commonly respond to each others' alarm signals.

There are two main ways of gaining specificity in pheromone signals. One is by the evolution of a large unique molecule. Peptide pheromones, using the 20 coded amino acids available in eukaryotic systems, offer an extraordinary variety of unique sequences; with a five- amino-acid polypeptide there are $20^5 = 3.2$ million (Browne *et al.* 1998). For example, two related species of the newt *Cynops* have species-specific decapeptide pheromones which differ by just two amino acids (Yamamoto *et al.* 2000) (Section 1.5.2). Among insects, a very few species use a unique complex molecule as a single-component pheromone; for example, periplanone-B is the sex pheromone of the American cockroach (*Periplaneta americana*) (Roelofs 1995).

More commonly, specificity is gained largely by using a unique blend of relatively simple compounds as a *multi-component* pheromone. For example, female sex pheromones in moths usually consist of five to six fatty acids or their derivatives. Vertebrates may also have multi-component pheromones. For example the mouse pheromone, which elicits aggression in other males, consists of two compounds, each of which is inactive alone (Section 2.6). Similarly, in the goldfish, while each of two female prostaglandin pheromones, F2α (PGF2α) and 15-keto-PGF2α, have similar effects on male behaviour when presented singly, both are needed together to stimulate a gonadotropin surge in males (Sorensen & Stacey 1999). It is possible that other components add species specificity.

A particular specificity open to biological systems is to use stereoisomers (molecules that have the same atoms connected in the same order but differ in the arrangement of atoms in space, see Appendix A2). Pheromone synthesis is catalysed by enzymes and pheromones are detected by receptors. Enzymes and receptors are proteins and recognise their substrates and signal molecules (ligands), respectively, by shape, so that different stereoisomers will be treated differently. Stereoisomers are very important in pheromone signalling for both invertebrates and vertebrates (Silverstein 1979; Seybold 1993; Mori 1996). Some molecules are **enantiomers**, mirror images of each other (said to be chiral, from the Greek meaning hand). Some pairs of species gain specificity by using different enantiomers of the same compound; for example, among sympatric scarab beetles in Japan (Leal 1999), the Japanese beetle *Popilla japonica* uses (S)-japonilure as its female sex pheromone whereas the Osaka beetle (*Anomala osakana*) uses (R)-japonilure (see Appendix A2 for notation).

1.7 | Composite signals: pheromones working in concert with other modalities

Pheromones are often just one of the sensory channels (modalities) involved in communicating a signal. The signals in parallel sensory channels can be important in three different ways. First, to ensure that the message gets through by having it repeated in each of the sensory channels; this is termed signal 'redundancy' (as for example when we write a cheque in both words and figures). Second, there may be modulation of the signal intensity by addition of other signals. Third, signals in each of the communication channels may be necessary for the message.

An example is given by the black-tailed deer, in which alarm signals are transmitted not only as an odour signal, but also as sounds and visual signals (Chapter 8). Any one of these may be effective in alerting other deer in

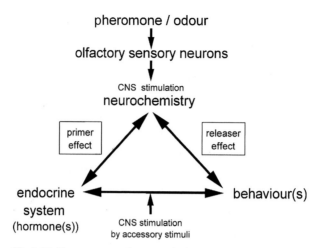

Fig. 1.10. Pheromones can be a stimulus leading to a prompt behavioural response by nerve impulses from the brain (CNS, central nervous system) (**releaser effect**) or can act indirectly by stimulation of hormone secretion resulting in physiological changes, 'priming' the animal for a different behavioural repertoire (**primer effect**). A given pheromone can of course act as a primer or releaser at same time or in different contexts. The distinction between primer and releaser pheromones has become blurred as we understand more about the links, interactions, and feedback loops in the sequence from odour to behavioural and endocrine effects. Hormonal effects can be rapid, and memories, sometimes facilitated by local neurochemistry changes, can be long lasting (see Fig 9.10 for an example using maternal learning of the odours of her lamb by a sheep). Diagram brings together ideas from figures in Wilson & Bossert (1963) and Sachs (1999).

the group. This redundancy in signal can make dissecting the role of pheromones much more difficult (Chapter 2). While this complexity is perhaps characteristic, and perhaps even the rule, of many signals in mammals, there are also many invertebrate examples of modulation of pheromone signals by other stimuli. For example in the desert ant *Novomessor*, recruitment of nestmates to a new food source is faster when release of pheromone is accompanied by stridulation sounds from scouts (Chapter 7). In the snapping shrimp *Alpheus heterochaelis*, male responses to visual threat signals are changed if these are accompanied by female pheromones (Hughes 1996). Successful courtship by male *Drosophila melanogaster* fruit flies requires a combination of chemical and visual stimuli from the female; pheromones are necessary but not sufficient alone (Chapter 3) (Greenspan & Ferveur 2000).

1.8 | Primer and releaser pheromones

Most of the phenomena described so far have been of **releaser pheromones**, with immediate effects on the behaviour of the receiver. An equally important group of pheromones, **primer pheromones**, have longer term physiological

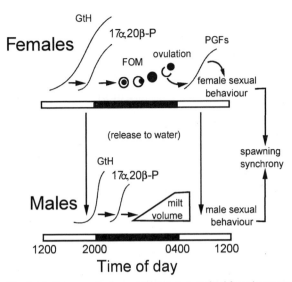

Fig. 1.11. Primer and releaser pheromones, evolved from hormones, coordinate reproduction in the goldfish, *Carassius auratus* (Sorensen & Stacey 1999).

Preovulation primer pheromones: In the evening while the female matures her eggs before release, rising levels of the steroid 17,20β-P in her blood leak into the water. The male's olfactory system detects the pheromone and stimulates his brain to release male gonadotropic hormone (GtH) which in turn stimulates testicular 17,20β-P synthesis and promotes increased milt (sperm and seminal fluid) production.

Postovulatory releaser pheromone: The next morning at ovulation, eggs within the reproductive tract stimulate the production of prostaglandins (PGFs), which act as hormones to stimulate female sexual behaviour. Metabolites of PGFs leak out and function as a releaser pheromone stimulating male sexual behaviour (Dulka 1993).

   The sensitivity to the primer pheromones benefits male reproductive success (Sorensen & Stacey 1999). Males primed overnight are more aggressive in chasing other males away, spawn almost three times more often, and fertilise more eggs than males who were not primed. Figure redrawn from Dulka (1993).

effects on the receiver, in the original definition by stimulating olfactory sensory neurons that send signals to the brain to release the hormones of the endocrine system (Fig. 1.10) (Wilson & Bossert 1963). Studying primer pheromones is a challenge because the timescale for primer effects can be long and it may be difficult to associate a gradual physiological change with an earlier contact with a chemical signal.

   Examples of primer pheromones include stimulation of sperm production in fish (see Fig. 1.11), termite caste determination, locust development rates, and menstrual cycles in humans and other mammals (Chapters 4, 6, 9 and 13).

   The division into releaser and primer pheromone is only a rough classification, first, because many pheromones have both roles. For example, honeybee (*Apis mellifera*) queen mandibular pheromone attracts males on her nuptial flight but also has primer effects on worker reproduction (Chapter 6). Second, response to some endocrine-mediated pheromone signals can be rapid. For example, pheromones of oestrous females cause release of hormones into the blood in sex-experienced male rats, which give them erections and elicit sexual behaviours within minutes (Sachs 1999). Third, the mutual influences and feedback loops between hormones, olfactory centres and endocrine cells in the brain, are complex. For example, long-term behavioural effects can come from neuronal memories for odours, themselves modulated by hormones (e.g. sheep maternal behaviour, Chapter 9). Sachs comments that it may be more useful to view primer and releaser pheromones as acting on a continuum of physiological and behavioural pheromone-induced effects occurring earlier or later in a sequence (Fig. 1.10).

While the physiologies of mammals and insects are very different, there are parallels between the modes of action of primer pheromones in both. For example, in mammals, dominance hierarchies are reflected in blood gonadal hormone concentrations. In social insects JH (juvenile hormone) is often important in pheromone-mediated effects (Chapter 9).

It is sometimes argued that mammals do not have many releaser pheromones that elicit a particular behaviour. However, some have been identified and more are likely to be discovered. One example is the 'standing behaviour' of an oestrous pig (*Sus scrofa*) female in response to the steroid sex pheromones, 3α-androstenol and 5α-androstenone, of the male pig in his saliva (Chapter 12). Another is the rabbit nipple pheromone, which allows the young rabbit pup to find the nipple rapidly (Chapter 9). A third example is the vaginal pheromones attracting males and stimulating mounting in the golden hamster (*Mesocricetus auratus*) (Chapter 9).

You will have noticed that one group missing from the discussions so far are the birds (see Box 1.2).

1.9 | Cost of signalling

The metabolic cost of most pheromone signalling is low compared with that of other signals (Thornhill & Alcock 1983), in part because the quantities of material needed are so small. For example, the lifetime cost to a male boll weevil beetle (*Anthonomus grandis*) to produce its monoterpene sex pheromone is estimated at only 0.2% of its body weight (Hedin *et al.* 1974). In contrast, male crickets devote over half their daily respiratory budget to acoustic signalling (Prestwich 1994). Some of the advantages of chemical communication over sound relate to scale effects of signal production and the size of the organism: smaller organisms cannot make the lower frequency sound signals which travel further (Dusenbery 1992) (see Chapter 10). The costs of production are

---

Box 1.2 | Birds and pheromones?

Birds are not famous for their pheromone communication ... yet. Just as studies of mate choice in birds were transformed by the recent discovery that, unlike us, birds see ultraviolet (e.g. Bennett et al. 1997), pheromones could be important for birds. Most birds have a good sense of smell but we have been so attracted by bird songs and visual displays that the role of olfaction in reproduction and recognition behaviour has been little studied (see review in Roper 1999). All but a few species of bird have a uropygidial gland, at the base of the tail, which secretes lipids used in preening the feathers. The secretions differ between species. Darwin observed that sexually selected characters will be expressed differently in the sexes, at sexual maturity, and during the breeding season (Chapter 3). The uropygidial secretion in the female mallard duck (*Anas platyrhynchos*) meets all these predictions. It changes during the breeding season but during the rest of the year it is the same as that of males and ducklings (Balthazart & Schoffeniels 1979; Jacob et al. 1979). The sexual behaviour of male mallard ducks was reduced if their olfactory nerves were cut, although their other behaviour was apparently normal. Other potential secretory glands in birds include the sebaceous and anal glands.

Olfaction may play an important role in the navigation of adult procellariiform seabirds – 'tube-nosed' birds which include petrels and albatross, to colony sites and nesting burrows, marked in many species by notoriously smelly stomach-oil secretions (nestlings of the British storm petrel *Hydrobates pelagicus* recognise their nest by smell, Minguez 1997). Almost all the Procellariiformes have very large olfactory bulbs in the brain, although these may have evolved primarily for the location of patchy food resources in the ocean (see Chapter 10). In the Antarctic, petrels and albatrosses appeared within minutes at oil slicks scented with food odours (Nevitt 1999b).

Pheromone-mediated behaviour in birds is just as likely to be found in species with small olfactory lobes. The relative size of the olfactory lobe is not necessarily an indication of its importance: if the pheromones are used at close range there need not be large or specialised olfactory structures (female moths respond to male courtship pheromones despite having tiny antennae compared with the male – Chapter 3). Olfactory cues in courtship in ducks and other birds would be well worth investigating; this is another dimension of the world of birds to be explored.

---

probably similarly low for vertebrates (Chapter 3). For example, just 40 nanograms of the peptide pheromone of the magnificent tree frog *Litoria splendida* released into the water one metre from a female, will attract her to the source in minutes (Wabnitz *et al.* 1999). Nonetheless, pheromone signals in some mammals, such as mice, may be costly, as they secrete significant amounts of protein in their urine to make long-life signals (Chapter 5).

However, the costs of signalling in most animals are not limited to the simple cost of production. For example, in territorial animals, the time taken to maintain scent marks may be much more important (Chapter 5). Similarly, there are time and energy costs, in species which do not synthesise their pheromones themselves, of collecting plant materials used as pheromones or pheromone precursors (Section 1.4).

1.10 | Pheromones in humans?

Sight and hearing are our most important senses, which makes us very untypical mammals. There are many exciting things going on around us just beyond our noses (although even the animals that do use pheromones extensively are usually only sensitive to their own; other odours pass them by). However, one of the intriguing possibilities emerging in recent research on human senses is that olfactory signals may be more important to us than supposed, at both conscious and unconscious levels. This is explored in Chapters 9 and 13.

1.11 | Conclusion

Across the animal kingdom, more interactions are mediated by pheromones than by any other kind of signal. A wide variety of compounds are used as pheromones but there are many examples of the same compounds being used by different species. The design of the olfactory system makes evolution of pheromones very likely because there is selection for any odour cue that increases reproductive success or survival. There is less difference between vertebrates and invertebrates, in both the pheromones produced and in the range of behaviours that pheromones influence, than is often thought. Pheromones perhaps provide the supreme honest signals. Given the ubiquity of chemical communication among animals, chemical cues are likely to emerge as one of the key criteria animals use for choice of mate.

1.12 | Further reading

For a general introduction to pheromones, written at a more popular level, there are two excellent books: Agosta (1992), in English, and Brossut (1996), in French. A stimulating discussion of the definition of pheromones in both invertebrates and vertebrates is given by Johnston (2000) (see also Müller-Schwarze 1999). Dusenbery (1992) offers a comprehensive and very readable discussion of sensory ecology. For pheromones in particular taxonomic groups see Johnston *et al.* (1999) and Marchlewska-Koj *et al.* (2001) on vertebrates, Hölldobler & Wilson (1990) on ants, Cardé and Minks (1997) largely on moths, and Hardie & Minks (1999) and Cardé & Bell (1995) for other insects.

In this book I have not attempted a review of animal communication as this has been done excellently by Bradbury & Vehrencamp (1998).

Chapter 2

Discovering pheromones

2.1 | Introduction

Until recently, our own poor sense of smell and the relative difficulty of studying pheromones has left the subject lagging behind the study of visual and auditory signals. Apart from the behavioural complexities of studying signals of any kind, pheromones present particular problems. First, the chemical messages are complex in many species, especially in mammals and social insects. Second, the small amounts available to the chemist are often at the limit of current analytical techniques. Third, whereas sound signals such as cricket song can be played back from a tape or indeed 'synthesised' or manipulated experimentally by computer, 'playback' for pheromones is much more difficult given the limits of chemical synthesis and the sensitivity of olfactory receptors to the precise nature of the pheromones.

It is difficult to imagine now, when the role of pheromones in behaviour is so well established, that not so long ago it was hard to show that chemicals *were* involved, let alone identify the compounds. In the nineteenth century, the French naturalist Jean-Henri Fabré found that a virgin female emperor moth (*Saturnia pavonia*) could attract dozens of males into his study. The males were still attracted to a female hidden under a cover so long as it was not airtight. These observations convinced him that the attraction must be chemical even though he could not smell anything (Fabré 1911, p.137). We know now that humans cannot smell most of the chemicals important to other animals.

Despite these and many other early clues that chemical messages were important to animals, the first chemical identification of a pheromone was not until the late 1950s, after almost two decades of work by a German team led by Adolf Butenandt (Hecker & Butenandt 1984). They chose the domesticated

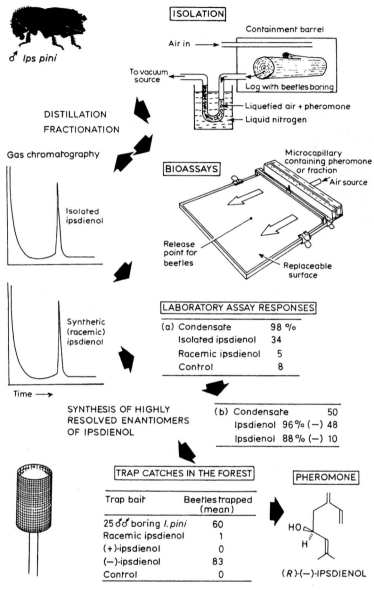

Fig. 2.1. Bioassays are essential at every stage of isolating and identifying pheromones. Here the stages of collection, purification, and identification (along with confirmatory synthesis and field testing) are illustrated by the procedures leading to the identification of (R)-(−)-ipsdienol as one of the components of the aggregation pheromone of the bark beetle *Ips pini* in California. The volatiles produced by male beetles boring in a log were entrained using a cold trap cooled with liquid nitrogen. The resulting concentrate was fractionated by gas chromatography (GC). Bioassays of the response of walking beetles showed attraction to only one isolated compound, ipsdienol. However, synthetic (racemic-mixed (+)- and (−)-ipsdienol) ispdienol did not attract the beetles (see Appendix A2). Purified synthetic 96% (−)-ipsdienol was attractive in the laboratory whereas the (+)-ipsdienol interrupted (stopped) the response, which explained the earlier lack of response to racemic ipsdienol. In the field, synthetic R-(−)-ipsdienol was as attractive to flying beetles as real male beetles boring in a log. Caption and figure adapted from Birch (1984).

silk moth (*Bombyx mori*) which could be reared in the enormous numbers needed. They extracted the female pheromone gland in solvent and then fractionated the complex mixture. Biological activity in the extracts was monitored with a simple laboratory bioassay to see which fractions excited the males into fanning their wings (Section 2.3). The silk moth pheromone, which they named 'bombykol', was eventually identified as a long chain alcohol with 16 carbons, E-10,Z-12-hexadecandien-1-ol (the chemical terminology is explained in Appendices A1 and A2). More than half a million moths were required to obtain some 6 mg of material. Today, powerful new tools, particularly for chemists, have revolutionised the study of pheromones so that now much of the task could be attempted with a single female insect.

It was not until the mid-1960s that the first vertebrate pheromone was identified. It was then – and still is – a much more difficult task than work on insect pheromones. The reasons include a greater complexity of odour, greater variation between individuals and greater complexity of response. Interestingly, some of these characteristics are shared with the colony-recognition pheromones of social insects (Chapter 6) (Hölldobler & Carlin 1987).

2.2 | How are pheromones identified?

The chemical methods used to track down pheromones depend, of course, on the properties of the compounds thought to be involved. For example, there are big differences between the techniques used to investigate volatile hydrocarbon pheromones of insects and the water soluble peptides of some marine animals. Nonetheless, there are similarities in the stages involved (collection, purification, identification) and at every stage the results need to be tested with a good behavioural or physiological bioassay (Fig. 2.1).

2.3 | Bioassays

The first step is the observation of a behaviour that appears to be mediated or influenced by chemicals. The behaviour forms the basis of a bioassay, a repeatable experiment for measuring response to a stimulus – in this case a potential chemical signal. Such a behaviour might be time spent sniffing and marking (Fig. 2.2), settlement by marine planktonic larvae (Fig. 2.3), or longer term physiological and behavioural changes such as becoming sexually mature. There are as many potential bioassays as there are animal species and behaviours to study.

A key feature of any bioassay is that it should be a reliable measure of the behaviour you ultimately want to assess: a laboratory bioassay, for example,

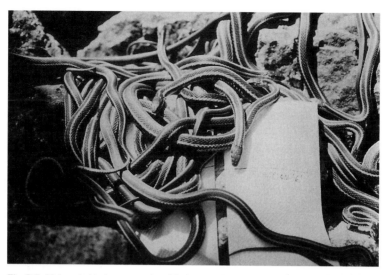

Fig. 2.2. Male red-sided garter snakes (*Thamnophis sirtalis parietalis*) chin-rub female contact pheromones presented on a paper towel in the field. This bioassay was used to identify the pheromones (see Chapter 3). Figure from Mason (1993).

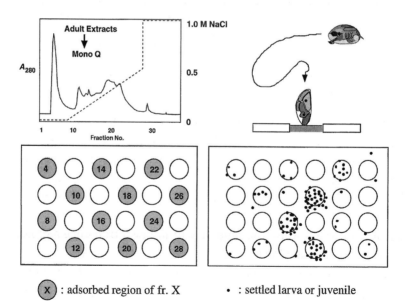

Fig. 2.3. The planktonic cyprid larval stages of marine barnacles search for chemical evidence of other barnacles so they can settle close to them. This bioassay separated the barnacle proteins by high performance liquid chromatography (top left) and then dropped the different fractions onto defined spots on a nitrocellulose membrane (lower left). The dosed plate was then exposed to searching cyprids in seawater for 48 h to see which spots would cause them to settle and metamorphose (right). In this case, fractions 16–22 were judged to be active proteins. (See Box 4.2 for the barnacle life cycle; Chapter 12 for applied uses.) Figure from Matsumura *et al.* (1998b).

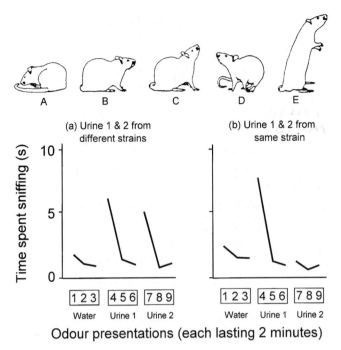

Fig. 2.4. Bioassays can use habituation as a tool. (Top) When a resting rat is presented with a new odour from above, he rears up and sniffs. This response can be used to investigate which mixtures/individual odours the test animal treats as the same. The time he spends sniffing when he is repeatedly presented with the same odour declines rapidly (habituation), but increases again if a new odour is offered (dishabituation). Figure from Schellinck & Brown (1994).
(Bottom) Urine odours from rats from different major histocompatibility complex (MHC, Chapter 3) strains are treated as different at trial 7 out of nine trials (a) whereas (b) urine from a second rat of the same strain does not cause dishabituation and is treated as if it was the same as the first. Figure redrawn from Brown et al. (1989).

should be able to identify a pheromone that works in the field. Another important feature is ease of use. Simple bioassays such as the wing fanning response by male moths are much easier to run than flight tests in a wind tunnel. Careful design may be needed to tease out the different behaviours involved. For example, in *Solenopsis* fire ants, different trail pheromone components have different roles, some being involved in recruitment (alerting) and others in orientation (the tendency to follow a trail once alerted) (Chapter 7) (Vander Meer & Alonso 1998).

Different bioassay methods can give different results, perhaps because they are testing different things. For example, studies of discrimination of odours by rats found different results depending on whether the study used a trained discrimination task (with a reward for choice of one odour over another) or used a habituation task (which measures what the animals naturally notice on their own) (Fig. 2.4) (Schellinck et al. 1995).

Bioassays for primer pheromones measure longer-term physiological effects. For example, scientists tracked down the identity of steroid pheromones released by females of the goldfish *Carassius auratus* by exposing males to candidate compounds overnight and measuring the volume of milt (semen) production the next morning (Chapter 1) (Stacey & Sorensen 1986).

Sometimes, bioassay results can be misleading. For example, some termites will follow lines drawn with ink from a ballpoint pen even though these compounds are not found in their trail pheromones (Becker & Mannesmann 1968 in Pasteels & Bordereau 1998). Similarly, for many years it was thought that most termite species had the same trail pheromone. Only when offered a choice *between* two trails, was evidence found that conspecific trails were more attractive and thus we can conclude that there *are* differences in trail pheromones between termite species (Chapter 7).

An important consideration in running bioassays is to ensure that conditions are standardised – so (for example) that animals are tested at the same time in their day. Apart from external, environmental variations, there may also be internal conditions such as hormonal states which need to be standardised as well as obvious factors such as age and sex. In addition, experience may change behaviour in subtle ways. Learning is being shown to be more and more important in insect behaviour (Papaj & Lewis 1993). Studies of vertebrates routinely require this factor to be taken into account. Sometimes, an animal receiving a signal will make no visible response at the time; for example, the female mouse learns the odour of her mate during copulation but gives no sign of this until later tested with the odours of different males (see Chapters 6 and 9).

2.3.1 Appropriate concentrations

For compounds to be identified as a pheromone they should be active at the concentrations released under natural circumstances. In studies of ant alarm pheromones, it is possible to get an alarm response from almost any volatile compound extracted from ants if the concentrations used are high enough (Hölldobler & Wilson 1990, p. 261). However, pheromones often have a lower sensitivity threshold than similar compounds: for example, goldfish males are many times more sensitive to their own pheromones than to related steroids.

2.3.2 Redundancy

The classic test of a possible role of a stimulus is to remove it and see the effect. Unfortunately, many animals can use not only a multiplicity of stimuli but use one sense when another is not available. For example, testing alarm pheromones in black-tailed deer is complicated because the alarm reaction also includes visual (erected tail and anal hair, cocked ears, stiff gait) and auditory stimuli (hissing, stamping the forefoot on the ground, and the rhythmical impact of feet in the 'stotting' gait) (Chapter 8) (Müller-Schwarze 1971). Such redundancy allows an animal to function despite loss of input from one of its senses through old age, injury, or vagaries of the environment. The advantage to the animal is clear but since most experiments involve either depriving an animal of one sense at a time or providing limited controlled stimuli, drawing conclusions can be difficult. If the animal can continue to respond despite the loss of a particular cue, it is not proof that the animal does not normally use these cues. Redundancy can also occur with pheromone blends, such as male mouse pheromones, in which it seems that some components can substitute for each other (Chapter 9). Conversely, many animal behaviours require *combined* inputs from many senses. Multifactorial tests may show interactive effects of stimuli that would be missed had each been tested alone (Harris & Foster 1995). The challenge for the experimenter dealing with multiple stimuli is to devise testable, simpler questions to help understand a more complex real world.

2.3.3 Using the animal's sense organs as physiological detectors

A powerful technique developed for studying the sensory biology of insects and many other animals, including vertebrates, is to record the electrical impulses from the chemosensory cells as they are stimulated by pheromones. The technique uses the animal's own olfactory system, which of course is finely tuned to the chemicals important to it.

The electroantennogram (EAG) detects the stimulation of the male moth antenna by female sex pheromone (Box 2.1) (Chapter 9). Electrodes are placed at each end of the antenna and connected to a DC amplifier. The rapid depolarisation when the sensory cells of the antenna are stimulated by pheromone is dose dependent and varies between chemicals depending on the number of sensory cells on the antenna which are sensitive to them. Portable EAG equipment can now be used to measure pheromone concentrations in the air in the field (van der Pers & Minks 1998).

However, an EAG will not detect pheromone components for which there are few receptors. Single cell recording (SCR) is needed to find cells sensitive to these components. Techniques similar to the EAG have been developed for vertebrates, such as fish, to give an electroolfactogram (EOG) (Hara 1994).

These electrophysiological techniques only indicate sensitivity to a compound but not how the animal would behave in response. For example, males of the noctuid moth *Helicoverpa zea* have antennal sensory cells which detect a compound characteristic of the female sex pheromone of another, sympatric, species. The *H. zea* males turn back if they detect this compound (see Chapters 9 and 10).

2.4 | Collection of chemical signals

If animals deposit enough material, such as the orbital gland marks left on twigs by male antelope (Chapter 5), the scent marks themselves can be collected.

Box 2.1 | Gas-liquid chromatography and pheromone research

Gas–liquid chromatography (GLC, often abbreviated to GC) has revolutionised pheromone research by allowing separation of mixtures of chemicals; the process is illustrated in the figure below. Capillary GC enables repeatable work with very small quantities of material; the peaks on a GC trace can represent as little as a nanogram (10^{-9} g) of a compound and even picograms (10^{-12} g) are detectable (Howse et al. 1998, p. 188). The process is effectively the same as the chromatography of coloured inks separating in water on filter paper.

In the GC, the **mobile phase** is an inert gas (usually nitrogen or helium) which blows through a fine tube (up to 50 m long). The **stationary phase** is an involatile liquid held on an inert porous support inside the tube. The compounds in the mixture are separated by differences in vapour pressure (boiling points) and relative solubility in the stationary phase. The column is placed in an oven and raising its temperature in a controlled way during the run helps to speed the passage of less volatile compounds.

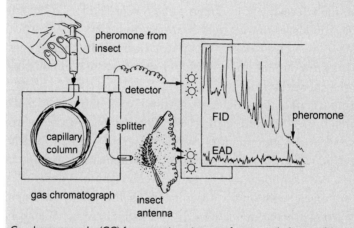

Gas chromatography (GC) for separating mixtures of compounds. A powerful way of identifying possible pheromone components in a female insect extract is GC combined with a male antenna as a biological detector. The GC separates the mixture of compounds into a series of peaks, which are detected by the FID. Part of the GC output is diverted over the electroantennographic detector (EAD, a male antenna of that species). The FID and EAD traces are lined up so that the positive EAD responses can be matched up with the GC peaks. For more details see Howse et al. (1998). Figure from Löfstedt (1986). Artist Jan van der Pers.

All chromatography systems need a method for detecting the compounds as they emerge in turn from the end of the column. The commonest detector for insect pheromones is the **flame ionisation detector** (FID) in which the output of the column leads into a flame of pure hydrogen and air. The length of time a compound takes to reach the end of the column is its **retention time,** which is characteristic and repeatable given the particular conditions (including oven temperature, column characteristics, and gas flow). Different column packings (the stationary phase) can be used to achieve the separations desired – for example, chiral columns enable the separation of enantiomers (see Appendix A2).

Part of the output of the column can be tested on an electroantennogram preparation, the results of which are illustrated in the figure (trace EAD). Behavioural bioassays to assess fractions as they come direct from a GC have also been developed (e.g. Leal et al. 1992), and human armpit-odour testing, Chapter 13). However, looking for interactions between compounds in different fractions is difficult because they come separately from the GC column during the course of a run.

The GC enables chemical separation, but determination of chemical structures themselves is typically done by sending the GC output to a mass spectrometer (MS).

Usually, pheromones are collected by wiping potential glands or collecting whole scent glands. However, one of the disadvantages of making crude extracts or washes of pheromone glands is that many non-pheromone compounds come too, giving quite a 'dirty' and complex mixture. In addition, the process may extract precursors in the gland rather than the active pheromone.

Entrainment methods collect and concentrate the pheromones released into the air or water around an animal. One method is to pass cleaned air over the subject animal and 'trap' the odours from the exhaust air on a chemical 'sponge' such as Poropak-Q ™ or Tenax ™, special synthetic porous polymer resins that will adsorb organic molecules and then release them when either heated or washed with organic solvents such as pentane. The same principles work for concentrating aquatic pheromones using different resins. Another method is to condense the volatiles, along with the water vapour, by passing the exhaust air through a cold trap, cooled to $-195\ ^{\circ}C$ with liquid nitrogen (Fig. 2.1).

Washings or entrainment tend to produce a relatively weak solution and when evaporating a large volume to a few microlitres before injection into the gas chromatograph (GC) (Box 2.1) for analysis, important volatiles may be lost. An alternative method is the recently developed Solid Phase MicroExtraction (SPME) technique which allows solvent-free sampling and sample preparation (Pawliszyn 1997). In SPME, an inert fibre coated with a thin layer of polymer (similar to the polymers used inside GC columns) is exposed to the sample (such as the air over elephant urine, Chapter 1). Compounds migrate into the polymer until an equilibrium is reached (which occurs in minutes or less as the polymer layer is so thin). The fibre, just a millimetre or so in diameter, is mounted in a device like a syringe for ease of use (Fig. 2.5). When retracted, the fibre retains the molecules until introduced into the GC injector port where the compounds are desorbed by heat and carried by the gas into the column.

2.5 | Separating the chemicals, finding the active components, and identifying them

Having collected the potential chemical signals, the next task is to find the active components in a complex mixture. Traditionally, the approach has

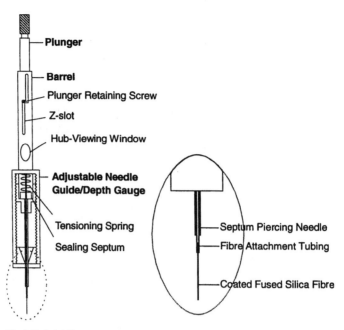

Fig. 2.5. Solid Phase MicroExtraction (SPME) uses an inert silica fibre coated with a polymer similar to those used inside gas chromatography columns. Odours are adsorbed into the coating of the fibre when it is exposed. Later, the fibre can be placed in a GC input port and the trace run. The diagram shows the first commercial SPME device made by Supelco. Figure from Pawliszyn (1997).

been 'fractionation'– separating the components by exploiting their chemical or physical characteristics such as solubility in different solvents. The dual aim of this process is to both purify and concentrate the pheromone. Classically, the sample would be shaken with a water and organic solvent mixture. After the two phases, water (polar) and organic solvent (non-polar), had separated, each would be tested for activity in the bioassay to trace which fractions contain biological activity; had the active compounds moved into one phase or other, into the water phase say, rather than the non-polar organic solvent?

Provided the sample is clean enough, it is now often possible to separate and identify the active components without solvent fractionation. The techniques which have changed this are gas–liquid chromatography (GC), linked GC–electroantennogram/single cell recording, and linked GC–mass spectroscopy (GCMS) (Box 2.1).

A method for identifying likely components, particularly when the mixture of odours is complex or entrainment is used and there is a heavy background from other odours, is to line up the GC traces of (say) an alpha female ant and a sterile worker (Fig. 2.6) and look for differences. This comparative approach has helped with identification of mammal pheromones by exploiting the dependence of male sex pheromone secretion on testosterone. Comparison of the GC traces of urine odours from castrated and

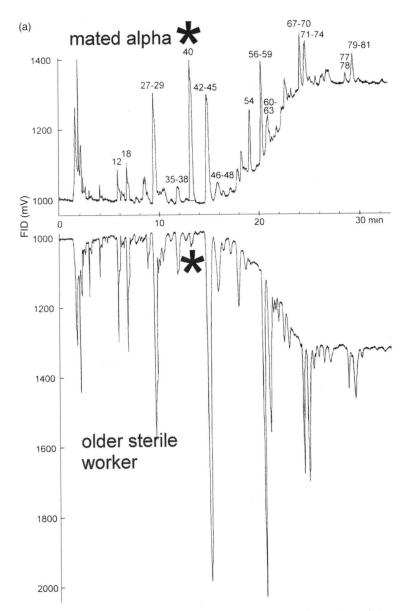

Fig. 2.6. In social insects such as the 'queenless' ant, *Dinoponera quadriceps*, the egg-laying status of individuals is reflected in particular hydrocarbons on their body surface.
(a) The SPME–GC trace above, of the cuticle of a mated alpha female (who dominates the colony), and the trace below, of a sterile worker, show the differences in their cuticular hydrocarbons; note the difference for peak 40 (starred) which is 9-hentriacontene, characteristic of alpha females.

(b)

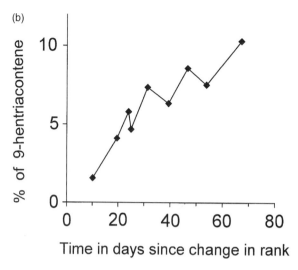

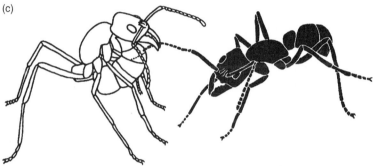

(c)

Fig 2.6. *(cont.)*
(b) SPME allowed changes in the hydrocarbons (data are shown here for 9-hentriacontene) of the same individual to be followed as its dominance position changed because taking SPME samples did not harm the animal. Figures from Peeters *et al.* (1999).
(c) The way the SPME sample was taken is similar to the way the dominant worker (shown here in white) rubs the tip of the antenna of a subordinate (shown in black) against its intersegmental membranes; each ant is 3 cm long. Figure from Monnin *et al.* (1998).

intact male house mice, and castrated males treated with testosterone, reveals the peaks corresponding to pheromones that are lost if the male is castrated (Novotny *et al.* 1999b).

2.6 | Multi-component pheromones and synergy

The early work on moths in the 1960s suggested that each species has a single, different, compound as its pheromone. These identifications were largely based on simple laboratory bioassays and most of the compounds failed to attract moths in field trials; often the purer the compound, the more disappointing the results (Silverstein 1977). The answer was found almost by accident in work on bark beetle pheromones. Silverstein and D. L. Wood were having no success in getting responses to individual chemicals isolated from male *Ips paraconfusus* beetles. After a relaxing lunch they came back to the laboratory, tried the bioassays again and found that the beetles responded. They later realised that this was because in doing the experiments less carefully, by mistake they had created mixtures of compounds which were attractive to the beetles. The male pheromone, which attracts both males and females, consists of three components, none active in the field by itself. Silverstein and Wood had demonstrated **synergy** between the compounds: that is, the combined effects were greater than the sum of the individual activities (Silverstein 1977). Most insect sex pheromones have been found to be multi-component, consisting of a number of different components acting in synergy when combined in a particular ratio.

There are examples of synergy in vertebrate pheromones too. Two compounds isolated from the urine of male mice (*Mus musculus*) provoke aggressive behaviour in conspecific males: dehydro-*exo*-brevicomin and 2-*sec*-butyl-4,5-dihydrothiazole (Novotny *et al.* 1999b). For this effect, both compounds have to be present together and in addition they need to be presented in mouse urine. Simply dissolved in water they have no effect (Chapter 9).

In contrast, complex odours that serve in recognition or discrimination of conspecifics in mammals and social insects are not set combinations of defined compounds (see Chapter 6). Instead, the differences in these complex odours are the message.

2.6.1 Subtractive approach

Pheromones composed of many components acting together synergistically, but inactive singly, present problems for behavioural analysis. This is because sub-fractions of an extract may show no behavioural activity unless other synergists are also present. As the number of major fractions rises, the complexity of bioassaying the potential combinations increases in a power function. The solution may be a subtractive approach: fractions containing active synergists are revealed when removal of the fraction reduces the activity of the whole extract (see Byers 1992).

2.7 | New tools in pheromone research

Techniques exploiting the opportunities offered by mutant studies and the revolution in genomics are likely to become increasingly important. For example, our understanding of *Drosophila* spp. sex pheromones and speciation has been transformed by studies of genetically engineered *Drosophila melanogaster* females (Chapter 3) (Savarit *et al.* 1999). The roles of the female- and male-produced pheromones in *D. melanogaster* courtship behaviour by males have been investigated recently with genetically manipulated individuals, gynandromorphs. Such individuals produce different mixtures of female and male pheromones on their cuticle or express female or male genes in different parts of the brain.

These genetic techniques are likely to be centred on well characterised model systems such as *Drosophila* (Carlson 1996) and the nematode *Caenorhabditis elegans* (Mori 1999) but genetic techniques have also been used, for example, to explore pheromone production or reception in moths (e.g. Haynes 1997; Cossé *et al.* 1995) and honeybees (e.g. Pankiw *et al.* 1995) and are likely to be used increasingly with vertebrates.

The new approaches coming from, for example, molecular biology and brain imaging, are likely to increase the need for collaboration between different branches of biology. Some of the breakthroughs such collaboration could bring are illustrated in Chapter 9. Underlying everything, however, will be a need for a good understanding of the behaviour of the animal concerned and how it evolved.

2.8 | Conclusion

The challenges presented by pheromones have meant that some of the most productive research teams have combined the expertise of chemists and biologists. The collaboration is not always easy as each group of specialists has its own ways of working and, most troubling, each often thinks that the work of the other is easier or should be more straightforward than it often is (biologists assume chemists will always find the answers, however small or contaminated the sample; chemists sometimes do not appreciate how much the behaviour of animals, even laboratory cultured insects, can vary unpredictably from day to day) (Albone 1984).

2.9 | Further reading

Silverstein (1977) provides an entertaining and incisive account of early work on insect pheromones. For an excellent introduction to the chemistry of insect pheromones see the chapters by Stevens in Howse *et al.* (1998). Other chapters cover extraction, identification, and synthesis of insect pheromones. For comprehensive information on the best current techniques in chemical ecology see Millar & Haynes (1998) and Haynes & Millar (1998).

Chapter 3

Sex pheromones:
Finding and choosing mates

3.1 | Introduction

Darwin proposed sexual selection to explain why the males of so many species have conspicuous bright colours or long tail feathers even though such extravagant features or behaviour might reduce survival by natural selection (Darwin 1874). He included chemical signals alongside visual and sound signals, describing the distinctive odours of breeding male crocodiles and many mammals, as well as male moths. He concluded that the development of elaborate odour glands in male mammals is 'intelligible through sexual selection, if the most odoriferous males are the most successful in winning the females, and in leaving offspring to inherit their gradually perfected glands and odours' (Darwin 1874, p. 809). The notorious odours of male goats in the breeding season, for example, might be as spectacular as a peacock's tail, if only we had the nose to appreciate them.

Darwin suggested that sexually selected signals in animals would have many features in common, including elaboration or expression of the signals in only one sex, development only in adults, often only in the breeding season, and use primarily or exclusively in mating (Darwin 1874, p. 807). Chemical signals, and the scent glands that create them, commonly show all these characteristics (in humans too, Chapter 13). After long neglect, studies of chemical signals in sexual selection are now providing some of the most exciting results of all.

Sexual selection is shorthand for selection that arises through competition over mates or for matings. It is a subcategory of natural selection and operates in fundamentally the same way, by hereditary variation among individuals which translates into different numbers of offspring produced by each individual (Alcock 1998). Sexual selection can take many forms (see reviews by Andersson 1994, Andersson & Iwasa 1996). Best studied are **mate choice**, in which members of one sex display to be chosen by the other sex, and **same-sex** 'contests' in which members of one sex fight each other to be the top animal. However, other important mechanisms of sexual selection include aspects of **sexual competition** such as scrambles to be first to arrive at a potential mate, sperm competition, and tactics later in the reproductive cycle, such as the induction of embryo-implantation failure (e.g. the Bruce effect, Chapters 6 and 9). Any character influencing success in these competitions for fertilisations, before or after mating, can be sexually selected. All these mechanisms can be influenced or mediated by chemical communication.

Sexual selection also includes basic factors affecting reproductive success including choosing a mate of the right species, sex, and reproductive stage (Andersson 1994; Sherman *et al.* 1997). Many pheromones traditionally seen as having releaser or species recognition roles may have evolved by sexual selection through female choice.

Given the ubiquity of chemical communication among animals, chemical cues are likely to emerge as one of the key criteria animals use for mate choice.

3.2 | Mate choice and sexual selection

Typically, males compete for matings with females because males in most species invest less in their offspring. As a result of investing less, males tend to have greater potential reproductive rates than females and so compete for the limiting, choosier sex (Clutton-Brock & Vincent 1991; Clutton-Brock & Parker 1992). Thus there are usually more males seeking females than the other way round, leading to the ratio of males to reproductively available females at a given time and place (the operational sex ratio, OSR) being greater than 1:1. Breeding systems are also influenced by the ecology of feeding and resource distribution, among other factors (Reynolds 1996).

While there is much evidence that females choose between different males, that choice favours conspicuous male traits, and that females benefit from these choices, one or more of three evolutionary mechanisms could be involved (Table 3.1) (Andersson 1994; Johnstone 1995; Ryan 1997).

The first mechanism is a **direct benefit** to females, for example by choosing males with better nuptial gifts (Section 3.6.1) or avoiding contagious disease by using odour cues to choose healthy males as mates (Section 3.6.3).

The second mechanism is 'runaway sexual selection' or 'Fisherian' ('sexy sons') selection in which male attractiveness is itself heritable (so choosy females get the indirect benefit of producing attractive sons). The original female preference can be for indicators of genuine quality or an exploitation of sensory bias (Chapter 1). In the later 'runaway' phases, 'sexy sons' are not necessarily better at surviving but have a high reproductive value simply because they are attractive to females in that population. Runaway elaboration

Table 3.1. *A comparison of the key features of three theories for the evolution of extreme sexual ornaments and displays by courting males which may apply in lekking[a] and non-lekking species*

Benefit to the female can be either direct or via offspring reproductive success or survival.

Evolutionary mechanisms (not mutually exclusive)	Females prefer a trait that is:	Primary adaptive value to choosy females	Odour based examples
Direct benefits			
Healthy mate selection	Indicative of male health	Females (and offspring) may avoid contagious diseases/parasites	Female house mice choose urine of uninfected males (Kavaliers & Colwell 1995
Resources	A signal of greater resources	Resources for egg laying or protection of young	Female tiger moths (*Utetheisa ornatrix*) choose males with most pheromone and thus greatest givers of protective alkaloids (Iyengar & Eisner 1999b)
Runaway sexual selection ('sexy sons', Fisherian: indirect benefits)	Sexually attractive	Sons inherit the trait that made their fathers attractive to females; daughters inherit the majority mate preference	Female tiger moths (*Utetheisa ornatrix*) choose males with most alkaloids, which are larger, and whose offspring are large and attractive (Iyengar & Eisner 1999a)
			Female mice prefer odours of dominant mice, and the sons of dominant mice (could be good genes?) (Drickamer 1992)
Indicator ('good genes': indirect benefits)	Indictive of male survival chances	Sons and daughters may inherit the viability advantages of their fathers	Female tiger moths (*Utetheisa ornatrix*) choose the pheromones of larger males and the daughters inherit this larger size and thus lay more eggs (Iyengar & Eisner 1999a)

[a]See Section 3.7.
After Alcock (1998) and many authors.

of odour might be selection for stronger longer-lasting odours, by analogy with visual and acoustic signals, which females prefer stronger and more rapidly repeated (Ryan & Keddy-Hector 1992).

The third mechanism, the **indicator** mechanism, also offers an indirect benefit. If males display a trait, for example strength of odour (Section 3.6), in proportion to their health and viability, and the viability is heritable, then females choosing males with stronger odours will gain genes for high viability to pass to their offspring (Andersson 1994, p. 53). The idea is the basis of many models (including 'handicap', 'indicator' and 'good genes'). However, the mechanism could work even if the benefits are direct rather than inherited (Grafen 1990b).

In practice, a number of sexual selection mechanisms can act together in the same species. For example, both male–male contest and female choice occur in the African cockroach *Nauphoeta cinerea* and are mediated by the same sex pheromones, produced by males (Moore & Moore 1999). Males form linear dominance hierarchies characterised by odour badges (Section 3.5.2) and females use these olfactory cues to choose males, leading them to mate with dominant males. In these experiments, female choice (measured as time to start mating) can be separated from male–male interactions. Male dominants typically produce dominant sons. Female mating preference is also inherited: daughters of dominant males were more likely to prefer dominant males as mates.

A second selective mechanism behind female choice in *N. cinerea* may be 'good genes'. Matings with attractive males chosen by females give rise to offspring that develop faster. The female benefits, first, directly from her own increased reproductive success because she can mate and produce another brood sooner and, second, indirectly via her offspring's reproductive success, because they also reach maturity sooner, which gives reproductive advantages in a growing population.

While models of sexual selection tend to assume a central role of 'genetically determined' mate preference, there is growing evidence that in some species mate preferences are partly learned through imprinting in early experience (Owens *et al.* 1999) (Chapters 6 and 9). This is particularly relevant to olfactory imprinting, for example to odours associated with the major histocompatibility complex (MHC) (Section 3.6.4.1), colony odours, and for species-specific odours. Imprinting could lead to choosing partners like the parents (as with species-specific odours) and if there were a preference for sexual ornaments more elaborate than the parents, there could be runaway sexual selection.

3.3 | Which sex should call?

Who should signal to attract mates? In most animals the display is not made by the limiting sex so there are few female frogs croaking in chorus or brightly coloured female birds. The argument that the limiting sex should put least effort or risk into display and attraction of mates makes it seem paradoxical that the main example of long-distance calling with pheromones should be female moths which attract males. This is in sharp contrast to other insects: these use visual or acoustic signals, almost always produced by the male. Why the exception with moths? The reason is likely to be the low energy cost of advertising with pheromones compared with sound (Chapter 1) and possibly the relatively small risk of predation to the caller (compared with the larger number of predators responding to sound or visual signals). The major costs are borne by the males in expensive flight and the risk of predation on the wing (Greenfield 1981).

Only a minority of insects produce male pheromones that are long distance attractants and the majority of these are beetles. With the exception of lekking species (Section 3.7), male insects producing long-range pheromone signals tend to be associated with patchy resources needed by the female (Landolt 1997). Males find the resource and then call. Females can get both mating and resources at the same place. For example, males of the scorpion fly *Panorpa* pheromone-call when they have an insect food item to offer as a gift (Section 3.6.2) (Thornhill 1979) and male papaya fruit flies call from fruits which are oviposition sites (Landolt *et al.* 1992). Male plant-feeding insect pheromones are usually synergised by plant volatiles (Landolt 1997). When male insects pheromone-call they often attract males as well as females; the pheromones concerned are called aggregation pheromones even if their original function was sex attraction (Chapter 4).

The divide into 'females signal, males respond' is not true for all moths. For example, in the cabbage looper moth (*Trichoplusia ni*) both males and females produce pheromones (Landolt *et al.* 1996). The male pheromone, which is released from the hair pencils around dusk, attracts males and females. The female pheromone, released in the latter half of the night, attracts males.

There are perhaps fewer examples of pheromones used to attract mates from a distance among vertebrates as a whole (Drickamer 1999). There are, however, many examples among fish: male lampreys release sex pheromone to attract females, and both visual and pheromone signals are used by territory-holding male gobies to attract females full of eggs (Stacey & Sorensen 1999). A male pheromone has now been described for frogs (Chapter 1) (Wabnitz *et al.* 1999). In sheep, ewes are reported running towards the ram in response to his long-range attractant odours (Cohen-Tannoudji *et al.* 1994). There are female signals too, with female hamsters leaving trails of vaginal secretions (Chapter 9) and female dogs produce long-range attractant pheromones, as dog owners will know.

3.4 | External fertilisation and chemical duets

Animals such as marine worms and sea urchins, which rely on external fertilisation by releasing eggs or sperm or both into the water, must precisely

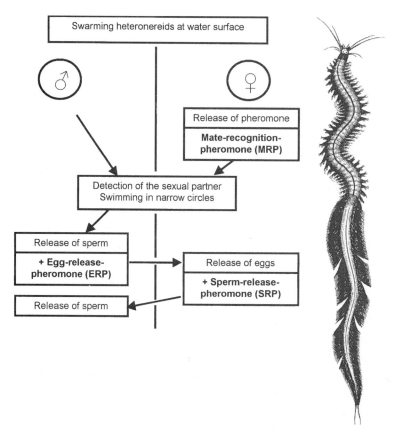

Fig. 3.1. Animals that release their eggs and sperm into the sea must ensure release is coordinated. (Left) The marine polychaete worm *Nereis succinea* spawns in a 'nuptial dance' timed by the phases of the moon, initiated by sundown, and choreographed by a chemical duet. The mature heteronereids leave their burrows and swarm near the surface of the water. The female swims slowly in circles, releasing a volatile sex attractant pheromone (the mate-recognition-pheromone in this figure). Her sex pheromone attracts many males who start to swim fast around her, releasing small quantities of sperm and coelomic (body cavity) fluids that contain the egg release pheromone (inosine, L-glutamic acid and L-glutamine). On sensing the egg release pheromone, the female swims fast in narrow circles (the nuptial dance), and spawns, discharging a massive cloud of eggs and coelomic fluid (which contains the sperm-release-pheromone (SRP) L-cysteine-glutathione disulfide). Males respond immediately to sperm-release-pheromone by swimming through the egg cloud and releasing large clouds of sperm (Zeeck *et al.* 1998a, b). Diagram from Müller *et al.* (1999).

(Right) Heteronereis reproductive stage of a nereid worm. The posterior part of the body is full of gametes. Figure from Harter (1979).

synchronise gamete release. Sperm do not live long and are quickly diluted by the huge volume of the sea, so the sexes need to be close together at gamete release (which is one reason many marine organisms form aggregations, see Chapter 4) (Levitan & Petersen 1995). Responses to cues such as tides, lunar cycles, temperature, and day length enable all members of a species to reach sexual maturity together for an annual synchronised spawning event. The exact moment of spawning is coordinated by a chemical duet of pheromones released into the water by males and females (Fig. 3.1) (Zeeck *et al.* 1996; Hardege *et al.* 1998). Synchronised spawning maximises the chances of gamete contact and fertilisation. Getting the timing right is especially important for animals such as marine nereid worms which have only one chance to reproduce, as they die after the event. It is likely that most externally fertilising species, including many fish (Chapter 1), have coordination by pheromones.

3.5 | Sexual selection, scramble and contest

3.5.1 Scramble competition

Sexual selection can act on the ability to scramble, to get to a mate ahead of rivals. In the great majority of moths, males make long-distance flights to find females emitting minute quantities of pheromones, in pico- or nanograms per hour. Williams (1992, p. 111) doubted that female pheromones exist. However, pheromone communication by moths is one of best established of all animal communication systems (Phelan 1997). Both female pheromone signaller and male receiver are as highly evolved and specially tuned as the sound signal systems of cricket mate location, with specific receptors, neural circuitry, morphology and hormonal control systems for production and release in the signaller (see Section 3.3). A low pheromone release rate selects for strong, effective searchers (Lloyd 1979; Greenfield 1981). Females mating with the first arrival(s) will produce winning sons in the next generation if they have similar features. The intense selection to be first has led to selection for great sensitivity to detect pheromones, reflected in the huge feathery antennae of males, and in neurosensory adaptations enabling faster flight or greater ability to track a complex pheromone plume (Chapters 9 and 10). Once the males have arrived, there can be female choice between the chemical displays of males (Section 3.11.1.2).

As soon as a female red-sided garter snake (*Thamnophis sirtalis parietalis*) emerges from the overwintering den in the spring, she is surrounded by a 'mating ball' of up to 100 competing males (Mason 1993). Males tongue-flick every snake they contact (see Fig. 2.2). They ignore other males, but respond to a female by rapid tongue-flicking and chin-rubbing up and down her back. The males are responding, via their vomeronasal organ (Chapter 9), to a female contact sex pheromone consisting of specific skin lipids - (Chapter 2). Males are ignored because their skin lipids include a male recognition pheromone, squalene. As soon as the winning male starts to copulate with the female, he releases a pheromone which causes the other males to disperse rapidly, to wait for the next female (see also Section 3.9.2).

The pheromone-controlled mating system in the male goldfish *Carassius auratus* (Chapter 1) provides another example of scramble competition. Males are attracted by the pheromone released by a female ready to spawn and then compete to be the one closest to her, trying to push other males away (Sorensen & Stacey 1999).

3.5.2 Contests

Sexual selection by contest, involving males fighting each other (for access to females or for ownership of resources needed by females), may lead to marked sexual dimorphism in traits which improve fighting ability, such as large antlers in male red deer. In social mammals, the urine of the dominant individuals can inhibit reproduction by subordinates of the same sex (Chapter 6).

Dominant individuals in hierarchies may display 'badges' of high status, for example a dark chest patch in birds, but debate about the honesty of such signals continues. The pheromones associated with dominant males and their role as 'badges' of status have been neatly investigated in an African cockroach, *Nauphoeta cinerea* (Moore *et al.* 1997). Dominance appears to be based entirely on odour cues. Females prefer to mate with the dominant male, identified by his odour, and there is strong competition between males to be top of the dominance hierarchy. Moore *et al.* investigated the honesty of the signal by adding different components of the male pheromone (used both for male–male interactions and female choice) to a filter paper disc on a test male's pronotum and observing the interaction with a control male marked with solvent alone (Fig. 3.2, Left). The male pheromone has three main components: 3-hydroxy-2-butanone, 2-methyl-thiazolidine, 4-ethyl-2-methoxyphenol (Sreng 1990; Sirugue *et al.* 1992). Adding 3-hydroxy-2-butanone caused the male to be treated as, and also act as, a subordinate. Adding either, or particularly both, of the other components caused him to be treated as, and act as, a dominant. If all three were added together, the effects cancelled each other out, leading to the conclusion that the compounds are acting independently (Fig. 3.2, Right). The dual action of the pheromone, in male–male interactions and female choice, allows the signals to be honest (Dawkins & Krebs 1978; Johnstone & Norris 1993). Males might be selected to boost the compounds characteristic of dominance in male–male interactions (as females choose dominant males) and reduce 3-hydroxy-2-butanone, characteristic of subordinates. However, it is the combination of all three compounds which is attractive to females so the male cannot eliminate the 3-hydroxy-2-butanone from his secretion (Moore & Moore 1999).

Males of the lobster *Homarus americanus* fight to establish dominance and thus ownership of preferred large holes (Atema 1995). Females only choose dominant males with such holes (Section 3.10.1). Aggressive males release pulses of urine during a fight, losers stop urine release as soon as they have lost. The urine is blown towards opponents in the strong exhalent

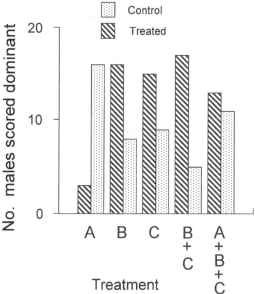

Fig. 3.2. Honest badges confer odour status in the cockroach *Nauphoeta cinerea*. (Left) Each of a pair of males wore a filter paper 'hat' dosed with either the test compound(s) (A, as here, or B, C) or the acetone solvent alone as control. Photograph by A. J. Moore.

(Right) Males with increased 3-hydroxybutanone (A) were more likely to be treated as, and act as, subordinates. Males treated with either 2-methylthiazolidine (B), 4-ethyl-2-methoxyphenol (C), or both (B + C), were treated and acted as dominant. Adding all three components (A + B + C) had no effect on status. Figure from Moore *et al.* (1997).

gill current, which jets 1–2 m in front of the lobster. During encounters in the week following an aggressive encounter, opponents seem to remember each other and almost no fights take place as the former loser tends to avoid the winner, who is recognised by individual odour cues (Karavanich & Atema 1998a,b).

Male red-backed salamanders (*Plethodon cinereus*) compete with each other for the highest quality feeding territories and these are usually occupied by larger males (Mathis 1990b). The territories are centred on stones or logs, which offer shelter and a reliable food source of small invertebrates during dry periods. Males mark their territories with faeces (together with pheromone secretions) (Chapter 5). In the breeding season, females choose faeces that indicate a high quality diet – the 'sexy faeces' hypothesis (Jaeger *et al.* 1995). The higher quality diet consists mostly of termites, which are easier to digest than ants.

While these examples show benefits for females choosing dominant males, this may not be the case in all species (Qvarnström & Forsgren 1998).

3.6 | Sexual selection, mate quality and courtship

In many species, the most important role for chemical communication occurs in courtship, once the male and female are close to each other. Short-range pheromones may be important first in species and sex recognition. Females then choose a male using odour as cues to his intrinsic quality which reflect factors, in mammals, such as his social status, the quality of nutrients he has consumed, his reproductive state, his immunological genotype or other genetic compatibility, and his health. Signs related to quality of resource or male quality are likely to reflect overall condition, rather than be related to one disease type or nutritional state.

In vertebrates, an important internal measure integrating many of these effects, including dominance status, is the level of androgen hormones such as testosterone in the blood. Testosterone stimulates many of the glands important in producing pheromones or other signal odours such as the leg (femoral) gland in the male lizard *Iguana iguana*, the temporal glands in elephants (below) and many rodents. Females of the bank vole (*Clethrionomys glareolus*) prefer the odours of a dominant male to those of a subordinate (Kruczek 1997). In rodents, the androgens affect the glands secreting into urine, so urine scent marks reflect hormonal and dominance states (see Chapters 5 and 9). Testosterone implants in castrated male meadow voles (*Microtus pennsylvanicus*) show that the male scent glands respond in a graded way to testosterone levels (Ferkin *et al.* 1994). In *M. pennsylvanicus*, males on a higher protein diet produced more attractive smells in their urine marks, perhaps via effects on their physiological condition (Ferkin *et al.* 1997).

African (*Loxodonta africana*) and Asian (*Elephas maximus*) elephant males have an annual period of heightened aggressiveness and highly elevated testosterone concentrations known as 'musth' (Rasmussen & Schulte 1998; Schulte & Rasmussen 1999). During musth, mature males secrete copiously from the temporal gland on the head and dribble strong smelling urine (Fig. 3.3). Young males produce a different set of temporal gland secretions (Rasmussen *et al.* 2002). Only males in good physiological condition enter musth and females prefer to mate with them rather than with non-musth males. Oestrous females can detect the physiological state of males from chemical cues in male urine.

Females of the tobacco moth (*Ephestia elutella*) are more likely to mate with larger males, which produce more than twice as much wing-gland pheromone as smaller ones (Phelan & Baker 1986). Matings with these larger males produced a greater number of offspring that survived to the pupal stage and these pupae were larger. Both Fisherian 'sexy sons' and 'good genes' effects are suggested by these data.

3.6.1 Paternal investment and female choice

There is a close relationship between paternal investment and pheromone signal in some moth, butterfly and beetle species in which males offer protective anti-predator compounds to the female and eggs. All have convergent courtship behaviours that allow females to assess the value of potential gift from each male. The value of his gift is indicated by an honest signal with a pheromone derived from the defensive precursor.

The North American tiger moth (*Utetheisa ornatrix*) uses plant alkaloids for defence and the male produces an honest signal based on his store of alkaloids; this has a decisive role in sexual selection and a selective advantage for survivorship of its offspring (Eisner & Meinwald 1995). When the male mates, he transfers plant pyrrolizidine alkaloids (PA) to the female which supplement her protection and provide alkaloids to protect the eggs from predators such as coccinellid beetles (Gonzalez *et al.* 1999). Both sexes, as larvae, sequester PAs as they feed on their natural diet of *Crotalaria* plants. The poisonous PAs protect the larvae and are passed on from the caterpillar stage through the pupal moult to protect the adults from predators such as spiders.

The mating system is conventional for a moth, the female calling at dusk with a conventional female sex pheromone which attracts males. When the male arrives he hovers around her, everting his scent brushes containing the male pheromone, hydroxydanaidal (HD) (Fig. 3.4) derived from PAs in the diet. The female can use the amount of pheromone on the coremata to assess the potential gift from a male since his HD pheromone content correlates with his PA levels and the quantity of alkaloid he would

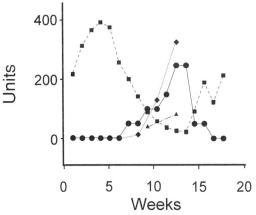

Fig.3.3. Asian elephant (*Elephas maximus*) males and females communicate with chemical signals. (Left) When the male is about to enter mating readiness, his annual musth, this is signalled to other males by volatiles, including 2-butanone, frontalin and ketones, released in a stream of liquid from his temporal gland (arrowed) (Rasmussen 1999; Schulte & Rasmussen 1999). Females can detect the musth signal. Photograph by B. Rasmussen.

(Right) Males respond to female urine with flehmen (responses/h) (●) when females are entering oestrus. Males are responding to the level of female urinary pheromone (Z)-7,12-dodecyl acetate (concentration g/ml $\times$ 10^{-1}) (♦) which rises as oestrus approaches. At the same time female serum progesterone (pg/ml) (■) falls and luteinising hormone (ng/ml $\times$ 10^{-1}) (▲) rises. Figure from Rasmussen (1998).

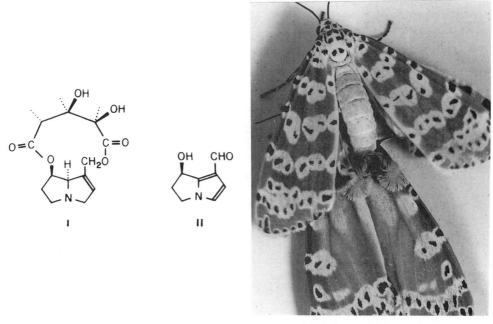

Fig. 3.4. (Left) Male tiger moths (*Utetheisa ornatrix*) sequester the monocrotaline (I) pyrrolizidine alkaloids from their host plants, convert it to hydroxydanaidal (HD) (II), and (Right) present it to females from their everted coremata during courtship. Females choose males with more HD pheromone. Figures from Eisner & Meinwald (1995).

transfer (Fig. 3.5) (Dussourd *et al.* 1991). In nature, male PA content is proportional to male size – larger males have more PAs (Lamunyon & Eisner 1993). So by selecting males with high PAs, the female also selects big males, which transfer larger spermatophores, providing more nutrients for her eggs as well as more PA (Iyengar & Eisner 1999b). Body mass is heritable in both sexes, so by choosing larger males, females obtain genetic benefits for their offspring as well as direct benefits as larger sons and daughters are, respectively, more successful in courtship and more fecund (Iyengar & Eisner 1999a).

3.6.2 Fluctuating asymmetry

Females may use body symmetry as a sensitive indicator of male quality (see review in Møller and Thornhill 1998). More symmetrical males are said to have low fluctuating asymmetry (FA) which reflects the ability of the individual to develop perfect bilateral symmetry despite environmental perturbations such as food availability. However, the role of FA in sexual selection or indicator of quality is still debated (see Andersson & Iwasa 1996, and Clarke 1998). Females of the scorpion fly *Panorpa japonica* chose the pheromone of males which have more symmetrical wings (Thornhill 1992). In a choice test, with males hidden behind screens, females choose the more symmetrical male, even if the difference in symmetry was very small.

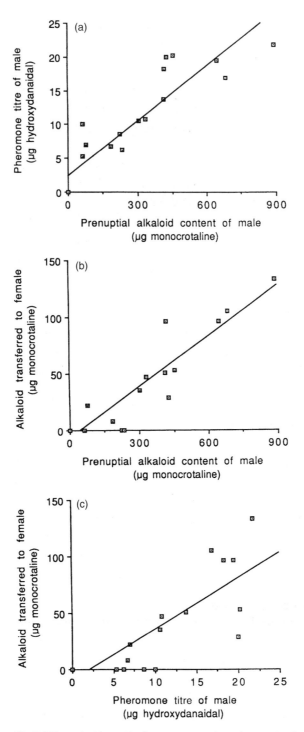

Fig. 3.5 The male tiger moth pheromone provides an honest signal of the value of his alkaloid store that he will transfer to the female during mating. Graphs (a), (b) and (c) show the various relationships between the prenuptial adult male (*Utetheisa ornatrix*) alkaloid titre (monocrotaline), the adult male pheromone hydroxydanaidal (HD) titre, and quantity of alkaloid transferred to the female at mating. Figures from Dussourd *et al.* (1991).

Females showed no discrimination between hidden males of different symmetry if each male's gland was glued up to prevent pheromone release.

3.6.3 Mate choice and parasites

The chemical signals produced by a male might advertise health and genetic quality to mates, including his resistance to parasites and pathogens (see review in Penn & Potts 1998a). Numerous studies have shown that a male's secondary sexual displays, such as the brightness of his colours, honestly reveal his parasite load. However, until recently, chemical signals have been largely ignored.

Female house mice are attracted to male scent marks (Chapter 6). Females show a preference for the odours of uninfected mice (Kavaliers & Colwell 1995) and aversion to urine from mice experimentally infected with either a parasitic protozoan or nematode. An analgesic (pain-killing) response to the odour of parasitised males might be the cause of female aversion and thus adaptive, for example in avoiding infected males, but it is probably too early to say if this is an evolved response (Zuk et al. 1997).

More recent experiments using a virus infection rather than a parasite showed that the attractiveness of male urine was eliminated, not made aversive, during infection and that females chose to settle in nest boxes scent marked by uninfected males (Penn et al. 1998). It seems that females can use olfactory cues to choose healthy mates. The most likely cues are the direct or indirect change in androgen levels in illness and olfactory cues via androgen-dependent glands, including pheromone glands, together with changes in expression of major histocompatibility complex (MHC) proteins (and the odours linked to them, see below). It is possible that androgens are involved in controlling trade-offs of resource allocation between reproductive functions and immunological defences. Infected males may have to reduce androgen-related signals to devote more resources to fighting infection (Wedekind & Folstad 1994; Smith, F. V. et al. 1996). (See Westneat & Birkhead 1998, for alternative hypotheses for immunocompetence effects.) Manipulations of host androgen levels may be made by the parasites themselves (see Penn et al. 1998).

3.6.4 Mate choice for optimal outbreeding

Some mate choice is for a mate that is genetically different or complimentary to the chooser but not necessarily 'better' than other potential mates. Females and males in many species use odour cues to select mates for optimal outbreeding. For example, sweatbee (Lasioglossum zephyrum) males use smell to choose mates that are not too closely related to themselves (Smith & Breed 1995), and female crickets (Gryllus bimaculatus) reject males that have similar cuticular pheromones to their own (Simmons 1989). More remarkable are the powerful roles that odours play in the selection of genetically different mates in mice, and possibly in humans (Chapter 13). Learning of odours plays a central part of these recognition systems (Chapter 6).

3.6.4.1 The major histocompatibility complex and mate choice in mammals

Some of the odours important in mate choice are associated with the enormously variable 'major histocompatibility complex' (MHC) genes underlying the mammalian immune system (Brown & Eklund 1994; Edwards & Hedrick 1998; Eggert *et al.* 1999; Penn 2002). Mice choose mates that are different from themselves at the MHC loci. This phenomenon was discovered by accident when technicians observed that two strains of laboratory mice, specially inbred so they were identical apart from one locus in the MHC, preferred to mate with mice of the other strain (Beauchamp *et al.* 1985). Odour is the key: mice can be trained to distinguish the body odours and urine of mouse strains differing at only one gene in the MHC but cannot tell apart genetically identical inbred mice. Incidentally, rats, dogs and even people can also be trained to distinguish the smells of these mice, which differ only at the MHC. Importantly, the distinction is not limited to mice tested in laboratory mazes: it can be made by untrained, wild-derived mice and influences the mate choice of mice in semi-natural enclosures (Potts *et al.* 1991; Penn & Potts 1998d). The odours are used as a marker for the degree of kinship between two individuals.

The MHC codes for cell surface proteins that are involved in self and nonself recognition at the centre of the vertebrate immune system for defence against disease organisms. In humans the MHC (known as the HLA – Human Leucocyte Antigen) region consists of at least 82 genes on the short arm of chromosome 6. The HLA genes are extremely polymorphic with many alleles (50–70 for some genes), giving billions of individual combinations and making each individual almost unique unless one of identical twins. The alleles for a given gene occur in similar frequencies, unlike most other genes in which a few alleles dominate (Brown & Eklund 1994; Edwards & Hedrick 1998).

The extreme genetic diversity coupled with the ability to discriminate MHC genotypes by smell makes it a potentially useful system for avoiding mating with kin (Box 3.1) (Fig. 3.6). For example, by avoiding mating with prospective mates who carry one or more alleles identical to those of its parents, a mouse will avoid all matings with full and half sibs and 50% of matings with cousins (Potts & Wakeland 1993).

Box 3.1 | What odours are associated with the MHC?

Behavioural choices by male and female mice associated with the major histocompatibility complex (MHC) are well established, but what odours are involved? The MHC genes are the most significant contributor to the odours, but they are not the only factors, which also include other genes, diet, hormonal state and parasites (see figure below) (Zavazava & Eggert 1997).

How the MHC proteins influence odours is still not clear (Penn & Potts 1998b; Penn 2002). The proteins themselves are too large to be volatile, but they may affect the bacterial flora and thus bacterial odours. The MHC proteins may also act as specific carriers for smaller

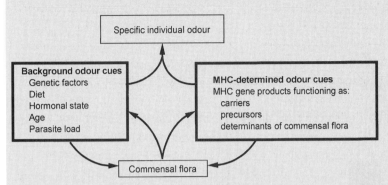

The specific individual odour is the sum of the interplay of MHC-determined products, odours produced by the genetic background of the individual and its commensal microbial flora. Figure from Zavazava & Eggert (1997).

molecules perhaps derived from bacteria or the host animal (Chapters 1 and 13). In mice, the differences in MHC type are reflected in the relative concentrations of volatile carboxylic acids in the urine (Singer *et al.* 1997). Eggert *et al.* (1996) found that some different odour compounds were associated with different MHC types.

(1) The MHC molecule hypothesis

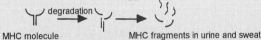

(2) The peptide hypothesis

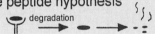

(3) The microflora hypothesis

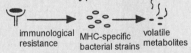

(4) The carrier hypothesis

(5) The peptide–microbe hypothesis

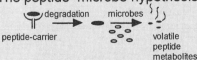

As the MHC proteins themselves are too big to be volatile, how do MHC genes influence individual odour? Among the possible mechanisms are: (1) fragments of MHC molecules in urine and sweat provide the odourants (the MHC molecule hypothesis); (2) MHC molecules may alter the pool of peptides in urine whose metabolites provide the odorants (the peptide hypothesis); (3) MHC genes may alter odour by shaping allele-specific populations of commensal microbes (the microflora hypothesis); (4) MHC molecules could carry volatile aromatics, including volatiles produced by bacteria (the carrier hypothesis); and (5) MHC molecules may alter odour by changing the peptides that are available to commensal microbes (the peptide-microflora hypothesis). Figure from Penn & Potts (1998b).

(a) House Mice

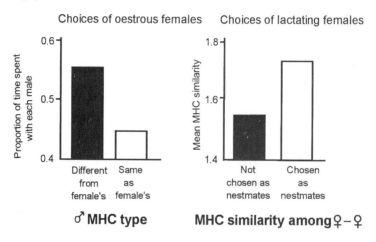

(b) Humans

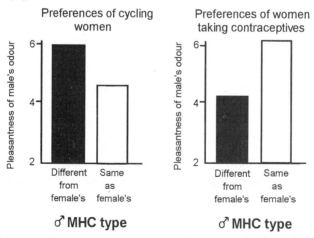

Fig. 3.6. MHC and mate choice in mice and humans. (a) Sexually receptive female mice avoid males with similar MHC genotypes (thereby promoting outbreeding), whereas pregnant and lactating females are attracted to females with similar MHC genotypes, with whom they form nepotistic communal nesting associations (data from Egid & Brown 1989, and Manning et al. 1992).
(b) Women find odours of males with a different MHC genotype more pleasant than odours of males with the same genotype as themselves. However, women taking oral contraceptives prefer males with a similar genotype to themselves (data from Wedekind et al. 1995). Figures redrawn from Sherman et al. (1997).

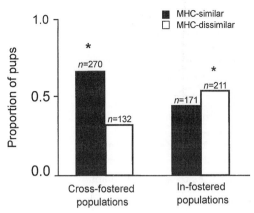

Fig. 3.7. House mice (*Mus musculus domesticus*) use odour cues to avoid mating with individuals that are genetically similar at the major histocompatibility complex (MHC) (Fig. 3.6). When young they learn the odours of their close kin (familial imprinting) and when adult they compare this with the MHC identity of potential mates. Wild-derived female mouse pups cross-fostered into MHC-dissimilar families reversed their mating preferences compared with in-fostered controls. Genetic typing of their progeny indicated that females avoided mating with males carrying the MHC genes of their foster family (left), thus supporting the familial imprinting hypothesis. Stars indicate significant differences. Figure redrawn from Penn & Potts (1998c).

The mouse learns the MHC-associated odours of its parents in the same way that many types of animals learn the odours of nest mates for kin recognition (Box 3.1) (Fig. 3.7). The MHC preferences of mice can be experimentally reversed by cross-fostering when newly born (Penn & Potts 1998c).

The high diversity of MHC alleles may be maintained by at least three selection pressures (Fig. 3.8) (Penn & Potts 1999; Penn 2002). The first selective advantage of disassortative mating for MHC genotype may be the increased diversity of MHC alleles in offspring. This gives greater protection in the immunological arms race against parasites and pathogens. Second, the same behavioural choice would also reduce the chance of inbreeding depression, with effects across all genes. Third, there are higher implantation rates in mammals if the foetus and mother differ in MHC (possibly shown in human populations, Chapter 13). For some species, such as the stickleback (*Gasterosteus aculeatus*), the potential mate's MHC diversity may be more important to females than the MHC difference from themselves (Reusch *et al.* 2001).

The same MHC-related odours used to avoid mating with kin can also be used to recognise kin for cooperation. While mice choose mates with dissimilar MHC odours, lactating females choose to nest with sisters or closely related females with similar MHC (Manning *et al.* 1992). The main reason for communal nesting seems to be cooperative defence against conspecific infanticide, but females nesting together suckle all the young indiscriminately and kin selection would allow this apparent altruism (Chapter 6). The change in MHC-odour linked behaviour is presumably due to changes in

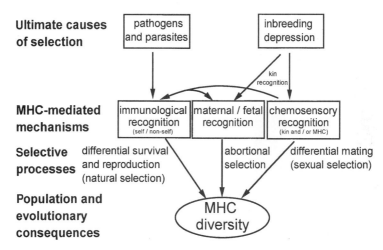

Fig. 3.8. The nature of selection on MHC genes. Figure from Apanius *et al.* (1997).

hormonal state in pregnancy. Human MHC preferences may also change with hormonal state (Chapter 13).

MHC-associated odours are involved in other behavioural and physiological effects, including pregnancy block in female mice in response to the odour of strange males (the Bruce effect, Chapters 6 and 9).

MHC multigene systems are found throughout the vertebrates and it is highly likely that MHC-based mate choice is important in many of these. A role for the MHC in kin recognition has been demonstrated in salmonid fish (Olsen 1999), and study of birds and reptiles is beginning (Wittzell *et al.* 1999).

As researchers extend studies to wild populations of other species, MHC-based mate choice is not always found. One reason may be the species' mating system. For example, Soay sheep do not show disassortative mating by MHC type, perhaps because their harem mating system, based on male–male fighting, gives little opportunity for female mate choice (Paterson & Pemberton 1997). Some human populations appear to demonstrate MHC-based mate choice, but others do not (Chapter 13).

3.6.4.2 Mate choice to avoid lethal alleles

Female mice also use odour cues to avoid mating with males carrying the same potentially lethal alleles as they do (Jemiolo *et al.* 1991; Lenington *et al.* 1994). The alleles are of a set of genes called the 't-complex'; this is highly polymorphic with some 15 recessive t-alleles which kill the developing embryo if it is homozygous (t/t) for the same t-allele (embryos with a pair of different t-alleles at the t-locus may survive but all such males are sterile). Many wild house mice are heterozygous (+/t). Female mate preferences match the likely success of matings. Heterozygous (+/t) females show a strong preference for wild type (+/+) males. Offered a choice of heterozygous males, heterozygous females prefer males carrying a t-allele different from their own. In a more extensive choice of +/t males, they choose a male with the dissimilar t-allele least deleterious in combination with their own (Lenington *et al.* 1994). The t-allele preference, unlike mate choice in mice involving the MHC-system, is not affected by rearing experience. It may be either genetically determined or based on self-inspection ('avoiding mates that smell like you') (Box 6.1).

3.6.5 The coolidge effect

The Coolidge effect describes the way that an animal which appears to have lost interest in sex with an animal it has already mated with becomes sexually active again when offered a new mate (see Agosta 1992, for a nice account of the story behind the term). Males that show this behaviour will potentially leave more offspring as it will prompt them to mate with more partners. Females which mate with more than one male can also gain a variety of benefits including the material benefits from spermatophores and ensuring that at least one male is fertile (Newcomer *et al.* 1999). For example, female snakes (adders, *Vipera berus*) mate up to eight times per season and avoid mating with the same male twice (Madsen *et al.* 1992). The proximate cue in many kinds of animals, including adders, is likely to be recognition based on odour cues. The effect has been demonstrated frequently in male rodents (e.g. Rodriguez-Manzo 1999). Female crickets, (*Gryllus bimaculatus*) preferred to mate with a novel partner over a partner with which they had mated previously (Bateman 1998), probably recognised by cuticular pheromones.

3.7 | Leks

Leks are arenas where displaying males apparently offer only sperm to attending females so these places offer an opportunity to test the fitness benefits of female choice, through direct benefits, 'sexy sons', or 'good genes' mechanisms, without the complication of paternal investment (Högland & Alatalo 1995). Mate choice is particularly strong on leks and typically a few males make most of the matings.

Females may come close to a number of potential mates before choosing one to mate with (Högland & Alatalo 1995). For example, female carpenter bees (*Xylocopa fimbriata*) fly just downwind of several territories (males deposit a scent-marking pheromone attractive to the female), briefly entering each territory, before choosing a mate (Vinson & Frankie 1990).

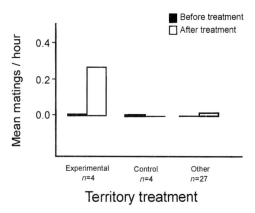

Fig. 3.9. Uganda kob antelope (*Kobus kob*) females use smell to identify the most desirable male territories on the lek. These are the territories most often visited by other females. The urine deposited by the previous females impregnates the soil and so moving the soil from successful territories to previously unpopular ones (experimental), greatly increases visits by females and mating rates in contrast to controls (with the earth only disturbed). 'Other' indicates unmanipulated previously unsuccessful territories. Figure redrawn from Deutsch & Nefdt (1992).

In two African antelope species, the Uganda kob (*Kobus kob*) and Kafue lechwe (*Kobus leche*), the territories preferred by oestrous females were the ones most often visited previously by other females (Deutsch & Nefdt 1992). Successful territories remained popular for months even though the male owners changed almost daily. Females sniff the ground to identify the territories most often urine marked by earlier females while soliciting the owner (Fig. 3.9). Males compete for these heavily marked territories. Females choosing these territories would find the most vigorous and dominant males as mating partners.

Among lekking insects, pheromones are commonly used by males to attract females. Shelly & Whittier (1997) found lekking species in some 52 genera in a broad range of insect taxa including Diptera, Hymenoptera, Lepidoptera, Homoptera, Coleoptera and Orthoptera. Within these taxa, some groups were characterised by male pheromones: notably tephritid fruit flies, bumble bees, anthophorid bees, ants, sphecid wasps and vespid wasps.

Males of the tephritid Mediterranean fruit fly (or medfly, *Ceratitis capitata*) each defend a single leaf in leks. Once the male has a territory, he releases pheromone as a long-range attractant to females (predators are also attracted, Chapter 11). In this species male–male competition is not important and territories are only weakly defended (Whittier *et al.* 1994). In other species, lekking spots may be fiercely contested (see Shelly & Whittier 1997).

Insects have been used to test the theoretical benefits of female choice on leks as it is possible to create leks artificially in some species. In the South American sandfly *Lutzomyia longipalpis*, male pheromones attract females to leks on or near the host, a bird or mammal (Kelly & Dye 1997). In a laboratory study of lekking in this species, Jones *et al.* (1998) demonstrated that the most important benefit to females of choosing attractive males was the Fisherian one of 'sexy sons': rearing the next generation showed that the sons of attractive fathers tended to be attractive when they in turn formed leks. Direct benefits to the female or good gene benefits to the larva, reared in sib groups in the laboratory, were not shown.

In the medfly (*Ceratitis capitata*) despite non-random mating, no benefits were associated with female choice (either in increased fecundity or as 'sexy sons') (Whittier & Kaneshiro 1995).

3.8 | Conflict between the sexes revealed in signalling

The conflict between the sexes can be significant at the point of signalling for mates when the attraction of additional mates is in the interests of only one party. The burying beetle *Nicrophorus* rears its brood on a dead mouse or other small carcass (Eggert & Müller 1997). Such carcasses represent a rich but rare resource. Males use pheromones to attract females, with a 'sterzeln' calling behaviour with the abdomen in the air. If a male has no carcass, attracted females mate immediately and then leave. This means that most females arriving at a carcass are already fertilised; her behaviour reflects the ability to rear brood alone. If females are attracted to a pheromone-releasing male with a carcass, they do not mate immediately, but delay until taken to the carcass. Why is the female coy? One suggestion is that the delay is to keep him occupied so that he will not call for further females who would compete with her for the carcass. The conflict continues even after the carcass is buried and the first female lays her eggs. If the carcass is big enough to support more offspring than a single female can lay, the male would still benefit by attracting another female, but competition from the second female's brood would reduce the survival or growth rate of the first female's larvae. If the male emits pheromone in the presence of the female she tries to interfere by pushing him off his perch or biting the tip of his abdomen (Eggert & Sakaluk 1995). This behaviour reduces the time that the male can emit pheromone, whereas if the female is tethered in an experiment he is able to call for much longer. This may be an example of female-enforced monogamy. On small carcasses, none of these behaviours are observed, presumably because male and female interests converge as only the eggs of one female can be supported.

A converse example is the conflict between the female spider *Linyphia litogiosa,* who might benefit from additional males attracted to her pheromone, and the first male arriving, who would benefit from reducing the competition from other males. The spider male neatly bundles up her web to prevent further release of pheromone from the silk (Watson 1986).

3.9 | Alternative mating strategies

Staying 'silent' to avoid predators drawn to mating calls or pretending to be a female are both well known alternative strategies used by some males in a great variety of species to obtain matings without the risks of advertisement or direct confrontation with other males (see also Chapter 11).

3.9.1 Silent satellite males

Males of the cockroach *Nauphoeta cinerea*, which has male pheromones attractive to the female (Section 3.5.2), are able to control release of pheromone by either exposing the gland or not. Subordinate male cockroaches do not release pheromone but instead stay near calling males, as silent satellites, sometimes gaining sneaky matings (Moore *et al.* 1995). Moore *et al.* wondered if there might be parallels with subordinate cowbird (*Molothrus ater*) males who are prevented from using the songs females most prefer by the threats of dominant males (West *et al.* 1981). Similarly, there may be parallels in mouse social groups, with physiological suppression of key urine components in subordinate males because of the challenge from dominants (Chapter 5).

3.9.2 'She-males' and chemical camouflage

Males in many species steal matings by mimicking female morphology, behaviour or both (Gross 1996). Males in species of snakes and beetles may camouflage themselves with female sex pheromones to confuse other males.

Males of the red-sided garter snake (*Thamnophis sirtalis parietalis*) form 'mating balls' of up to 100 males surrounding a female (Section 3.5.1). However, in about 15% of mating balls, there is no female at the centre, instead one male, a 'she-male', is being courted by the other males (Mason & Crews 1985). The skin lipid pheromone of the she-males matches the pheromone of the female and importantly, like females, she-males lack squalene, the 'male signal'. In some experiments, she-males outcompeted ordinary males for mates (Mason & Crews 1985). However, recent work (Shine *et al.* 2000) suggests that she-maleness is not an alternative mating strategy of a subset of males and that most, if not all, males may go through a temporary stage of being a she-male immediately after emerging from hibernation. During this time, while restoring physiological functions, they are slow moving and cannot compete. As suggested by Mason & Crews, one benefit of she-maleness may be to confuse other males and waste their energy and time.

Chemical mimicry is also found in invertebrates. The carrion-feeding rove beetle *Aleochara curtula* has intense male–male competition for females (Peschke 1990). In combat, males push against each other with their heads and mandibles, and in a chemical contest they also drum the end of their upcurved abdomen on the body of the other beetle, at the same time releasing poisonous chemicals such as benzoquinones from their tergal glands. Larger males usually win. The female sex pheromone of *A. curtula* is a mixture of (Z)-7-heneicosene and (Z)-7-tricosene, found in her cuticular hydrocarbons. Immature or starved males, or multiply mated males stimulate mating attempts from other males because they have the female sex pheromone on their cuticle. It seems that by having these components on their cuticle, these she-males avoid male–male fights and gain access to potential sneaky matings and to the food. However, females actively repulse males with these compounds on their cuticle – so there are costs.

3.10 | Sperm competition and mate guarding

In many species, females can only be fertilised at particular times and males compete to be there at the right time. However, copulation may not be enough to ensure fertilisation because of the many mechanisms that females can use to manipulate sperm of different males in 'cryptic female choice'.

3.10.1 Precopulatory mate guarding

Recognising when a female is about to be fertilisable is most often done by chemical cues. In many insects, males emerge before females (possibly as a result of sexual selection among males to be first), and males then attempt to find the pupae of females about to emerge, which they guard, mating with the female as she emerges (Thornhill & Alcock 1983).

Male crustaceans and mammals share the problem that their females can only be fertilised at the moult or oestrus, respectively (events controlled by hormones). In both types of animal, urine cues advertise hormonal state. Males attempt to monopolise access by precopulatory mate guarding (Ridley 1983; Christy 1987).

Males of the lobster *Homarus americanus* create odour currents from their dens, thus 'singing a chemical song' (Fig. 3.10) (Bushmann & Atema 1997). Females choose the locally dominant male by his urine signals. Males accept premoult females of all sizes, recognised by chemical cues that stimulate a high-on-legs response in the male. However, the male only accepts one female at a time. While with the male, the female moults and mates. Apart from any indirect benefits, the guarded female benefits by protection from predation and cannibalism during her vulnerable soft-shell post-moult condition.

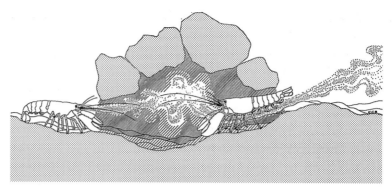

Fig. 3.10. A female lobster (left), attracted by a male's chemical 'song', jets her own urine towards a male in his shelter. He responds by retreating to the opposite entrance and fanning his pleopods. The female will join the male until her moult and copulation. Figure from Atema (1986).

Many mammals show oestrus testing and precopulatory mate guarding by dominant males, sometimes called 'consorting'. For example, male kangaroos sniff the pouch and cloacal opening of females and follow them if they are near oestrus (Fig. 3.11) (Russell 1985). Male Asian elephants (*Elephas maximus*) can detect pheromones in the urine of females coming into oestrus and about to ovulate (Chapter 1) (Rasmussen *et al.* 1996; Rasmussen & Schulte 1998). Dominant males of the African elephant (*Loxodonta africana*) only monopolise females during the phase of oestrus when it matters. At other times in the cycle, females may mate with subordinate males, but few matings lead to offspring.

Fig. 3.11. Male mammals use smell to identify females approaching oestrus and then stay close to them, guarding them from rivals. Here, an adult male kangaroo smells the pouch and cloaca of a female. Figure from Russell (1985). Drawing by Priscilla Barrett.

3.10.2 Cryptic female choice – arms races between males and females

Females can influence the likelihood that sperm from any given copulation will fertilise their eggs – sexual selection by the female, using a wide range of mechanisms, may continue long after copulation ends. Called cryptic choice, because the male need not know, it is also hidden from the human observer (Eberhard 1996, 1997).

For example, female flour beetles (*Tribolium castaneum*) find the pheromones of some males more attractive than others. After they were allowed to mate with a number of males, females skewed the likelihood of fertilisation of their eggs towards the sperm of males with more attractive pheromones (Lewis & Austad 1994).

Each additional mating gives the female arctiid moth *Utetheisa ornatrix* nutrients and protective alkaloids in each male's spermatophore (Section 3.6.1), increasing her egg output by 15% each time. However, after the matings she cryptically controls which sperm fertilise the eggs and is able to choose the sperm of the largest male, whatever the mating order (Lamunyon & Eisner 1994).

There will be selection pressure on males to influence cryptic female choice if they can. One way is to use 'chemical genitalia', secretions passed to the female with the sperm that potentially can affect many aspects of female physiology and behaviour including oviposition, resistance to further mating, and sperm transport (Eberhard & Cordero 1995; Eberhard 1996). The invasive action of male secretions could go direct to target organs such as the female nervous system or ovaries, bypassing the sense organs. In turn females would be expected to evolve traits that reduce the impact of the male.

For example, the rejection behaviour of recently fertilised female fruit flies (*Drosophila melanogaster*) to further males seems to be triggered by semen, which contains peptides from the first male's accessory glands (see review by Yamamoto *et al.* 1997). The male peptides have a number of effects on the female that benefit the male in sperm competition by inducing rejection of other males by the female, stimulating ovulation, and destroying pre-existing sperm. However, the male peptides also have physiological side effects that reduce the female lifespan (Chapman *et al.* 1995). Support for an evolutionary arms race between male and female over the effects of these peptides is suggested by work with *D. melanogaster*: if the evolution of competitive characteristics in the seminal fluid of males was allowed to occur in a special experimental design that prevented females from co-evolving with the males, then increases in the fitness of males were at the expense of the females (Rice 1996).

Effects mediated by seminal products are also found in vertebrates. Almost immediately after mating, the female red-sided garter snake

(*Thamnophis sirtalis parietalis*) becomes both unreceptive to courtship and unattractive to males (Section 3.5.1) (Mason 1993). The change in her behaviour may be due to a surge in the prostaglandin hormone $PGF_{2\alpha}$ in her blood minutes after mating. Some of the prostaglandins are the female's own, but a significant proportion is deposited by the male during copulation; the prostaglandins stimulate her ovaries to secrete oestrogens which stimulate the transport of recently deposited sperm and also cause her to evacuate sperm stored in her oviduct from the previous year.

The loss of attractiveness of the female garter snake is not caused by loss of her skin pheromone (skin lipid extracts of mated females remain attractive). Instead it may be due to a pheromone left by the male in his gelatinous mating plug or other secretion (see review by Mason 1994).

Like garter snakes, many male insects include hormones in their seminal secretions, for example, juvenile hormone, which closes down female calling behaviour in the saturnid moth *Hyalophora cecropia* (Eberhard 1997). Human sperm also contain many hormones. It is interesting to note here that an early definition of pheromone was 'ectohormone' (a hormone acting outside the sender).

3.11 | Sex pheromones and speciation

Sex pheromones are important in many taxa for species recognition, and like other signals favoured by sexual selection, they may have also played important roles in speciation. Choosing genetically compatible individuals of the right sex and reproductive state is an important precondition for reproductive success, along with other important but finer distinctions between potential mates (Andersson 1994). For example, selection acting on females to avoid hybrid mating with sympatric species (species living in the same place) may have been a factor in the evolution of male moth pheromones (Section 3.11.1) (Phelan & Baker 1987).

New species develop when previously interbreeding populations no longer interbreed successfully. There are two main alternative explanations for speciation (Orr & Smith 1998). In allopatric speciation, populations become separated by a geographical barrier. While apart, many characters diverge by drift or by different selective conditions, so that the two populations do not interbreed if they come back into contact.

Sympatric divergence of mate recognition systems, either in populations that come back into contact or in populations that have always lived in sympatry, could occur if genetic divergence has reached a stage where hybrid offspring are less fit than within-population offspring. This could either be because hybrids are less viable, or by sexual selection, if hybrid offspring are less successful in getting matings than pure-bred individuals (see Andersson

1994, p. 214). In either case, selection favours an increase in assortative mating (non-random mating resulting from a preference for similar partners, Butlin & Tregenza 1997). Assortative mating will increase reproductive isolation, further reducing the exchange of genes between the populations. This mechanism leading to divergence in mating signals or preference is known as reinforcement (Butlin 1987).

If different sexual selection pressures operate in the two populations, runaway selection could lead to increased divergence in mating recognition traits and faster speciation (a process that could also occur in allopatry). Conversely, Fisherian or indicator processes might start with traits favoured initially to avoid the risk of hybridisation (Andersson 1994).

Empirical support for reinforcement has grown recently (Butlin & Tregenza 1997), particularly from studies of chemically mediated systems, such as cuticular sex pheromones in *Drosophila* spp. A comparison of more than 120 pairs of closely related *Drosophila* spp. worldwide suggests that there has been faster evolution of mating discrimination (prezygotic barriers to mating) in pairs of species that were in sympatry than in pairs in allopatry (Coyne & Orr 1997).

The demonstrations of the strong influence of sexual selection on mating signals suggest that their involvement in reproductive isolation is not incidental. This is in contrast to the concept of the Specific-Mate Recognition System in which mating signals are maintained by stabilising selection so long as the environment does not change (Paterson 1993). (For a good discussion of this, using moth pheromones as examples, see Linn & Roelofs 1995.)

Evidence is growing that divergence in sympatry can be maintained by assortative mating despite gene flow between populations (for example, in the corn borer moth, see Section 3.11.1.1). However, reinforcement, with stronger separations of sexually selected characters in areas of sympatry, may not be seen if sexually selected traits can spread through the wider population outside the zone of sympatry as suggested by genetic models (Lande 1982); this may have occurred with male moth pheromones.

A separate process from reinforcement can lead to divergence in mating recognition signals in species, perhaps from completely different taxa, which share the same communication channel (Butlin 1989). This 'reproductive character displacement' is comparable with ecological character displacements such as changes in beak shape in different bird species sharing the same niche (Butlin 1987). For example, moths in different families may share some pheromone components and selection for changes in reception or signal will be expected to reduce interference in communication (e.g. through wasting male efforts in false pursuit) (Löfstedt 1991; Linn & Roelofs 1995).

3.11.1 Species isolation and evolution in moths

While there many examples of species discrimination by odour by vertebrates, for example in salamanders (Dawley 1987), moths provide the best understood examples. Much of the early literature on moth pheromones emphasised their role in species isolation (see reviews by Cardé & Baker 1984, and Linn & Roelofs 1995). The pheromones of almost all moth species are multi-component blends of chemicals, typically composed of one abundant 'major' component and lesser amounts of several 'minor' components. A feature of the pheromones of many species is the use of a major component and its geometrical isomer in a particular ratio (Appendix A2). Within closely related taxonomic groups (families or subfamilies), species often use the same or similar components, as a result of sharing a common biosynthetic pathway. Closely related species within a family gain specificity by using unique ratios or combinations of these components (Linn & Roelofs 1995). Niche separation in the imaginary 'sex communication channel' (Greenfield & Karandinos 1979) can come from such blend differences, from calling at different times of the day (or season), from host plants, or from a combination of these.

A study of nine sympatric species of European small ermine moths (*Yponomeuta*) shows the use of all of these dimensions (Löfstedt *et al.* 1991). Six of the species use a mix of (E)-11- and (Z)-11-tetradecenyl acetate in different ratios as primary components of their pheromone. Three species which have different host plants from each other appear to share very similar pheromone blends and were attracted to each other's pheromone in wind-tunnel tests, but not in their natural habitats (Fig. 3.12) (Hendrikse 1986). However, three species which shared a host plant, and thus would be likely to meet in the field, showed no overlap in pheromone blend and did not cross attract in the wind tunnel. A phylogenetic analysis suggests that the present pattern of blends has occurred by loss and gain of minor components and shifts in ratio. In addition, two species with similar blends call at different times in the night. The pheromone components of some species act as behavioural antagonists (see below) to males of other species.

The use of pheromone blends by female moths affects selection on male signal reception. On theoretical grounds, perception of the whole blend would dramatically enhance the signal-to-noise ratio against background chemical noise and thus would enable a male to detect a conspecific female reliably from further away (Cardé & Baker 1984). Important support for this came from the discovery that male moths are most sensitive, at the greatest distance, to the whole blend produced by the female, and not to the major component by itself as had once been thought (Fig. 3.13). (Linn *et al.* 1987; Linn & Roelofs 1989) The minor components might be less than 10% of the concentration of the major component, but they give the signal its

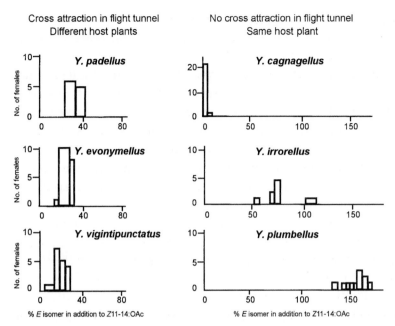

Fig. 3.12. Species isolation and mating site. (Right) Three of six sympatric species of small ermine moths (*Yponomeuta*) share the same host plant and so have evolved different pheromone blends: males are only attracted to the blend of their own species of female (there is no cross attraction in a flight tunnel).

(Left) The other three species have different host plants from each other and have similar pheromone blends which show significant cross attraction in the flight tunnel but *not* in the field. The data show the frequencies of (*E*)- to (*Z*)-11-tetradecenyl acetate blends produced by individual females (incidentally, showing the *intra*specific range of blends). Figure redrawn from Löfstedt *et al.* 1991.

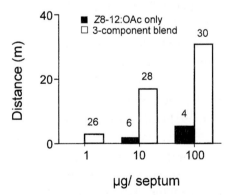

Fig. 3.13. Far downwind, male oriental fruit moths (*Grapholitha molesta*) are more sensitive to the whole female pheromone blend than to the major component alone. Wing-fanning activation of male moths was tested downwind from pheromone sources: major female pheromone component alone (Z8-12:OAc) or the three-component blend (Z8-12:OAc + 6% E8-12:OAc + 3% Z8-12:OH) at three doses. The experiment was done in the field and soap bubbles were used to track the path of pheromone plumes in the wind. Values above bars are the number of males of the 30 tested that activated. Figure redrawn from Linn & Roelofs (1989), data from Linn *et al.* (1987).

specificity. Because of the way pheromone plumes travel downwind as filaments of pheromone-rich air (Chapter 10), all the components are present in the ratio as released, even far downwind.

A responding male would benefit from early rejection of a female not from his species. He would then save the energy cost of flight, predation risks and the time opportunity cost, as he is not likely to be accepted by the female when he reaches her, or if she accepts, the mating would be infertile. Males have specialised olfactory sensilla for these 'behavioural antagonists' – key components that distinguish the blends of other species similar to their own – and they turn back if they detect antagonists (Chapters 9 and 10) (Linn & Roelofs 1989).

3.11.1.1 Speciation and signal shift: are signals and receptors linked?

One of the key questions in the evolution of sexual communication is how signal and response are coordinated for species specificity, but change during speciation. It is particularly challenging because the characters for signal and receiver are only rarely genetically linked (Butlin & Ritchie 1989; Boake 1991). For example, in the insects so far studied, pheromone production in females and response in males are controlled independently on different chromosomes (see end of this section).

If the signal and receiver are not linked, how might change occur? In the case of the long-distance pheromone of the female moth, the changes in mating signals are often treated as an equal co-evolution between male and female. However, Phelan and others point out that because females are a limiting resource to males, there is more selection pressure on individual males to have a broad response spectrum than there is on females to produce a particular pheromone blend (Fig. 3.14). In isolated populations, there may be chance changes in the female blend but the wide response spectrum of each male means that males will still respond and evolve towards this new blend; Phelan (1997) terms this 'asymmetric tracking'. If the change in blend is large then there may be disruptive selection towards the extremes of male response leading to polymorphism, assortative mating and possibly speciation.

One example of how major shifts in pheromone blends can occur comes from a chance observation of change of pheromone blend produced by mutant females in a laboratory population of cabbage looper moths (*Trichoplusia ni*) (Haynes 1997). Mutant females produced a blend with 50 times more of one of the minor components. This resulted in a 40-fold reduction in the number of conspecific males attracted in the field, but *some* did arrive. In the laboratory, crossing experiments on the mutant population showed

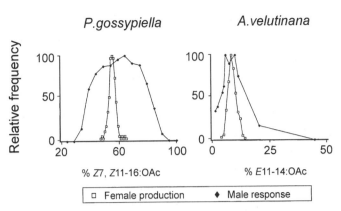

Fig. 3.14. Male moths have much wider pheromone response windows than the range of variation in female pheromone production. This is shown in two species of moths: *Pectinophora gossypiella* and *Argyortaenia velutinana*. Frequencies within each class are calculated relative to most common type of female production or male response respectively (taken as 100 by definition). Figure redrawn from Löfstedt (1990).

that genetic control of the female signal and male response were independent. The change in the female was due to a single recessive autosomal mutation at one locus. The males from the mutant colony had the same proportion of antennal receptors sensitive to each of the components as normal males. In flight tests to pheromone lures, the mutant males, like normal males, were initially much more responsive to the normal blend but after 49 generations the male response spectrum broadened to include the new blend.

The European corn borer moth (*Ostrinia nubilalis*) has become a model species for the study of pheromone genetics (Löfstedt 1993; Linn & Roelofs 1995). Two pheromone races of the species use opposite ratios of the Z and E isomers of the main component 11-14:acetate: the Z race uses a 97:3 mix and the E race a 1:99 mix of Z/E 11-14:OAc. An extensive field study of *O. nubilalis* in Maryland, USA, where both races are sympatric, showed strong assortative mating (Klun & Huettel 1988; see also Bengtsson & Löfstedt 1990). However, some hybrid individuals are found in the field, and in the laboratory the two pheromone races will produce viable and fertile hybrids: F_1 females produce an intermediate 63:35 ratio and the F_1 male response is also intermediate. The pheromone signalling polymorphism is controlled by pairs of alleles on three different chromosomes: the female pheromone production locus is autosomal, the male behavioural response locus is sex linked, and a third locus (autosomal but not linked to the female production locus) controls the male pheromone-sensitive sensilla (Roelofs *et al.* 1987; Cossé *et al.* 1995).

3.11.1.2 Sexual selection of male moth pheromones and speciation

In some moth species, such as *Utetheisa ornatrix* (Section 3.6.1), when males find the female, an elaborate courtship follows as the male releases his own pheromones. The male pheromones, often called scents, are released from special structures such as hair pencils, brush organs, and coremata (see, for example, Fig. 1.5) (Birch *et al.* 1990; Phelan 1992, 1997). Courtship heromones are sometimes called 'aphrodisiac pheromones'.

The male pheromones are chemically diverse and the scent organs show enormous variety, reflecting multiple independent evolution across the taxon. This is in contrast to the pheromones of female moths which are largely similar compounds across the group based on Δ11-desaturation of fatty acid precursors and produced from roughly homologous glands, reflecting evolution early in the lineage (Löfstedt 1991; Phelan 1997). One strong selective force for the evolution of male pheromones and courtship displays may be selection by the female in order to avoid hybrid matings. Phelan & Baker (1987) hypothesised that if male pheromone-emitting structures have arisen via selection by females to avoid hybrid matings, then they would be more likely to be found in closely related species with the greatest chance of making mating mistakes, for example species pairs sharing the same host plant. In each of five different moth families, all having some species with male scent-emitting organs, significantly more pairs of species sharing a host plant had male scent-emitting structures than not. When the data were pooled, 53% of 396 species sharing a host plant had male pheromone structures compared with only 28% of 419 species without a host plant in common.

3.11.2 *Drosophila*

Sex pheromones in Diptera are mostly cuticular hydrocarbons, relatively involatile molecules only detected at very short range or on contact during courtship (see review by Greenspan & Ferveur 2000). These cuticular hydrocarbons play a fundamental role in isolation between the fruit fly (*Drosophila melanogaster*) and its sibling species (Fig. 3.15).

The courtship behaviour of *Drosophila melanogaster* tends to follow a stereotyped sequence. First, visual cues orient the male towards the female. After following her, he approaches and taps her with his front legs (receiving contact pheromone signals from the female via receptors on his tarsi), and if the female is from his species, vibrates his wings in a species-specific courtship song. Her response to his song is to protrude her genitalia, which he licks, presumably gaining further chemical information, possibly on her mating state. If accepted by the female, the male will mount and copulate.

Studies of *Drosophila melanogaster* females genetically engineered to produce almost *no* cuticular hydrocarbons on their surface suggest that four *Drosophila* sibling species all share a common set of ancestral cuticular

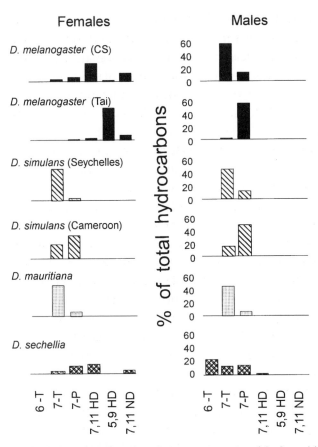

Fig. 3.15. Cuticular hydrocarbon pheromones in species of the *Drosophila* complex (plotted as percentages of total hydrocarbons). Two strains of *D. melanogaster* and *D. simulans* show the geographical variation within each of the species. Key to hydrocarbons: 6-T: 6-tricosene; 7-T: 7-tricosene; 7-P: 7-pentacosene; 7,11-HD: 7,11 heptacosadiene; 5.9 HD: 5, 9-heptacosadiene; 7,11 ND: 7,11-nonacosadiene. Tai: Ivory Coast; CS: Canton-S laboratory strain. Figure after Cobb & Ferveur (1996).

pheromones which stimulate mating in males of all species (Savarit *et al.* 1999). The ancestral pheromones stimulate mating interest when detected by the male but he only continues the courtship if the correct cuticular hydrocarbon pheromones for females of his species, and not others, are present. Thus the *Drosophila* cuticular pheromones so far described are not the main stimulus but instead modulate male responses and, combined with behavioural responses by the female, prevent interspecific matings. The male cuticular pheromone in *D. melanogaster*, 7-tricosene (7-T), inhibits male–male courtship by masking or inhibiting the action of the ancestral female pheromones. *D. simulans* females have 7-T as a major component and hence *D. melanogaster* males do not court them.

The genetics of species-specific female hydrocarbon pheromones in two sibling *Drosophila* species has been examined in detail (Coyne &

Charlesworth 1997). The main female component of *D. sechellia* is 7,11 heptacosadiene (7,11-HD) whereas *D. mauritiana* uses 7-tricosene (7-T) (Fig. 3.15). These compounds seem to play a large role in the sexual isolation between *D. mauritiana* males and *D. sechellia* females by inhibiting *D. mauritiana* male courtship, while sexual isolation in the reciprocal hybridisation results more from differences in female behaviour.

The cuticular pheromones of males and females in the cosmopolitan species *D. melanogaster* differ greatly around the world (Ferveur *et al.* 1996). Females from sub-Saharan Africa and the Caribbean have a low percentage of 7,11-HD and high quantities of 5,9-HD whereas females everywhere else have the opposite ratio. Male pheromones show a clear latitudinal cline: equatorial strains have a high concentration of 7-P and low 7-T, whereas males from high latitudes have the reverse. Ferveur *et al.* suggest that it is possible that sexual selection and/or a tendency to assortative mating based on these pheromone differences may have contributed to the development or maintenance of genetic differences between the sub-Saharan strains and those elsewhere in the world.

3.11.3 Hawaiian *Drosophila*

Sexual selection may have played a major part in the explosive speciation of *Drosophila* fruit flies on the Hawaiian islands, leading to more than 800 endemic species (unique to the islands), characterised by complex courtship behaviour and elaborate secondary sexual features (Kaneshiro & Boake 1987). Rapid speciation may be encouraged by the relaxation in sexual selection during the founding phase of new populations: with few males around, less choosy females will have a selective advantage as they can be surer of finding a mate (Kaneshiro 1989). Less choosy females will allow male sexual signals to expand in variety. By the time that populations increase or other species invade, returning an advantage to choosier females, the male signals will have changed. There is limited evidence that newer species have females that are less choosy than their 'ancestral' species. Many of the Hawaiian *Drosophila* lek (see Section 3.7), males produce pheromones used in advertisement and courtship, and cuticular pheromones on the females enable species recognition by males, as for example in three species of the *Drosophila adiastola* subgroup (Tompkins *et al.* 1993).

3.12 | Conclusion

Chemical signals are sexually selected in the courtship behaviour of many species of invertebrates and vertebrates. There are examples of pheromone mediated mating systems that illustrate Fisherian ('sexy sons'), 'indicator' ('good genes'), and direct benefit mechanisms for the evolution of female choice. Male moths, goldfish and garter snakes respond to pheromones in scramble mating systems. Alternative mating strategies include 'silent' satellite males and deception with cuticular pheromones in beetles and garter snakes. Assessment of male value by honest signals is shown in the cockroach *Nauphoeta cinerea* and the tiger moth *Utetheisa ornatrix*. Assessment of mates by olfactory cues is taken to the height of sophistication by mammals, ranging from the species, sex, and reproductive state to the genotype and health of a potential mate.

Pheromone signals are used by males in many insect species that lek. Odours are important in female choice of territories in some ungulate leks. Pheromones act in cryptic mate choice and may offer 'chemical mate guarding'.

Speciation mechanisms have been investigated in moths and the fruit fly *Drosophila*. Pheromones in male moths are consistent with multiple independent evolution to avoid mating mistakes in sympatric species. Worldwide comparisons of pairs of *Drosophila* species provide evidence for reinforcement of characters in sympatry. Insect pheromone systems in particular offer the advantages of tractable genetics and laboratory populations for the study of speciation and sexual selection.

3.13 | Further reading

Andersson (1994) gives a readable and comprehensive account of sexual selection. Butlin & Ritchie (1994) describe the mechanisms involved in speciation and the role of mate choice. A special issue of *Genetica* (**104** (3)) is devoted to the MHC. Many chapters in Johnston *et al.* (1999) and Marchlewska-Koj *et al.* (2001) cover pheromones and reproduction in vertebrates. Articles in *Trends in Ecology & Evolution* regularly cover advances in sexual selection.

Chapter 4

Coming together
and keeping apart:
aggregation and host-marking
pheromones

4.1 | Introduction

The attraction of thousands of bark beetles an hour in response to the aggregation pheromones released by the first arrivals on a suitable tree is one of the most impressive demonstrations of the power of pheromones. More subtle are the pheromone marks, left by a female parasitic insect on its host, which seem to deter conspecific females from laying in that host.

But why do the first beetles call and why do conspecifics respond so spectacularly to these pheromones? Why do female parasitic insects mark hosts and why should other females take notice? Until relatively recently, many authors suggested that aggregation and host-marking pheromones enabled the most efficient use of resources by a species. However, it is hard to see how such behaviours could evolve unless they benefit individual reproductive success. This chapter investigates the individual advantages for signallers and responders in these pheromone systems.

4.2 | Aggregation pheromones and Allee effects, the advantages of group living

Aggregation pheromones lead to the formation of animal groups near the pheromone source, either by attracting animals from a distance or stopping ('arresting') passing conspecifics (Chapter 10). In contrast to sex pheromones (which attract only the opposite sex), aggregation pheromones by definition attract both sexes (and/or, possibly, larvae). However, the benefits of aggregation to individuals may be complex and two rather different mechanisms may apply: first, individuals may be aggregating for the benefits of living in a group. Second, what we call aggregation pheromones may be the response of eavesdropping conspecifics to sex pheromones released by the same sex (Section 4.2.5) (Chapters 3 and 11).

While the negative effects of overcrowding are often obvious, it is also possible to be 'undercrowded', to have a population density that is too low. For example, a lone animal may be more vulnerable to predators. The benefits of conspecifics are broadly referred to as Allee effects (named after the pioneering work of W. C. Allee (1931)). Allee effects are often revealed by lower reproductive rates per individual at low numbers or densities of conspecifics (Courchamp *et al.* 1999; Stephens & Sutherland 1999). The classic beneficial effects of group living include influences on one or more of the following: predation, reproduction, feeding and local environment. Allee effects may explain the evolution of many strong aggregation responses to pheromones. Pheromones also play an important part in the life of desert locusts, one of the most famous aggregating animals (Box 4.1).

4.2.1 Aggregation in space for defence against predation

California spiny lobsters (*Panulirus interruptus*) form groups of up to 30 animals in dens during the day. The aggregations are caused by attraction to chemical cues from conspecifics of either sex (Zimmer-Faust *et al.* 1985). Although vulnerable as individuals, as a group the lobsters wave their robust spiny antennae from the den opening to deter predatory fishes.

Many species of insect with strong chemical defences and bright warning colours (aposematic colours) form conspicuous aggregations mediated by pheromones. The aggregation pheromone of the seven-spot ladybird beetle (*Coccinella septempunctata*) is an alkylmethoxypyrazine (Al Abassi *et al.* 1998). This class of compounds acts as an olfactory warning signal in this and many other aposematic species – giving the olfactory equivalent of aposematic colouration (Guilford *et al.* 1987; Moore *et al.* 1990). The ladybird pyrazines are persistent and may provide the chemical cues leading to the traditional overwintering sites used year after year by large aggregations (Aldrich 1999). Whether aggregation or conspicuous warning signals came first in the evolution of aposematism is not yet clear (Krebs & Davies 1993, p. 89).

Two common benefits of safety in numbers come from the dilution of risk (any individual animal is less likely to be eaten) and the swamping of predators (when there are locally too many prey for predators to eat). Female insects such as mosquitoes and blackfly (Simuliidae) may gain such benefits for their offspring by laying their eggs close to others, particularly since their aquatic larvae usually have plentiful food so are not in competition with each other (McCall & Cameron 1995). *Culex* mosquito females are attracted to pheromones produced by egg rafts laid by congeneric females (Pickett & Woodcock 1996). The pheromone, (5*R*,6*S*)-6-acetoxy-5-hexadecanolide, found in a droplet at the top of egg, is not produced for the first 24 hours so only

Box 4.1 | Aggregation pheromones in the life of the desert locust

Pheromones are important in the biology of the desert locust *Schistocerca gregaria* (see figure below) (see review by Hassanali & Torto 1999). *S. gregaria* can famously change between solitary and gregarious phases (forms) in response to population density. In its gregarious phase, the desert locust forms swarms of thousands of millions of insects that can affect 20% of the Earth's land surface. Animals on the extremes of the phase continuum are strikingly different in appearance, physiology and behaviour (and were once thought to be separate species). The change from solitary to gregarious starts subtly and is caused by crowding (see review by Simpson *et al.* 1999). The aggregation pheromones that bring the locusts together can thus have an important effect on gregarisation by starting a powerful behavioural positive-feedback loop.

Adults and nymphs have separate aggregation pheromone systems (see figure). One feature of the gregarious phase is maturation synchrony, mediated by primer pheromones, so all the adults will be able to swarm together (Hassanali & Torto 1999). The aggregation pheromone produced by gregarious adult males also acts as a primer that speeds up maturation of females. Nymphal aggregation pheromones have the primer effect of slowing down the development of adults of either sex. The combined effect of these primer pheromones is to synchronise the development of the entire population. Cuticular compounds produced by gregarious nymphs may gregarize solitary individuals (Heifetz *et al.* 1998).

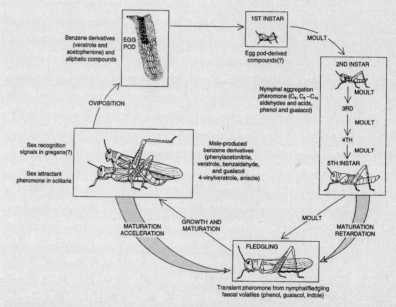

Life cycle of the desert locust, *Schistocerca gregaria*, showing the pheromones released and acting at different stages. Figure from Hassanali & Torto (1999).

Egg laying can affect gregarisation in the next generation in two ways. First, by the aggregation responses of laying females to 'oviposition pheromones' in the froth of eggs already laid in the soil (Rai *et al.* 1997; Torto *et al.* 1999). The young will emerge to a concentrated population. Second, crowd-reared, gregarious female locusts secrete a hydrophilic 'gregarising factor' in the foam on the eggs that promotes gregarisation of the hoppers (McCaffery *et al.* 1998).

surviving eggs will produce it. While the response to oviposition pheromone may have evolved as eavesdropping by the later females (Chapter 11), the 'signaller' also benefits from the same effects so this could also be an evolved signal. Oviposition pheromones may offer methods for control of these important insect vectors of disease (Chapter 12).

4.2.2 Synchronisation: aggregation in time

Aggregation can occur in time as well as space. Predators may be swamped if all prey larvae are released at the same time. In the mud crab (*Rhitropanopeus harrisii*), which lives in tidal creeks on the east coast of North America, all the larvae in the population are released on the same few nights, with the larval release signal coming from the eggs themselves (Forward 1987). The female carries the fertilised eggs glued to her abdomen until they are ready to hatch. Egg hatching is timed to occur at night, to reduce predation risk, and at the spring high tide, to avoid low-salinity stress to the larvae. When ready to hatch, the eggs swell and the outer egg membrane breaks, leaving a fragile inner egg membrane which parts easily to release a 'pumping pheromone'. The pheromone stimulates a stereotyped 'pumping' behaviour in the adult female, which stands on tiptoe, and vigorously flexes her abdomen; this helps to disrupt more egg membranes, releasing more pheromone, thus stimulating more pumping, and propels the larvae into the water column. The pheromone response results in a near-simultaneous release of her larvae, which benefit from 'safety in numbers'. The pheromone is a short peptide (which can be mimicked by amino acids and short synthetic polypeptides), with activity at concentrations as low as 10^{-21} M (Pettis *et al.* 1993).

In the crab, the pheromone signal is from the egg to the adult. In contrast, the egg hatching pheromone in the barnacle *Semibalanus balanoides* (syn. *Balanus balanoides*) is from the adult. The barnacle broods its eggs in its mantle cavity over the winter (see Box 4.2). The eggs are ready to hatch for some months, just waiting for the signal – the egg hatching pheromone released by the adult after feeding on the spring phytoplankton bloom (Clare 1995). The system ensures that the barnacle nauplii are released when the phytoplankton they need for food have arrived. The hatching pheromone consists of eicosanoids (called prostaglandins in vertebrates) which are C_{20} polyunsaturated fatty acids (PUFAs) (Clare 1997). Other eicosanoids are important aquatic pheromones for spawning in the abalone, a marine mollusc, and fish (see Chapters 1 and 3).

4.2.3 Settlement of marine invertebrates

Marine invertebrates living on the seabed or seashore show some of the most conspicuous aggregation behaviour known. Most of these organisms have a

Box 4.2 | Barnacle settlement behaviour

Settlement of barnacle larvae from the plankton has attracted great interest in the search for ways to stop them encrusting ships (Chapter 12). With the last larval moult, the barnacle nauplius larva changes into a cypris larva whose sole function is to find a suitable site to settle (see figure). Chemical signals from conspecific adults are paramount especially at distances of

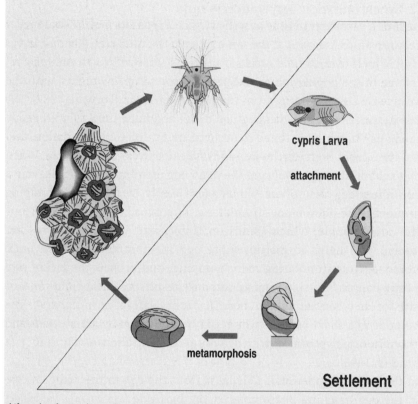

Life cycle of acorn barnacles, showing adults filter feeding and mating, and the planktonic nauplius larva which becomes a cypris larva. Before finally settling, the cypris larva searches potential surfaces for a metamorphosis site, in response to chemical cues from past or living conspecifics (see also bioassay, Fig. 2.3). Figure from Clare & Matsumura (2000) after Okano & Fusetani (1997).

centimetres or less. Two settlement pheromones are important (Clare & Matsumura 2000). First, a small peptide released into the water by adults causes the larvae to start searching hard surfaces. If the cypris contacts the second pheromone, the settlement-inducing protein complex (SIPC) which is adult glycoprotein (formerly called arthropodin) in the cuticle of a conspecific barnacle, or even in the base of a dead barnacle, its behaviour changes. It moves more slowly, with frequent changes in direction as it explores surface depressions before settling and extruding attachment cements to glue itself permanently to the substrate, and metamorphosing to the adult form (see also Fig. 2.3).

planktonic larval dispersal phase and it is crucial for their later survival and successful reproduction that the planktonic larvae settle in the right habitat, close to conspecifics.

Chemical cues produced by adult conspecifics or recently metamorphosed larvae are among the key stimuli (Box 4.2) (Zimmer & Butman 2000). For example, when searching for settlement sites, oyster larvae sink into the bottom boundary layer where they detect soluble cues from the mantle cavity of adult oysters. A larva has to make a decision to settle quickly as the current sweeps it past potential sites. A larva settling near conspecific adults has evidence that the site sustains that species – for example, a site with adequate water flow rates and food supply, and a reasonable exposure and temperature range if intertidal.

A wide range of other selfish advantages may come from settling close to conspecifics (Allee effects). The first are for reproduction. At maturity, adults need to be close enough to find or reach mates. This is especially important for internally fertilising animals such as barnacles that are permanently glued to the rock (Box 4.2). Size still matters, even though barnacles have the longest penis in relation to body size of any animal; for example that of *Elminius modestus* is up to 5 cm long, impressive for an animal of only 0.5 cm.

For externally fertilising animals such as sea urchins, which release their gametes (eggs and sperm) into the water, population density is still important. The ocean is big and quantities of gametes are small. Fertilisation rates are greatly reduced if adults releasing gametes are far apart so there are significant reproductive benefits to being close to conspecifics (Fig. 4.1) (Levitan 1996). Synchronisation of spawning is also important (Chapter 3).

Filter-feeding animals, such as sand dollars, byrozoans and barnacles, which exploit currents, may gain hydrodynamic advantages from living close to conspecifics (Fig. 4.1).

If larvae only settle in response to chemical cues from existing aggregations, how do such species colonise new, empty sites? The larvae of most marine invertebrates become less discriminating with age, so that if ideal sites are not found, larvae will 'settle for less', responding to suboptimal sites rather than dying unsettled. A different solution, evolved by the colonial tube-dwelling polychaete worm *Hydroides* is to have two kinds of larvae: one type responds to chemical cues from adult conspecifics, the other type consists of 'pioneer larvae' which colonize bare new substrate (Toonen & Pawlik 1994).

4.2.4 Ecophysiological benefits

One reason for forming some aggregations may be ecophysiological. For example, pheromone-mediated aggregation of conspecifics reduces water loss in ladybirds, cockroaches, house mites and ticks, and barnacles high on the

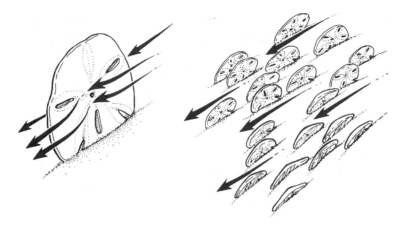

Fig. 4.1. The Pacific sand dollar (*Dendraster excentricus*) is a filter-feeding sea urchin, which 'stands on its head' in the sand. It gains hydrodynamic benefits for feeding by forming dense aggregations. (Left) The hydrofoil shape of the sand dollar creates lift which brings food particles to the feeding surface (O'Neill 1978; Vogel 1994). (Right) The sand dollars can gain hydrodynamic advantage by forming dense groups all oriented with the flow and each other to operate as a colonial multi-winged craft, bringing feeding currents closer to their oral surfaces. There are also reproductive advantages for fertilisation if the urchins are close together. Adult aggregation of the sand dollars is facilitated by the specific metamorphosis response of the planktonic larval stage triggered by adult odours on the sand (Burke 1984). Figure from Vogel (1994).

shore (Yoder & Grojean 1997; Bertness *et al.* 1999). The pheromonal basis of these aggregations has been explored in some cases. Cockroach aggregation in response to conspecific odours is well known (Rust *et al.* 1995). An assembly pheromone arrests wandering ticks, forming tight clusters of inactive ticks in sheltered locations where they sit out adverse conditions (Hamilton 1992). The tick assembly pheromone, guanine, is derived from excreted waste products of blood meal so places where ticks have successfully fed before, and to which hosts will return, will be marked most (Otieno *et al.* 1985). The pheromones are not species-specific because it benefits any tick to aggregate. The response is only induced when the relative humidity is very low, thus supporting a role as behaviour to reduce water loss (Hassanali *et al.* 1989).

The red-sided garter snake (*Thamnophis sirtalis parietalis*) overwinters in communal dens containing tens of thousands of individuals. Communal hibernation may be a crucial adaptation which allows it to extend its range further north than any other reptile in North America, allowing it to survive the bitter winters of the Great Plains and Canadian prairies where surface temperatures drop to −40°C. The migration to the dens at the end of the summer is partly mediated by orientation to skin-lipid pheromones, perceived through the vomeronasal system (see Chapters 2, 9 and 10) (Heller & Halpern 1982).

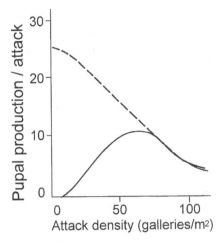

Fig. 4.2. Allee effects in bark beetles. Female *Dendroctonus montanus* attacking a living tree (solid line) have a higher individual reproductive success with increasing female density until competition starts to reduce success at densities greater than 60 beetle galleries/m². The effectiveness of the living tree's defences is revealed by the contrasting high reproductive success of females in dead logs even at low population density (dashed line). Figure redrawn from Raffa & Berryman (1983).

4.2.5 Beetle aggregation pheromones – calling for Allee benefits, or same-sex eavesdropping?

Epidemics of 'aggressive', tree-killing bark beetle species can produce a moving front of killed trees turning from green to brown, looking as if a fire is sweeping through the forest. Colonising a living tree is only possible by mass attack of thousands of beetles attracted by aggregation pheromones (Chapter 11). Only attack by large numbers can overcome the living tree's formidable defences, which include toxic chemicals such as terpenes and phenols as well as physical defences such as gums, latexes, and resins (Byers 1995). For the females of the bark beetle *Dendroctonus montanus*, these Allee effects explain why reproductive success on living trees rises with increased density until intraspecific competition for limited larval food space overtakes the benefits (Fig. 4.2) (Raffa & Berryman 1983).

On a smaller scale, there are some aggressive bark beetle species in which a single female attacks a live tree by herself (e.g. *Dendroctonus micans* and *D. valens*). Her larvae cope with the resin defences of the tree by aggregating, in response to larval pheromones, to form a feeding front of up to 50 larvae feeding side by side. Larvae reared in groups have higher survivorship and gain mass faster than isolated larvae (Storer *et al.* 1997).

Mass attack in most bark beetles starts with the arrival of the first beetles on a suitable tree. The pioneers start to bore into the bark and release long-range aggregation pheromones that attract conspecifics of both sexes. In most monogamous species, this is the female (e.g. *Dendroctonus* spp.); among polygamous species, the male (e.g. *Ips* spp.) (Vité & Francke 1976; Kirkendall *et al.* 1997).

By definition, aggregation pheromones attract both sexes. There has been much debate about the origin of these pheromones. Are they pheromones that invite fellow colonists of both sexes (for Allee benefits to the signaller)? Or instead, are these sex pheromones, usually released by the male to attract females, but eavesdropped by males hoping to intercept some of the females attracted (Chapter 11) (Greenfield 1981; Alcock 1982; Thornhill & Alcock 1983; Landolt 1997; Schlyter & Birgersson 1999)? The answer may depend on the lifestyle of the beetle species (Raffa *et al.* 1993; Kirkendall *et al.* 1997). In tree-killing *Dendroctonus* spp., same sex (female–female) responses dominate during colonisation. Only later during the attack do more males than females arrive. For less aggressive bark beetle species (the majority), which only attack dead trees, there may be little advantage in attracting fellow colonists of the same sex, and much to lose in competition. The shutting down of pheromone calling when the opposite sex arrives suggests that in these species eavesdropping is the reason for same-sex arrivals. For example, mating stops calling in the less aggressive *D. pseudotsugae* and there is little difference in the attractiveness of unmated and mated females, but pheromone calling continues after mating in the tree-killing southern pine beetle *D. frontalis* (Raffa *et al.* 1993).

If females are tuned to the male pheromone blend, it will be hard for calling males to change the blend they offer, even if other males start responding, because the cost of not attracting females would be too great. In the tropical beetle *Prostephanus truncatus*, the males produce a sex pheromone probably directed at females, but which is eavesdropped by conspecific males in search of mates. This is suggested by the way that the males shut off pheromone release when females arrive, like the bark beetles described above (Smith, J. L. *et al.* 1996; Hodges *et al.* 2000). This, incidentally, prevented the isolation of sufficient pheromone for many years until scientists tried collecting from males only, rather than from mixed-sex cultures.

Prostephanus truncatus is a stored-product beetle, a group of insects much studied because they infest human food stores. Stored-product beetles offer a similar range of lifestyles and pheromone strategies to bark beetles (Plarre & Vanderwell 1999). In many species, males produce pheromones that attract both sexes. Allee's original work (Allee 1931) cited studies of the flour beetle (*Tribolium castaneum*), which showed optimum individual reproductive success at intermediate population densities. This was either as a result of environmental amelioration because of concentrated beetle activity or perhaps because of the reproductive advantages for females through multiple mating. At high densities, cannibalism of eggs might reduce reproductive success.

Burkholder (1982) suggested that patterns of pheromone behaviour in stored-product insects might depend on the ecology of the adults. He generalised that stored-product beetle species with short-lived adults (<1 month) generally have female-produced sex pheromones, with a highly synchronised communication system, and usually do not feed as adults (for example *Trogoderma glabrum* and *Lasioderma serricorne*). In contrast, species with long-lived adults tend to have male-produced aggregation pheromones and need to feed to support reproduction (for example, *Prostephanus truncatus* and *Sitophilus*). Maximum production of aggregation pheromone in many of these species requires the presence of food and Burkholder proposed that aggregation pheromones bring both sexes together on significant food sources and that mating encounters then occur. However, Levinson & Levinson (1995) provide numerous examples of stored-product beetle species that break these predictions, sometimes in the same genera as ones that conform to them. For example, the Khapra beetle (*Trogoderma granarium*) is a short-lived and non-feeding species which has a dual-function female-produced sex pheromone that attracts both males and females, thus acting as an aggregation pheromone.

4.3 | Host-marking pheromones

After laying an egg inside a host, such as a small fruit or a caterpillar, many species of parasitic insects leave a pheromone mark on or in the host (host-marking pheromone, HMP). Other female conspecifics usually avoid laying in these marked hosts, a behaviour called 'host-discrimination' (van Lenteren 1981). Why should animals pheromone-mark hosts, particularly as the pheromone can attract predators (see Chapter 11), and why should conspecifics take any notice of the marks?

What host-marking species have in common is that they lay their eggs in hosts of limited size, which can only successfully support the development of a limited number of parasitoid larva (Roitberg & Prokopy 1987) (see Chapter 12 for the potential use of HMP in pest control); any further larvae will reduce the chance of the first surviving. HMPs have evolved independently across several orders of insects – in more than 200 species of parasitoid wasps (wasps which lay their eggs in other insects) (van Lenteren 1981) and more than 30 species of herbivorous insects across eight families of Diptera, Coleoptera and Lepidoptera (Fig. 4.3) (Table 4.1) (Roitberg & Prokopy 1987). I am not aware of animals apart from insects that have HMP, but the situations in which to look for it would be ones where the developing larva needs the whole of a finite resource.

Modelling suggests the primary reason HMP evolved is to avoid self-superparasitism (defined as laying again in the host by the same female), which would waste her second egg and risk the life of her first egg (Roitberg

Table 4.1. *Examples of host-marking pheromones in insects*

Insect	Example	Origin
Diptera Tephritidae *Rhagoletis* Fruit fly	$CH_3CH(CH_2)_6CH(CH_2)_6CONHCH_2CH_2SO_3H$ ... N - taurine	Midgut
Hymenoptera Ichneumonidae *Nemeritis* Parasitic wasp	Heneicosane	Dufour's gland
Lepidoptera Tortricidae *Cydia* Codling moth	Linoleic acid	Unknown
Pieridae *Pieris* Cabbage butterfly	Miriamide	Unknown
Coleoptera Anobiidae *Lasioderma* Cigarette beetle	Serricorone	Unknown

After Chapman (1998).

& Mangel 1988). Conspecific females gain from taking notice of the marks by not wasting an egg in an already parasitised host if the original female's offspring, which have a head start, would usually triumph. However, it may be adaptive for other females to superparasitise marked hosts when there are few unparasitised hosts available (Godfray 1994). Indeed, females wasps do adjust their rate of superparasitism according to the proportion of unparasitised hosts they encounter (Hubbard *et al.* 1999).

Some parasitoid species use only internal marks while others only mark externally and some use both mechanisms (Godfray 1994; Kainoh 1999). In many species Dufour's gland is the source of the mark and the variation in the chemical profile of gland secretion may allow distinction of self- vs. con-specific marks. Females of six out of nine parasitoid species could

Fig. 4.3. (Left) After an apple maggot fly (*Rhagolitis pomonella*) female has laid a single egg in a fruit, she deposits host-marking pheromone before leaving. Later, conspecific females walking on the fruit detect the pheromone and often, but not always, leave without laying eggs. Although *R. pomonella* is now a pest of apples, the original host was the hawthorn *Crataegus mollis*, whose small berries offer only enough resource for one larva each, so that larvae from later eggs rarely survive (Averill & Prokopy 1987). Photograph by R. Prokopy.
(Right) An adult female parasitoid wasp (*Trichogramma pretiosum*) in the process of parasitising an egg of the brown-tail moth. After laying her egg inside, she will mark the host egg with pheromone. Figure from Howard & Fiske (1911).

distinguish between hosts parasitised by themselves and those parasitised by conspecifics, perhaps by individual recognition of marks (van Djiken *et al.* 1992).

As well as marking hosts, some parasitoid wasps also pheromone-mark patches they have visited and spend more time searching unmarked patches (Hoffmeister & Roitberg 1997). Predatory insects behave in a similar way to the evidence that an area already has many conspecifics present. For example, ladybirds lay fewer eggs if they detect chemical tracks made by ladybird larvae (Merlin *et al.* 1996; Doumbia *et al.* 1998).

Nectar-feeding insects such as bees use marking pheromones to solve a different problem: how best to exploit a renewable resource by not returning to a flower too soon (see e.g. Giurfa & Nunez 1992; Goulson *et al.* 1998)?

4.4 | Conclusion

Chemical cues are among the most important cues used by animals to aggregate. While different taxa have been studied independently, for example marine barnacles and forest beetles, there are many common threads linking these studies. The renewed interest in Allee effects provides a useful focus on the individual benefits of aggregation. Host-marking pheromones can also be best explained by individual advantage. Chemical cues mediating responses to aggregate or disperse offer many possibilities for exploitation.

4.5 | Further reading

Chapters in Hardie & Minks (1999) discuss aggregation pheromones in bark beetles, stored-product beetles, and other insects. Godfray (1994) gives a good review of the evolution of marking behaviour in parasitoid wasps. Allee effects are well reviewed in a modern context by Courchamp *et al.* (1999) and Stephens & Sutherland (1999).

Chapter 5

Scent marking and territorial behaviour

5.1 | Introduction

Pheromones in the form of scent marks have a particularly important role in the territorial behaviour of mammals and other terrestrial vertebrates including lizards and salamanders. Territories can be broadly defined as defended areas (Stamps 1994). They are often for feeding, but can include other resources such as a den site, which is valuable for the owner or for attracting mates. Depending on the species, the territory might be owned by one male, such as a small antelope, or by a group, as in badgers. Many social insects are also fiercely territorial but they generally use scent marks in different ways from vertebrates (Box 5.1).

Pheromone or scent marking is one of the most conspicuous behaviours of many mammals. Mammals have an enormous variety of specialised scent glands but a common pattern of scent marking: glandular secretions, and often faeces and urine, are placed at conspicuous places in their home ranges or territories, often in lines along paths or boundaries (Fig. 5.1) (Gosling & Roberts 2001). Males tend to mark more than females, and dominant males or territory holders mark more than others. Scent marking forms a central part of many ritualised contests between territorial males or between competing groups, for example 'stink fights' between neighbouring groups of ring-tailed lemurs (Fig. 5.2).

One benefit of scent marks comes from the unique separation of signaller and signal: unlike other signals, pheromones 'shout' even when the animal is not there. However, while producing pheromones takes less energy than sound signals, scent marking nonetheless can involve significant costs in time and risk. For example, males of a territorial African antelope, the oribi (*Ourebia ourebi*), spend up to 35% of their time on marking or associated

Box 5.1 | Social insect territories

Many ant species fiercely defend foraging territories (Traniello & Robson 1995). Recruitment of major workers by alarm pheromones (Chapter 8) during territorial disputes is important in most species, but few ant species use pheromones to mark the territory boundaries in the way that vertebrates do (Hölldobler & Wilson 1990). African weaver ants (*Oecophylla longinoda*) are an exception as they mark their territory with colony-specific pheromones in faecal material (see figure below) (Hölldobler & Wilson 1978). When new 'territory' was offered to nests of *O. longinoda* in the laboratory, workers rapidly moved in and defecated systematically over the area (500 marks were made in the first hour in an area of 71 by 142 cm). When contests between two ant colonies were staged on arenas marked by one or neither of the colonies, ants walking over an arena already marked with their own colony's defecations were bolder, won more contests and took over

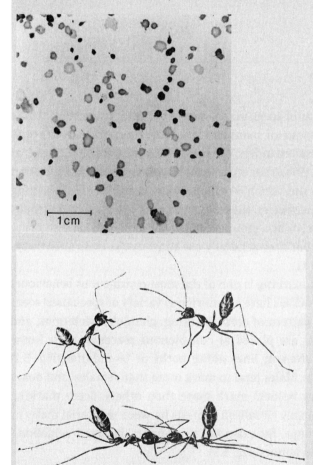

(Top) Chemical territorial marking by weaver ants (*Oecophylla longinoda*) shown by anal spots on a paper surface in a laboratory foraging arena. Experiments have shown that the spots contain a true territorial pheromone.

(Bottom) Two workers from different colonies in combat conducted by rearing up on their legs and mandibles in a threat display (above), dodging each other, and seizing one another with their mandibles (below). The ant fighting on an area marked by her own colony enjoys an 'owner advantage' typical in such cases. Figures from Hölldobler & Wilson (1978). Painting at bottom by Turid Hölldobler.

the whole arena in each of nine trials. If neither ant colony had marked the arena then contests were even (Hölldobler & Wilson 1978).

Social insect colonies have the advantage over mammals that they *can* be in many places at once when defending a territory (Hölldobler & Wilson 1990). Large colonies of *O. longinoda*, consisting of a single queen and up to 500 000 workers, can control exclusive three-dimensional territories. One Kenyan colony included the canopies of 17 trees (with a leaf surface area that would be equivalent, if the ant workers were scaled to human dimensions, to at least 100 km^2). In the related weaver ant, *O. smaragdina*, older workers dominate in barrack nests around the periphery of the territory, where territorial invasion is most likely. This led Hölldobler & Wilson (1994) to say that, while human societies send their young men to war, weaver ants send their old ladies. These ants are particularly aggressive towards members of other weaver ant colonies and the territories of different colonies are often separated by 'no-ant's-land' narrow zones into which few ants venture. In other habitats, the economics of territorial defence lead some ant species to defend trail trunk routes leading to valuable food resources rather than a wide area around the nest (Hölldobler & Wilson 1990).

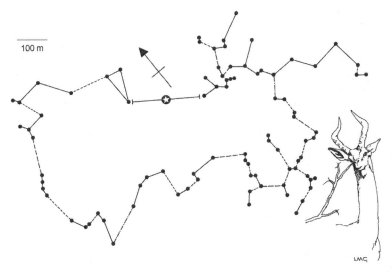

100 m

Fig. 5.1. Territorial male gerenuk (*Litocranius walleri*) mark the ends of projecting twigs with their large antorbital glands (inset). The pattern of marks is revealed by nearest neighbour mapping which 'joins up the dots' (left). Figures from Gosling (1981).

activities (Brashares & Arcese 1999b). Animals will go to some risk to investigate and overmark intruder's scent marks. Beavers, for example, will leave the safety of water to investigate foreign scent marks on the shore. Scent marks can be 'eavesdropped' by predators and parasites (Chapter 11).

Scent marks are honest signals, to an almost unique extent amongst animal signals (Gosling & Roberts 2001). This is because scent marks placed around a territory show that the marker is both successful in competition with other animals and has successfully held the territory long enough to mark it. Conversely, marks from a number of individuals indicate a group territory or an owner with only partial control of its territory. The

Fig. 5.2. The Madagascan ring-tailed lemur (*Lemur catta*) defends group territories by 'stink fights'. Males load their tails with secretions from their wrist and shoulder glands (right) and then wave their tails toward the opponents (left). The scent from these glands, which are testosterone dependent, identifies individual males and is used for intra-group dominance threats as well as inter-group territorial interactions. The olfactory threat is complemented by the conspicuous visual display of striking tail and face markings. Figure from Bradbury & Vehrencamp 1998, based on description and photographs in Jolly 1966.

components of pheromone marks are under endocrine control so their chemical composition will reflect the biological state of the marker, including its social status, health and nutrition (Chapters 3 and 9).

Even scent marks primarily for territorial marking may have many functions. For example, the yellow mongoose (*Cynictis penicillata*) in southern Africa, which lives in social groups of 10 or so animals, is unusual because most marking is done by subordinate adults of both sexes rather than by the dominants (Wenhold & Rasa 1994). The boundary marks may advertise the sexual maturity and identity of the subordinates, important because these animals go outside the territory to mate with neighbours.

5.2 | Why scent mark?

Despite their widespread use, how and why scent marks work is not clear. There are many different hypotheses (see Gosling & Roberts 2001). Among the three main theories are that scent marks (1) deter potential intruders (the **scent fence hypothesis** or chemical 'keep out' notice), (2) allow the intruder to recognise the owner, should he be met, by the match between his odours and those of the scent marks around the territory (the **scent matching hypothesis**) or (3) establish boundaries with major competitors (the **border maintenance hypothesis**). Scent marks could also provide information to intruding animals on the status of the resident through the intrinsic characteristics of the marks (for example, the compounds released in the urine of dominant mice, Chapter 9) or give the identity of the owner (allowing intruders to retreat before meeting the owner, if the intruder lost previously to that animal). Other suggested roles for marks within territories include the attraction or stimulation of mates (Chapters 3 and 6), and orientation of the resident.

The predictions for the different theories (which are discussed in Sections 5.3, 5.4 and 5.5) are not mutually exclusive, making it hard to assess them. The diversity of marking patterns among closely related species such as antelopes, or populations of the same species in different habitats, for example spotted hyenas (Section 5.6), shows the importance of ecological factors and suggests that it is probably a mistake to expect the selective forces for scent marking to be the same for all terrestrial vertebrates.

5.3 | Scent fence hypothesis

Scent marks could work as a chemical 'keep out sign' or 'fence' which would repel all individuals of a given category, for example, adult males. However, even though marks are often placed around the edge of the territory, field observations in almost all species studied so far show that territory intruders do not avoid marked areas (Gosling 1990). One exception is the European mole (*Talpa europaea*): males avoid the scent marks of other moles (whether familiar or unfamiliar), perhaps because escape is difficult in a confined burrow and the risk of injury in underground combat is high (Gorman & Stone 1990). There is some evidence that at least at low densities, beavers (*Castor canadensis*) will avoid marked sites (Sun & Müller-Schwarze 1999). However, in some species, marks paradoxically may make an area *more* attractive as they indicate that the habitat will support the species (Stamps 1994).

5.4 | Scent matching hypothesis

This hypothesis proposes that intruders learn the odour of the territory owner from the scent marks on the territory so that when they encounter another animal that matches the scent marks they recognise that this is the owner and can avoid fights they are likely to lose (Gosling & Roberts 2001). Scent matching helps to account for otherwise unexplained behaviour such as the way territory owners and other resource holders often mark their own bodies with the substances used for scent marking and make themselves available for inspection by sniffing at the start of many encounters (Gosling & Roberts 2001) (Fig. 5.3). Males of many species, for example the hyena *Crocuta crocuta*, evert their scent glands as they approach opponents (Kruuk 1972). Subordinates or intruders would be expected to search out scent marks, and this is the case in black-tailed deer (*Odocoileus hemionus columbianus*) and mice (see Section 5.4.1).

Fig. 5.3. The scent matching hypothesis would help explain two otherwise intriguing behaviours. First (left), self-marking by territory owners with the substances used to scent mark their territories: a territorial male Coke's hartebeest (*Alcelaphus buselaphus cokei*) rubs antorbital gland secretion onto its side. Second (right), presentation for inspection: in encounters with intruders, owners make the odours used to mark the territory available for olfactory inspection. This behaviour may allow intruders to identify the owner by matching the odour they detect with that of scent marks in the territory. A territorial male hartebeest stands immobile (far right) while an intruding male smells and nibbles the scent impregnated fur of its neck and shoulder. Drawings by L.M. Gosling from Gosling (1982).

The treatment of scent marks as sources of information (rather than as 'keep out' signs) by intruding males is selected for because they should not accept scent marks as indicating uncontestable ownership without meeting the owner and assessing his real capabilities. Owners are usually animals with high competitive ability and if they have more to gain by retaining their territory than an intruder has by taking it over (for example because familiar territory is more valuable than unfamiliar territory to an animal), it will pay the owner to escalate the fight (Gosling and Roberts 2001). Game theory predicts that if the identity of the territory holder is unambiguous, most intruders will retreat rather than become involved in an escalated contest that can result in serious injury. If scent marking and allowing inspection behaviour allows the territory owner to be identified unambiguously, then scent marks could reduce the cost of territorial defence for owners by avoiding costly (and risky) escalations. At stake are costs in time, energy and risk of injury or death. An intruder benefits from the knowledge that a particular individual it meets is the owner of that area (information gained by scent matching the smell of the owner with marks left over the territory) because the intruder can assess the long-term costs of interacting with that individual (Stamps & Krishnan 1999).

One major experimental difficulty in testing the scent matching hypothesis is that the intruder's response may be determined by information about the competitive ability of the owner received when they meet, for example from the resident's size and odour, rather than by matching scent to marks perceived earlier. Rather than part of a scent matching process, the resident allowing inspection could be advertisement of status by odours with dominance characteristics.

Does scent marking reduce escalated aggressive encounters between territory owners and intruders? The limited number of experimental tests of the hypothesis is explored in Sections 5.4.1, 5.4.2 and 5.4.3.

5.4.1 Mice

Dominant male house mice (*Mus musculus domesticus*) mark their territories extensively (Box 5.2). If males do assess potential opponents by comparing the opponent's odour with local scent marks, they should be less likely to fight if these odours are the same, indicating that the opponent is the resource holder (of territory or females). Gosling & McKay (1990) tested this by observing the behaviour of two unfamiliar males: an 'intruding' male faced with a 'resident' male on the other side of a mesh (Fig. 5.4). In some experiments the side of the intruding male was previously marked with urine from the 'resident', in others not. An air flow meant that the 'resident' could not detect the odours of the intruder. As predicted, intruding male mice were less likely to fight another male if the urine on its side matched

Box 5.2 | Marking behaviour of house mice

House mice (*Mus musculus domesticus*) are useful models for investigating the role of odour cues in structuring social groups. At high densities mice live in large groups and dominant males tolerate subordinate males on their territory. Urine marks are made all over the territory and all new objects are quickly marked. Dominant males mark at signal sites, which become 'urine posts' up to 25 mm high. Urine posts develop near resources such as food or nesting sites and along pathways. Dominant resident males and breeding females mark at much higher rates, up to 100 times an hour, than subordinates (Drickamer 1995). Within the group, low level marking by subordinates away from the urine posts and diffusely across the territory helps to ensure continued tolerance by the resident dominant male. (See Chapter 1 for discussion of urinary proteins and message life).

If an experimenter adds a small (mouse-sized) urine mark to a signal site, all males investigate these scent marks (Hurst 1993). If the urine was from a resident subordinate, the dominant male soon attacked that individual without warning. If the urine was from an unfamiliar subordinate, the resident dominant male investigated all subordinate males in his territory. If the artificial mark came from an unfamiliar dominant male, then there was much greater investigation of the urine post by the dominant male, more countermarking of it by him and a greater likelihood of the dominant male himself fleeing if challenged. A dominant neighbour's urine would almost be ignored (the dear-enemy phenomenon perhaps, see Section 5.7). It could be argued that some of the effects were due to the body odour of dominance rather than to reference to scent marking. Mice are able to recognise urine as coming from dominant males because of high levels of certain pheromone molecules (Chapter 9). However, the complex effects on the behaviour of responding males, depending on the source of the urine used to make the artificial urine mark (whether resident, dominant, or subordinate) suggest that the responses could depend on recognition of the individual odour of other mice in the group (see Section 9.6.1.1) rather than on the general characteristics of urine from subordinate or dominant males. Over-marking (see Section 5.8) may be used in mate choice by female mice (Rich & Hurst 1999). Females were more strongly attracted to approach territory owners that counter-marked the scent-mark challenges of competitors than those that had been countermarked.

the odour of the 'resident' male on the other side of the mesh. If assessment by the intrinsic characteristics of the marks was important, then differences in behaviour depending on whether the odour matched or not would not be predicted. In other experiments, Gosling and colleagues found that responses by subordinate mice to scent marks depended on their competitive ability: smaller mice tended to avoid marked areas, whereas relatively larger subordinates, with a greater chance of winning a contest with the owner, were more likely to investigate marked areas (Gosling *et al.* 1996). In experiments with larger more natural arenas (Hurst 1993), intruders were likely to flee if they encountered the dominant resident (recognised at a distance by sniffing) (see Box 5.2).

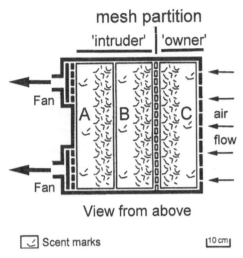

mesh partition

View from above

☑ Scent marks |10 cm|

Fig. 5.4. Testing the scent matching hypothesis is not easy because when the intruder compares scent marks and an individual met on a territory, the intruder gains information about the intrinsic abilities of the marker, not just the likelihood that the marker is matched to the owner of the territory. To control the flow of information, Gosling & McKay (1990) used an air flow to prevent the 'owner' mouse receiving odour cues from the intruder. The 'intruding' animal (left-hand side of mesh partition) had access to section A (on a substrate bearing its own scent marks) and access to section B (on the experimental substrate bearing either the marks of the owner or those of an unfamiliar male). The owner was confined to section C on substrate it had marked itself. The owner and intruder could interact through the mesh partition between B and C. Substrates were scent marked on the night before each experiment. Figure redrawn from Gosling & McKay (1990).

5.4.2 Salamanders

The scent matching hypothesis has also been tested in the terrestrial red-backed salamander (*Plethodon cinereus*) which lives in broadleaved forests in eastern North America. The adult salamander establishes a feeding territory under a stone or log. The territory is dotted with scent marks from the post-cloacal gland, which are placed on the substrate and on faecal pellets. Intruders repeatedly nose-tap these scent marks (nose-tapping allows access of molecules to the animal's vomeronasal organ, Chapter 9) (Jaeger & Gabor 1993). Consistent with the scent matching hypothesis, the scent marks themselves did not deter intruders. When intruder and resident met, both often nose-tapped each other, especially in the postcloacal gland region, giving an opportunity for scent matching. These initial contacts tended to reduce escalation of the contest to biting (Jaeger & Gabor 1993) as predicted by the theory. However, in an experiment following the design of Gosling & McKay (1990), escalated aggressive encounters between territory owners and intruders were not reduced when the intruder was on an area marked by the resident (Simons *et al.* 1997). Other information about the resident, such as his size, will also be given by faecal pellets, and may also be indicated by the pheromones left on the substrate. Even in the absence of the resident, intruders spent more time in a submissive posture when placed on substrate marked by a bigger resident male (Mathis 1990a).

5.4.3 Beavers

North American beavers (*Castor canadensis*) live in closed family groups centred on a lodge in a pond or dammed stream. Members of the group scent mark mud piles made along the shoreline (Sun & Müller-Schwarze 1999). Beavers have two main pheromone glands, which produce 'castoreum' and anal gland secretion respectively. The anal gland secretion differs between individuals, but family members have more similar profiles than unrelated animals; the anal gland secretion may thus be used for kin recognition (Chapter 6). Either gland may be used to scent mark although castoreum appears to have the more important role in territorial interactions between family groups (Sun & Müller-Schwarze 1998).

Castoreum is a strong-smelling secretion rich in phenolics. Beavers can distinguish between castoreum from family members, neighbours, and non-neighbours (Schulte 1998). When territory owners encounter scent marks made by non-group beavers they over-mark with their own secretion and often destroy the foreign scent mound. Sun & Müller-Schwarze (1998) reasoned that territory owners should habituate to repeated scent marks made by an unknown beaver so long as the intruding animal was not met (increasing scent signals alone would not be important if scent matching is the mechanism used). Conversely, if territorial marks are used as a scent fence (to deter entry into the territory) then the territory owners should increase the level of marking if intrusion by the strange beaver apparently continued. In the experiments carried out the behaviour of the beavers differed according to the gland secretion used to mark the experimental scent mounds: over six nights, there was no change in response to the anal gland secretion of the foreign beaver, whereas for castoreum, the territorial response declined or was level. Sun & Müller-Schwarze concluded that these results were consistent with the scent matching hypothesis, given other evidence that castoreum is the main territorial scent.

5.5 | Border maintenance hypothesis

In some species, males mark preferentially along borders adjacent to the most threatening rivals. Scent marks placed along territory borders may serve to form a 'property line' between neighbours, thereby preventing frequent and costly disputes between territory owners (Brashares & Arcese 1999a; Gosling & Roberts 2001). The oribi (*Ourebia ourebi*) which has some territories containing a dominant male and subordinate males, scent marks with both its preorbital gland secretion and faeces. Territorial males in the

Serengeti National Park, Tanzania, marked grass and shrub stems with their preorbital gland up to 45 times per hour (Brashares & Arcese 1999a). Almost all the marks were made along borders shared with other territorial males and males marked more often at borders shared with multi-male groups than at borders shared with a single male. This suggests that males perceived neighbouring male groups as a greater threat to territory ownership than neighbouring males that defended their territories without the aid of adult subordinates (almost half of takeovers of territories are by neighbours, Arcese 1999). The subordinate males contribute up to half the scent marks and help to defend the territory. In klipspringer antelope, males placed more marks on contested boundaries and on branch tips facing likely intruders (Roberts & Lowen 1997).

5.6 | Economics of scent marking patterns in territories

Marking strategy is an outcome of a trade-off between costs of producing and positioning scent marks and the benefits to the owner from the behaviour (Gosling & Roberts 2001). If a territory owner marked to ensure that any intruder would *always* encounter marks it would have to mark the whole perimeter. Instead, animals tend to produce a limited number of marks, at heights and places likely to be found, such as along paths (Fig. 5.1). There are many examples that suggest marking is limited – either by material (faeces, urine, gland marks) or by time; such economic limitations become more critical as the area of the territory increases. For example, even faeces, which may be 'free', are limited by the amount the animal can produce. The optimum position for a ring of scent marks may be some distance within the territory rather than at the actual boundary. This is found in klipspringer antelope and hyena latrines. Modelling suggested that this would maximise the probability of intruders encountering the marks and minimise the cost of intrusion (Roberts & Lowen 1997).

Most antelope marks are 'boundary' or 'perimeter' marks, towards the edge of the territory. Many carnivores also show this pattern. For example, like North American grey wolves (*Canis lupus*), Ethiopian wolves (*Canis simensis*) make the most urine marks along or near territory boundaries (Sillero-Zubiri & Macdonald 1998). Some species instead mark throughout the territory or towards its centre (hinterland marking), for example preorbital marking by dikdik (a small antelope). Among carnivores, which are noted for both interspecific and intraspecific variation in behavioural ecology (Macdonald 1985b), there may be great intraspecific variation in marking patterns depending on habitat. For example, spotted hyenas (*Crocuta crocuta*) living in large clans with small territories (in the Ngorongoro Crater, Tanzania) paste their marks along

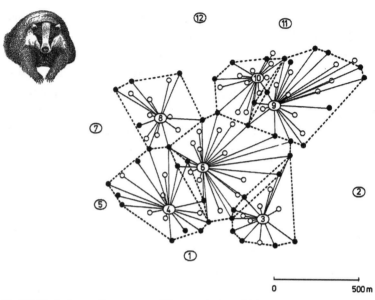

Fig. 5.5. The latrine marking patterns of European badgers (*Meles meles*) on the South Downs, an open landscape of chalk hills in southern England. Baiting each sett (den) with peanut butter containing pieces of a different bright coloured plastic marker revealed the patterns. Numbered ovals denote main setts; filled circles denote shared latrines; open circles denote unshared latrines in the territory hinterland. A line connects each latrine where bait markers were traced to the sett where the markers originated. Broken lines show territory boundaries drawn as minimum-area convex polygons around the outermost shared latrines belonging to each social group. Main figure from Roper *et al.* (1993). Badger from Cockburn (1991).

the perimeter. In contrast, in the Kalahari desert (southern Africa), where spotted hyenas live in small clans and very large territories, marks are clustered towards the centre of the territory (Gorman & Mills 1984). This may be because the territory is large in comparison with the number of defenders, making regular visits to the boundary uneconomical. A similar pattern is shown by four species of Hyaenidae: the aardwolf, spotted hyena, striped hyena and brown hyena. These species show hinterland marking behaviour when the length of the territory boundary divided by size of territory group is relatively large (Gorman & Mills 1984). This occurs both within and across species and indicates that the marking strategy is not species-specific but dependent on ecological factors that dictate the optimum distribution of marks.

European badgers (*Meles meles*) have both perimeter (border) and hinterland latrines (Roper *et al.* 1993) (Fig. 5.5). Within the social group the two sexes have different marking behaviour. Perimeter latrines are visited mostly by males, hinterland latrines by both sexes, but particularly by females. The early spring peak of perimeter marking by males during the main time for mating, suggests that male marking might be mate guarding or territory guarding for females rather than for food. Female marking on the other hand may be for protecting the den as much as for food. Marking and social behaviour are likely to be complex.

Long-lasting marks with a low volatility may be missed by animals attempting to find them by olfaction from a distance. Across the animal kingdom, many animals have ways of making marks more conspicuous (Alberts 1992). For example, the marks made by the desert iguana *Dipsosaurus dorsalis*, include molecules that strongly absorb long-wave ultraviolet light, which lizards see as dark patches (Chapter 1) (Alberts 1990). Antelope and hippopotamus dung heaps stand out visually. Many animals, including dogs, choose conspicuous objects such as lamp posts or trees to mark (Macdonald 1985b).

5.7 | Dear enemies

Many territorial animals respond more aggressively to intrusions by non-neighbours (strangers) than territorial neighbours; this is termed the 'dear-enemy phenomenon'. Several hypotheses propose that if animals learn to recognise and respect each other's territorial boundaries they can reduce the energy expended in territorial defence and more successfully focus defence against non-territorial floaters (strangers) (see review by Temeles 1994). However, territory owners are sometimes more aggressive to neighbours than floaters. Temeles suggests that owner responses are not based on the familiarity with neighbours or strangers but rather on the potential gains or losses in each interaction and the kind of resource being defended.

5.8 | Over-marking

Dogs sniffing the marks on lamp posts left by previous dogs and adding their own urine mark are perhaps the most familiar example of the common mammal behaviour of scent counter- or over-marking. Finding a mark from a stranger often prompts the over-marking behaviour, particularly if it is on the owner's territory. Adding more marks to a site could have three potential outcomes (1) scent blending may occur, (2) individual scents may remain distinct and (3) scent masking might occur (Johnston *et al.* 1994).

If scent blending occurred then a group odour might be created although this has not yet been established for carnivores (Macdonald 1985b). However, blending of marks on social insect nests is an important mechanism in formation of the colony odour of social wasps and bees (Chapter 6). Blending has been proposed for rabbits, mongooses, meerkats, beavers and marmosets.

If individual signatures in over-marks remained distinct, a chemical bulletin board or message centre would be created, where individuals could advertise themselves and find out about others in the area. This has been proposed for canids, spotted hyenas, and lions. For the marks to remain distinct, one might expect to have marks distributed over a relatively large marking site.

Finally, over-marking could 'mask' the previous marks, a hypothesis suggested for many mammal species that mark in territorial or home-area defence and/or advertisement of dominance, for example red foxes and mice.

One approach to disentangling these hypotheses is to ask what information animals receive from sniffing marks. In a fascinating series of experiments, Robert Johnston and colleagues have shown that male hamsters appear to ignore earlier marks and treat only the most recent (top) mark as 'familiar' when tested in a habituation experiment (Chapter 2) (Johnston *et al.* 1994). This was true even if the top mark was applied only 30 seconds after the bottom mark and if smaller in area and only partly covering the earlier one (Wilcox & Johnston 1995). The bottom of two scent marks was thus treated *as if* it was 'masked'.

5.9 | Scent marking in non-territorial mammals

Scent marking is also a significant behaviour of mammals that do not defend territories, including those with dominance and harem-defence mating systems such as bison and white-tailed deer (Gosling & Roberts 2001). Males may mark frequently as they move through their range and males in some species directly scent mark females, for example some species of antelope and rodents such as the South American mara (Fig. 5.6). Males in other species, such as goats, may pass on urine odours sprayed onto their underparts to females when they mount. Direct marking of females may be to advertise the costs to other males of any attempt to compete for the female. Could this be a form of chemical mate guarding (Chapter 3)? Scent matching

Fig. 5.6. Males in species with polygynous mating systems may mark females. For example, a male mara (*Dolichotis patagonum*, a South American rodent) marks a female with urine. Figure from Macdonald (1985a). Drawing by Priscilla Barrett.

could potentially be used by females to choose males as dominant males mark most frequently even in non-territorial species.

5.10 | Conclusion

Pheromone or scent marking is one of the most conspicuous behaviours of many mammals and other terrestrial vertebrates. Scent marks appear to be involved in defending territories. The three main theories are that scent marks (1) deter potential intruders (**scent fence hypothesis** or chemical 'keep out' notice), (2) allow the intruder to recognise the owner, should he be met, by the match between his odours and those of the scent marks around the territory (**scent matching hypothesis**) and (3) establish boundaries with major competitors (**border maintenance hypothesis**). The lack of unique predictions for any of the theories makes testing the hypotheses difficult. There are great opportunities for well designed field or laboratory experiments to explore scent marking behaviour further.

5.11 | Further reading

Scent marking in mammals is reviewed by Gosling & Roberts (2001). Chapters in Johnston *et al.* (1999) and Marchlewska-Koj *et al.* (2001) cover scent marking by particular species, including mice, beavers, and hamsters. Hölldobler & Wilson (1990) cover the behaviour of ants.

Chapter 6

Pheromones and social organisation

6.1 | Introduction

The most complex animal societies so far described are found among the social insects and mammals. The individuals in these societies interact via a complex web of semiochemical signals, including the odour signature that gives each member access to the group.

In both mammals and social insects, each individual has a great complex of odours produced by the animal and acquired from the environment and from conspecifics. The resulting chemical signatures of both mammals and social insects are complex and variable mixtures, giving a forest of peaks on a gas chromatograph trace, in contrast to the small number of defined peaks for the sex pheromones of moths and other insects (see Chapters 1 and 2). These complex mixtures reflect the overlaying of many different messages. For example, the saddle-back tamarin (*Saguinus fuscicollis*) a South American primate, produces chemical messages which identify species, subspecies, individual and gender, and may also contain information on social status (Epple *et al.* 1993). Social insects carry a chemical message on their cuticle that includes information about their species, colony, caste, age and gender. In both mammals and social insects the cues giving reproductive status, in particular ovarian status, may be the key to the role of pheromones in reproduction in social animals (Section 6.3.2). While signals of caste, gender, life stage or species may not vary much within the species and could thus be said to be anonymous (Hölldobler & Carlin 1987), the variability of colony and kin recognition chemical signatures is what gives them their specificity.

There are two main themes in this chapter: first, how chemical cues are used in colony and kin recognition (crucial for delimiting societies but also important for less social species); and, second, current debates about the role of primer pheromones in influencing reproduction in social groups – are they agents of control or signals?

6.2 | Colony, kin, family and individual recognition

Recognition of kin or fellow group members is central to social behaviour, whether a colony of millions or small family group. Throughout the animal kingdom almost all recognition relies on learning (Box 6.1) and chemical cues predominate.

6.2.1 Colony and kin recognition in social insects

The ability to recognise nestmates has been found in almost all social insects that have been studied. Typically, colony members will accept nestmates but exclude and possibly kill non-nestmate conspecifics (Breed 1998). This can be

---

Box 6.1 | Recognition mechanisms

How animals distinguish members of their group from non-members is a key behaviour allowing them to favour offspring and other relatives (kin) or fellow group members (see reviews by Sherman et al. 1997, Slater 1994, Pfennig & Sherman 1995). Kin recognition is also important for optimal outbreeding by avoiding close kin as mates (Chapter 3).

CUES AND MECHANISMS FOR RECOGNITION

Kin recognition cues may be any aspect of the phenotype that reliably signifies kinship (Sherman et al. 1997). Chemical cues are widely used for recognition signatures (labels), perhaps because even the earliest organisms had the receptor mechanisms for receiving and processing the information (Chapter 9) and perhaps also because of the enormous variety of compounds available, which allows an effectively unlimited number of possible combinations. In most cases, the metabolic costs of chemical labels are low because they are not produced especially for recognition purposes but are, instead, metabolic by-products, acquired from food or the nest, or have other primary functions.

The signature cues can be genetic, environmental, or a combination, and either endogenous (produced by the animal itself) or acquired from the environment. For example, odours implicated in parent–offspring recognition in some mammals, including humans, seem to be genetic and produced by the individual itself (see Section 6.2.2.1 and Chapter 13), whereas the hydrocarbon odours acquired from the nest by newly emerged *Polistes* paper wasp adults (which are secreted onto the nest by nestmates) are in part genetically determined, and partly of environmental origin (from plant material used in the nest, see Section 6.2.1.2). Tadpoles of some frog and toad species use acquired environmental cues around the spawn to recognise each other as siblings (Chapter 8). Thus chemical labels do not have to be genetically linked: a statistical association can be just as strong with environmentally acquired cues (Sherman et al. 1997).

Perhaps surprisingly, recognition cues are usually learned through behavioural 'rules', such as 'learn the odour of your nestmates'. There are three main potential mechanisms which animals use to recognise others as kin (see figure below): first, by learning the characteristics of surrounding individuals (by direct familiarisation with nestmates); second, by using this learning to allow phenotypic matching with unfamiliar kin; and third, by using self-inspection – the armpit effect (Dawkins 1982) – to allow phenotypic matching with unfamiliar kin. All three mechanisms rely on learning a memory template.

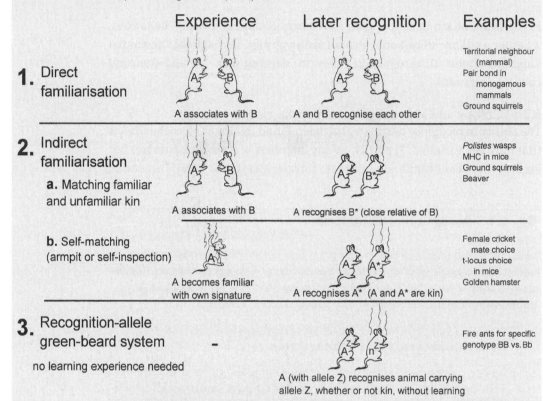

Kin recognition mechanisms in almost all animals, vertebrates and invertebrates, seem to involve learning a signature and then matching this template against the signature cues of other animals. (The diagram is somewhat anthropomorphic as mice do not have smelly armpits – but humans do.) Three mechanisms are represented:

(1) Direct* familiarisation, by learning the characteristics of nestmates and recognising these animals later.

(2) Indirect* familiarisation (phenotypic matching):
   (a) learning the characteristics of nestmates and using the template to allow phenotypic matching with unfamiliar kin; and
   (b) learning the odour of self to allow phenotypic matching of self with others (self-referent or 'armpit' phenotypic matching).

(3) Recognition allele ('green beard'). A (with allele Z) recognises an animal carrying allele Z, whether or not kin, without learning.

*Note: 'direct' and 'indirect' are used as by Porter & Blaustein (1989). The same words are used in a very different way by other authors who use 'indirect' for kin recognition rules using location e.g. 'any baby in the nest is treated as kin', compared with 'direct' for learning phenotypes, which would allow recognition away from the location, e.g. Pfennig & Sherman (1995) and Waldman et al. (1988).

Figure after Porter & Blaustein (1989) with modifications and additions.

For familiarisation and phenotypic matching there is usually a sensitive period, as with other imprinting, when chemical cues are learned. For example, in the Belding's ground squirrel and many other mammals, young animals learn the odours of their siblings, and mothers learn those of their offspring in the burrow, during the first few hours or days of life (Section 6.2.2). Self-matching seems to be used less than the learning of nestmate phenotypes. While in theory mice could compare their own major histocompatibility complex (MHC) odours with those of potential mates, instead they compare these with the MHC odours of nestmates imprinted early in life (Box 3.1). Similarly, *Polistes* wasps learn the odours of the nest rather than their own odour. Self-matching may be favoured in species where the young grow up alone (e.g. crickets, Chapter 3) and lacking contact with kin for learning, or if the available relatives in the nest would give error-prone templates, as for example when nestmates include full and half-siblings from multiple matings (Sherman *et al.* 1997). This second situation is the case of the golden hamster (*Mesocricetus auratus*), which mates multiply and produces multiply sired litters. Recently, recognition of kin by self-referent ('armpit') matching has been demonstrated in this species (Mateo & Johnston 2000). Hamsters that were reared only with non-kin since birth responded differently to the odours of unfamiliar relatives and non-relatives. Post-natal association with kin was not necessary for this discrimination. A theoretical means of recognition *without* learning is to have a system of three linked genes that code for a signal (visual or olfactory, etc.), *and* the genetic ability to recognise it in others, *and* a genetically determined appropriate response (Haig 1997) (see figure). This is the so-called 'green beard' phenomenon (Dawkins (1982) hypothesised linked genes that gave the owner a green beard *and* prompted the green-bearded person to look after others with green beards). There were no examples until recently, when fire ants were shown to demonstrate this behaviour via an odour cue 'green beard' (Keller & Ross 1998).

seen at the entrance to a honeybee hive. The guard bees use their antennae to touch all bees attempting to enter. Strangers are identified by their cuticular chemical signature and attacked. Guards are found in many species of termites, ants, social wasps and social bees. Guarding is the first line of colony defence and is important to prevent conspecifics from other colonies robbing food or brood and to prevent the entry of social parasite species (Chapter 11).

All mechanisms for nestmate recognition in social insects involve some form of learning (Box 6.1) (Smith & Breed 1995). The odour of nestmates is learnt, giving a 'template', which can be compared with the odour of other family members, and by phenotype matching with the template, nestmates and non-nestmates can be distinguished, effectively discriminating between kin and non-kin (Lacey & Sherman 1983).

6.2.1.1 Honeybees

The queen may mate with up to 20 males (Oldroyd *et al.* 1997) and so a colony consists of honeybee workers that are full sisters and half-sisters in different patrilines (with different fathers), yet guard bees accept all nestmates. What odours do the guard bees respond to?

The colony odour on the surface of the bee comes from cuticular pheromones secreted by the bee itself, together with compounds picked up from the comb wax in the nest (see review by Breed 1998). The compounds in the comb wax include pheromones produced by other workers and floral scents brought back to the nest during foraging. The cuticular pheromones are under genetic control. In laboratory experiments with small numbers of bees reared in isolation, away from the nest, bees can distinguish between full sisters, half-sisters or non-sisters, all reared in isolation (Getz 1991) (presumably the bees learn their own odour using 'self-inspection' or the 'armpit effect', see Box 6.1). However, laboratory-reared bees like these, emerging isolated from the nest, are rejected by their sisters when introduced to their parental hive at 5 days old so the bee's own cuticular pheromones are not enough for acceptance. The key cues are the compounds that the bees pick up from the comb wax. As little as 5 minutes contact with comb wax from a foreign nest is enough to contaminate a bee and cause attack by its sister guards (Breed et al. 1995). Cues are both picked up and learned from the comb: a bee exposed to comb from her own or another colony is rejected by unexposed bees reared in the laboratory, but will be accepted by laboratory-reared bees, whether nestmate or not, which have been exposed to the same comb.

Comb wax is an extraordinarily complex mixture of compounds from at least two glands of the bee – the wax glands plus mandibular gland secretions, which add hydrocarbons and fatty acids (Breed 1998). These pheromone secretions are under genetic control so workers from the different patrilines will be contributing their potentially different secretions to comb building, with the overall odour reflecting the mixture of genotypes in the nest. Different nests will, therefore, have different profiles. The fatty acids and hydrocarbons are moulded into the wax comb, then transferred to workers as they contact the comb. The result is a colony-level signature that is also passed on from insect to insect and differs little among workers in the colony. The fatty acids may prove to be key elements in nestmate recognition, as well as hydrocarbons such as hexadecane, octadecane and heneicosene (Breed 1998). In a cross-fostering experiment to partition genetic and environmental influences on guard acceptance, Breed et al. (1988) found that the environmental effects of hive exposure were so strong that no genetic effect was observed. It thus seems that in nestmate recognition cues from comb and colony environment have priority over cuticular pheromones produced by the bees although these can be used if other cues are absent.

The recognition priority given by bees to the overall colony odour may be the proximate mechanism for the low levels of conflict within the nest despite the many different patrilines. Whether workers from different patrilines show favouritism (nepotism) towards rearing queen larvae sharing their own father has been a controversial topic. Visscher (1998) concludes that queen-rearing nepotism does occur, but only weakly. One reason for the low levels of conflict is the potentially high cost of nepotistic queen rearing. Another could be the small benefits that result (possibly because workers are unable to discriminate full sisters and half-sisters accurately) (Bourke & Franks 1995; Seeley 1995, p. 226). Cohesion of the colony would suit the queen, in order to avoid having conflict between daughter workers fathered by different males. At colony-level selection this would also benefit workers. Alternatively, kin discrimination of different patrilines could be a non-adaptive by-product of the inter-colonial recognition adaptive for defence against the invasion of parasites and non-colony members (Grafen 1990a; Alonso 1998).

6.2.1.2 *Polistes* **and other wasps**

The primitively eusocial paper wasp (*Polistes fuscatus*) forms an excellent model system for investigating the development of kin recognition based on semiochemicals (reviewed Gamboa 1996, and Singer *et al.* 1998). After over-wintering, queens tend to found nests with close relatives, which they recognise by odour cues. There is nonetheless reproductive conflict as one of the queens lays most of the eggs (Section 6.3). The colony will accept orphan kin (for example, from nearby nests destroyed by birds) but they need to keep away non-kin, which may come to rob the nest. Aunts, nieces, and a minority of cousins are accepted.

Paper wasps and honeybees are strikingly similar in their use of nesting materials for transmitting recognition pheromones among colony members. Adult wasps (in particular the queen) deposit secretions, largely from the sternal glands, on to the nest surface. The secretions are hydrocarbons that can be extracted from the nest in hexane (Espelie *et al.* 1994). The nest will thus have the queen's chemical signature as well as compounds from environmental sources including the nest material.

When it emerges from its pupa, the young paper wasp rests on the natal comb and learns the odour template from the secretions laid down on the paper comb of the nest, not from its own cuticular chemistry. The young wasp recognises nestmates via the odours she learned from the nest surface (without this learning experience the young wasp is not able to distinguish nestmates from non-nestmates). The hydrocarbons are colony specific and heritable. Close non-nestmate kin (aunts and nieces but not first cousins) may have cuticular hydrocarbon profiles sufficiently similar to the resident's nest profile to allow them to be accepted. Gamboa *et al.* (1996) were able to use hydrocarbon profiles to predict if a given wasp would be accepted or rejected.

As in honeybees, the use of the nest like an artist's mixing palette for the chemical secretions of the paper wasp nest members allows the genetic profiles of all members of the colony to be accommodated. In the tropical paper wasp *Ropalidia marginata*, as in *P. fuscatus*, females discriminate nestmate from non-nestmate only after prior exposure to nest and nestmates (Venkataraman *et al.* 1988). The use of the nest odours as a template rather than genetic markers may have evolved because many of the wasps helping the colony are not genetically related (young strangers may join the colony and are accepted because the colony benefits) (Arathi *et al.* 1997).

6.2.1.3 Sweatbees

The guarding behaviour of a primitively eusocial sweatbee (*Lasioglossum zephyrum*) provided some of the first tests of kin recognition in Hymenoptera (see review by Smith & Breed 1995). This bee has small colonies consisting of a queen plus two to 20 daughters. Greenberg (1979) showed that the willingness of a guard bee to let in another bee is directly correlated with the relatedness between them. The recognition cues were learnt by frequent contact with nestmates, but closely related bees were allowed to enter even if they had not been met before, which suggests a genetic basis to some of the cues.

The identity of the recognition cues is not known, but candidate compounds include macrocyclic lactones and mono-unsaturated homologues from Dufour's gland. These have been shown to be correlated with species, familial and individual differences (Hefetz *et al.* 1986). Males use Dufour's gland odours to choose mates for outbreeding (Smith & Breed 1995) (Chapter 3).

6.2.1.4 Ants

Cuticular hydrocarbons are likely to be the main chemical cues for nestmate recognition in ants (Lenoir *et al.* 2001), although the role of hydrocarbons is still debated (see Vander Meer & Morel 1998). According to the 'gestalt model' (see Lenoir *et al.* 2001), the colony odour is gained by the sharing of these surface chemical cues. Ants may differ from bees and wasps since involvement of secretions deposited on nest materials seems limited (Smith & Breed 1995), and recognition cues produced by the queen and spread to individuals in the colony seem to be important in some genera, such as *Camponotus* (Carlin & Hölldobler 1987), but not in others, such as *Solenopsis* (Obin & Vander Meer 1989). Secretions of an exocrine gland in the head (the postpharyngeal gland) may be the source of colony hydrocarbons then transferred by grooming of nestmates (Soroker *et al.* 1998). The great diversity of social organisation in ants is likely to be matched in the variety of mechanisms used for nestmate recognition.

Newly eclosed workers (callows), after emerging from the pupa, have to pick up the colony odour from interactions with older workers. Until they have gained the colony odour they avoid attack by moving little and avoiding aggressive behaviours (see Chapter 11 for the exploitation of recognition rules by nest parasites) (Vander Meer & Morel 1998).

6.2.1.5 Termites

Much less is known about colony recognition in termites but it seems to be based, as in Hymenoptera, on blends of cuticular components; a similar role for genetically determined and acquired components in the blend is likely (Clément & Bagnères 1998). Termites are capable of kin distinctions as subtle as any hymenopteran: DNA fingerprinting suggests that in polygynous and polyandrous termite colonies, workers departing from the nest to their foraging areas tend to form working parties with their kin (Kaib *et al.* 1996).

6.2.2 Group, kin, family, and individual recognition in mammals

Mammal societies in species such as dwarf mongooses, ground squirrels, or hyenas form long-lasting, stable groups. Recognition of group members, which will often be kin, is important. They can then be favoured in altruistic behaviour, for example in alloparental care given by brown hyenas to young cousins, or sharing of limited food resources by spiny mouse siblings, or alliances by ground squirrel sisters to chase off territorial intruders (Porter & Blaustein 1989). The closest kin are parents and offspring: where confusion is likely, recognition of your own offspring is important.

6.2.2.1 Mother–infant recognition

A key role for olfaction in mother–infant recognition has been shown in many mammal species, including humans, and has been intensively studied in sheep. Many lambs are born to the flock in a short period so each mother (ewe) needs to recognise her offspring. She learns the odour of her lamb within the first few hours after birth (Chapter 9) (see review by Lévy *et al.* 1996). In this early period, but not later, an orphan lamb will be accepted and adopted, particularly if coated with amniotic fluid to encourage licking by the ewe (a method traditionally used by shepherds, see Chapter 12).

The mother uses phenotype matching (Box 6.1) to recognise her lamb, not odours produced by 'her genes' in the lamb or her secretions on it. This was shown by a neat study in which a surrogate mother sheep gave birth to identical twin lambs, from implanted monozygotic twin embryos (Romeyer *et al.* 1993). If allowed to learn the odour of one twin, she could recognise its identical twin, even though she was not related to either of them. This shows, first, that she does not use a comparison of the lamb's odour with her own for recognition and, second, some of the lambs' odour is genetic because, while she could recognise an identical twin, she could not recognise non-identical twins. Nonetheless, environmental odour cues acquired by the lamb are also important and some of these may come from the mother's saliva and milk, for example. Although these 'maternal labels' are not required for recognition, they are incorporated into the lamb's recognisable odour phenotype and learned by the mother (Lévy et al. 1996). Once the ewe has learnt the odour of her lamb, adding a foreign scent (whether vanilla or that of another lamb) does not confuse her. Therefore, it is her lamb's own complete odour that is important, not the absence of foreign scents.

Parents recognise their offspring but offspring also learn the odours of their parents. For example, human babies will turn their head towards their mother's odour, learnt soon after birth (Chapter 13).

6.2.2.2 Kin recognition

Many mammals use the rule of thumb 'learn the odour of your nestmates', as in the wild these are likely to be kin (Box 6.1). The mechanisms of learning odour templates can be studied in cross-fostering experiments. For example, young spiny mice (*Acomys cahirinus*) prefer to huddle with nestmates, which would normally be siblings, but if the experimenter creates mixed litters of non-sibs, they later treat the familiar but unrelated individuals as if they were sibs (Porter et al. 1981). Diet odours are an important source of cues used by the juveniles to recognise animals as familiar or not, and for offspring and parents to recognise each other as kin; non-kin fed on the same distinctive diet may be treated as kin (Porter et al. 1989). Nonetheless, genetic components to the odour profile are important.

Learning a combination of environmental and genetic odour cues is also implicated in kin recognition by Belding's ground squirrels (*Spermophilus beldingi*) (see review by Sherman et al. 1997). Mothers, daughters and sisters behave nepotistically, warning each other about predators and protecting their own and each other's pups against infanticide by establishing and defending a territory together. Nestmate females learn each others' odours just before they wean (when litters normally begin to mix) and later treat each other as siblings regardless of actual relatedness. This works because unrelated pups are rarely in the same burrow until weaning (and until then mothers will accept any offspring in the burrow – after weaning they discriminate by odour). Cross-fostering experiments show how odours of nestmates or parents are learnt during the sensitive period, so that as in spiny mice, ground squirrels are less aggressive towards familiar unrelated nestmates than unfamiliar unrelated ground squirrels. However, despite having shared the same nest, littermate full sisters and maternal half-sisters, resulting from multiple mating by the mother, seem to treat each other differently. In the field they were less aggressive towards full siblings than half-siblings, and more likely to share a territory with a full sister (Holmes & Sherman 1982). Phenotypic matching appears to allow them to behave differently to kin they have not previously met: in laboratory tests they are less aggressive to unfamiliar siblings than unfamiliar non-siblings (some genetic basis to the template odours is suggested because unfamiliar paternal half-siblings are preferred to unfamiliar non-siblings, Holmes 1986).

Relatively few field tests of phenotypic matching have been made but recent results from studies of beavers (*Castor canadensis*) support this mechanism (Sun & Müller-Schwarze 1997). As they disperse, young beavers might encounter unfamiliar siblings from litters dating from before they were born. The young beavers reacted less strongly towards anal gland secretions from unfamiliar siblings than those from unfamiliar non-relatives.

6.2.2.3 Clan or group odours and social mammals

Like beavers, many other social mammals such as ring-tailed lemurs, dwarf mongooses, hyenas, aardwolves, badgers, prairie voles, and mole rats are characterised by high levels of social behaviour and exclusion of strangers from the group, behaviour that relies on mutual recognition of group members. Odour cues dominate this behaviour in mammals.

Depending on the social structure and juvenile dispersal pattern of the species, in many mammal species newly adult members will join the social group at intervals. As these animals come from outside they will not be closely related to the rest of the group they join. So, while there may be kin-related distinctions in behaviour being made within the group, in so much as the group works as a social unit, animals will need to recognise as group members animals that are not kin.

In many social mammals there is a great deal of allomarking – marking of each other – often by the dominant individuals. For example, European badgers (*Meles meles*) squat mark each other with their subcaudal and anal glands (Kruuk *et al.* 1984; Kruuk 1989). All members of the clan then share a common odour identity – dominated by the odour of the dominant male. The fermenting bacterial flora of carnivore anal glands change over time (Albone 1984; Macdonald 1985b) leading to changes in odours, but the regular marking of each other keeps the clan odour 'up to date'.

Like badgers, brown hyenas can recognise foreign conspecifics by smell: they invariably paste mark on top of marks from other groups (Mills *et al.* 1980) (Chapter 5). However, rather than being a demonstration of a distinctive shared 'group' or clan odour, theoretically this response could be based on knowing all the individuals in the group and recognising the marks as coming from unfamiliar individuals (and thus outside the group) (Macdonald 1985b).

Eusocial naked mole rats (*Heterocephalus glaber*) are highly xenophobic, even to closely related conspecifics from outside their colony (O'Riain & Jarvis 1997). Like social insects, the principal mechanism of recognition appears to be distinct colony odour labels, contributed by each colony member and distributed among, and learned by, all colony members. Differences in the mixture of these odours may provide even genetically similar colonies with a unique odour label.

6.2.2.4 Individual recognition

Other recognition events take place throughout life, with animals learning odour identities and recognising individuals, so the perceiving animal can behave adaptively when it meets those individuals again. Recognition of individuals by odour alone has been demonstrated by training experiments in the dwarf mongoose (*Helogale undulata*) and Indian mongoose (*Herpestes auropunctatus*) (Rasa 1973; Gorman 1976) and for European badgers (Östborn 1976, cited in Macdonald 1985b). The mongooses could distinguish individual conspecifics by their anal gland secretions and badgers used subcaudal gland secretions.

In an interesting discussion of individual recognition, Halpin (1986) teases out the assumptions that underlie most investigations. Many would be better described as showing that animals can distinguish the odours from test subjects. Whether the subjects are recognised as individuals as opposed to simply being different from each other is usually unclear.

Many experiments on individual recognition rely on conditioning tests (for example using positive reinforcement) or habituation tests in which responses decrease as the same stimulus is presented repeatedly (Chapter 2). An individual's odour that is perceived as different from the previous stimulus will elicit a greater response. Demonstrations of individual recognition by pheromones are usually in the context of an experiment and their importance in the social behaviour of the animals is not always known.

In a more natural context, dominant mice seem to seek out particular subordinate individuals to attack if that subordinate (or experimenters on its behalf) urine mark a marking post reserved by the dominant male (Box 5.1) (Hurst 1993). Individual recognition is implicit in the 'dear-enemy' phenomenon (Chapter 5) as animals are presumably able to recognise particular neighbours by smell (so as to recognise them as familiar or not familiar) and they may be able to associate that individual with a particular expected location or direction. For example, when experimenters placed a scent mark from a neighbour on the territory of an aardwolf (*Proteles cristatus*) it responded by going to the appropriate boundary and marking (Sliwa & Richardson 1998).

Individual recognition by odour is important for pair-bonding in monogamous species such as prairie voles (Carter & Roberts 1997). In the monogamous lizard *Niveoscincus microlepidotus*, males follow the scent trails of their partners and this recognition may play a significant role in maintaining the prolonged partnership (Olsson & Shine 1998).

At a physiological level, the Bruce effect of pregnancy block caused by the odour of a strange male relies on individual recognition (Chapter 9), as does reproductive suppression, which is achieved in different ways in prairie voles and common marmosets (Section 6.3.3). An intriguing phenomenon is that within a species, some glands are more variable than others. For example, individual dwarf mongooses could be distinguished by their anal gland secretions but not by their cheek gland secretions (Rasa 1973). Beavers did not discriminate between castoreum gland samples from unfamiliar and familiar beavers, but they did distinguish between anal gland secretions (Section 6.2.2.2). In habituation tests, golden hamster males treated some odours (flank, ear, vaginal, urine, faeces) as different between individuals, but not others (for example saliva, secretions from the feet, behind the ear, gland-removed flanks) (Johnston *et al.* 1993). Some glands might be specialised for individual variability (see Chapter 3)

If individuals are actually recognised, animals should have integrated multifactorial representations of these individuals (Johnston & Jernigan 1994). A male golden hamster familiar with two females was habituated to vaginal odour from one female (A) then tested with flank odour from one or the other female. He treated the flank odour of the female (A) as if it were familiar (the habituation to her vaginal odour 'carried over'). This was not a result of 'generalisation' between the two odours, however, as this effect was only shown if the male was familiar with the female and had learned her various odours (vaginal and flank odours were not treated in this way if the female was a stranger).

6.3 | Pheromones and reproduction in social groups: control or signalling?

Pheromones are important for the many species of mammal that live in social groups with animals sharing and defending the same territory (Chapter 5) and for species of social insect that live in colonies (Box 6.2). In some societies of mammal and social insect, cooperation is taken a stage

Box 6.2 | Social insects and pheromones

The coordination and integration of colony activities, in particular recruitment for foraging and defence (see Chapters 7 and 8), has been an essential contribution to the success of social insects: the road to sociality was paved with pheromones (Blum 1974). Pheromones play a central role in these activities and in other functions such as recognition (of caste, sex, kin, colony, and species), caste determination, trophallaxis (mouth-to-mouth transfer of food), nest entrance marking and colony reproduction (Winston 1992). It is difficult to imagine how the integration of colony functions could be made without pheromones transferring chemical messages through the colony, as there may be several million individuals in colonies of some ant species (Wilson 1971, p. 432). Termites show convergent evolution of chemical signalling with the eusocial Hymenoptera.

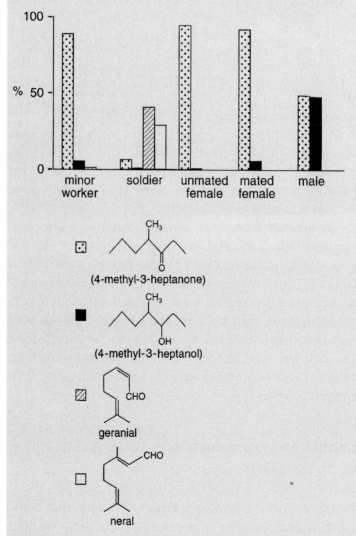

Caste differences in mandibular gland secretions of the leaf-cutting ant, *Atta*. Data from Donascimento *et al.* (1993) in figure from Chapman (1998).

There is an enormous diversity of glands and gland products in social insects (Fig 1.6). Different castes in the same species may produce different pheromones from the same gland, as for example in the mandibular gland in worker and queen honeybees and ants such as *Atta* (see figure earlier in box). Worker honeybee mandibular gland secretions have 10-HDA instead of the queen's 9-HDA, but produced by enzyme pathways that they largely share (see Pettis *et al.* 1999). Clues to the evolution of sociality and queen–worker signalling may come if we can deduce which pathway was primitive (Gadagkar 1997). Different responses to pheromones depending on the caste of the receiver, combined with context-specific responses, create an enormously complex chemical signal environment inside and outside the nest (see figure below). We are just beginning to decode these messages.

Multiple effects are common. For example in honeybees, the queen mandibular pheromone is a sex pheromone for reproductives and affects worker reproduction. It also has many other effects on worker behaviour, probably by affecting juvenile hormone levels. These include influences on swarming, stimulation of foraging behaviour, and an increase in the age workers start to move from working in the hive to foraging outside (see figure below) (Robinson & Huang 1998; Winston & Slessor 1998; Pettis *et al.* 1999). Many bee pheromones are now being exploited commercially (Chapter 12).

In honeybees, pheromones produced by the brood (eggs, larvae and pupae) help to coordinate activity to match the needs of the whole colony. Brood pheromones stimulate

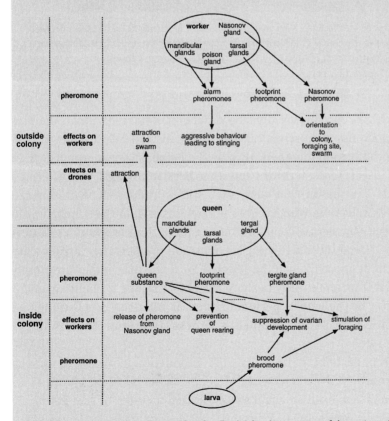

Pheromones of the honeybee, *Apis mellifera*. See Fig. 1.6 for the positions of the various glands. Figure from Chapman (1998).

worker brood-food glands, and promote larval feeding and cell capping (Le Conte *et al.* 1994). Cell capping is induced by fatty acid esters produced by older larvae and these esters are used as an attractant kairomone by the parasitic varroa mite (Trouiller *et al.* 1992).

A queen honeybee (*Apis mellifera*) surrounded by her retinue of attendants. The attendants collect queen mandibular pheromone (QMP) and then distribute it around their nestmates. Photograph courtesy of US Department of Agriculture, Carl Hayden Bee Research Center.

further, with a reproductive division of labour: some individuals (helpers or workers) do not themselves reproduce but instead help to rear the off-spring of other group members, usually the helpers' sisters or mothers. In the social insects (ants, termites and some bees and wasps), this is termed **eusociality**. The reason that such altruistic reproductive behaviour can persist is kin selection, which allows the helpers to gain their inclusive fitness indirectly by rearing copies of their genes in their brothers and sisters (Bourke 1997). Hamilton's rule for kin selection predicts that altruistic behaviour will be more likely to be selected for if the individuals are closely related (and thus more likely to share the helper's gene for helping) and if the decrease in the actor's personal fitness is relatively small compared with the increase in the recipient's fitness (Hamilton 1964; Keller & Chapuisat 1999).

One way of describing the sharing of reproduction in social groups is by the term **reproductive skew**, which describes how much the spread of reproduction differs from an equal share for each member of that sex in the group (Keller & Reeve 1994). Reproductive skew for males or females in a group ranges between 'zero', with equal shares (where all group members of a sex reproduce, for example female spotted hyenas), and 'one', in highly

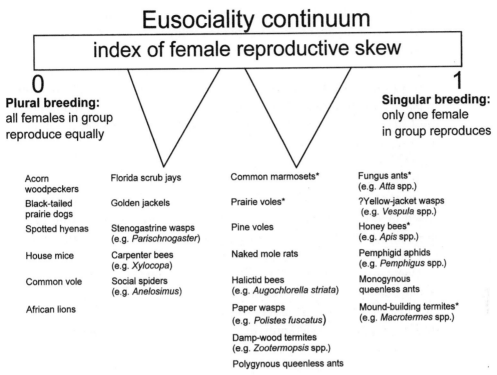

Fig. 6.1. Cooperatively breeding vertebrate and invertebrate species can be arranged along an axis, 'the eusociality continuum', which reflects how evenly reproduction is shared among females in a social group (Sherman *et al.* 1995). The axis ranges from a skew of '0' when female lifetime success is equal among group members to '1' when reproduction is limited to one female in a group. Systems where pheromones have been identified as the mechanism for interaction between the breeding female and subordinates are starred* (see text). After Abbott *et al.* (1998), itself modified from Sherman *et al.* (1995).

skewed animal societies in which effectively only one or a few members reproduce (for example, females in termites, ants, honeybees and the naked mole rat) (Fig. 6.1). Species with small colonies can be highly reproductively skewed, for example bumblebees, paper wasps, and the common marmoset.

Using reproductive skew as a measure, one can envisage social animals of all kinds placed on a eusociality continuum from no skew to high skew (Fig. 6.1) (Lacey & Sherman 1997, but see counterviews of Crespi & Yanega 1995, and Crespi & Choe 1997). At the high skew end, eusocial species show cooperative care of the brood, overlap of adult generations (with offspring helping parents) together with reproductive division of labour, with some individuals specialised for reproduction (called queens or kings in social insects), and other more or less sterile individuals showing reproductive altruism (Wilson 1971). In addition to the well-known highly eusocial insects among the Hymenoptera (wasps, bees, and ants) and Isoptera (termites), and among mammals (naked mole rats), there is a growing list of other eusocial animals, with species of eusocial spiders, aphids (Chapter 8), gall thrips and coral reef shrimps all now recorded.

Eusocial societies have cooperative broodcare but in even the most co-operative societies, genetic conflicts of interest are inevitable (Emlen 1997). In particular, group members will compete over who gets to reproduce. In most mammalian societies and in those social insect species in which al-most all individuals could potentially reproduce, fierce fighting deter-mines who reproduces. Perhaps surprisingly at first sight, in some of the most skewed societies, with the greatest morphological differences be-tween the queen and workers, and in some mammals, pheromones pro-duced by the dominant female 'settle the dispute', by appearing to stop subordinate females from reproducing. In social insects this phenomenon was traditionally viewed as pheromone control by the queen. An alterna-tive view is gaining ground: that the queen's pheromones are cooperative signals, not control by a form of chemical aggression. The proposals were outlined first for social insects so I describe these first, but the same or similar points probably apply to cooperatively breeding (eusocial) mam-mals (Section 6.3.3).

6.3.1 Social insect queen pheromones

Eusocial insect colonies are characteristically divided into two **castes**, repro-ductives and workers: a few individuals, queens or kings, are reproductive and workers reproduce little or not at all. The kin conflict between workers and the queen within social insect colonies can be over the level and timing of resources put into rearing reproductives, their sex ratio, and egg laying by workers (Keller & Reeve 1999).

However, despite the conflict, dominance with open aggression by the queen is virtually absent in more advanced insect societies, which have effectively sterile, morphologically distinct, worker castes (Wilson 1971, p. 302). As colonies increase in size, it is hard to see how physical domination could work for more than a few tens of animals let alone the 500 000 indi-viduals in a weaver-ant colony, controlled by a single queen (Hölldobler & Wilson 1977). Instead, in advanced insect societies, pheromones take the place of fights. The phenomenon of queen pheromone influence within social insect colonies is clear in advanced ants, wasps, and bees. Pheromones play a similar role in termites (Box 6.3).

The queen's primer pheromones affect the colony production of repro-ductives by influencing the behaviour of workers; this is important both for maximising her reproductive fitness and that of the colony (Winston 1992). Her pheromones also appear to cause the workers not to develop their ovaries.

News of the health of the queen is continually spread throughout a colony of social Hymenoptera, mediated by the queen's pheromones passed from one colony member to another. These queen pheromone effects can be

---

Box 6.3 | Primer pheromones in termites

Pheromone coordination of reproduction and the production of different castes in termites parallels that of social Hymenoptera, but, like their eusociality, it has developed quite independently. All termite species are eusocial. They differ from Hymenoptera in many ways: both sexes are diploid, nymphs are workers, and there are equal numbers of both sexes in the colony. Most colonies are effectively extended families that start with a monogamous pair of winged sexuals, the queen and king.

Termites, which are closely related to cockroaches, are divided into two groups, the higher termites which are all in the single large family Termitidae, and the lower termites which include all other termite families. In the lower termites, nymphs (termed pseudogates = false workers) form the functional equivalent of workers. These nymphs are capable of moulting into winged imagoes (which can become kings and queens of new colonies), replacement reproductives to replace the colony's king and queen, or soldiers. In the higher termites, which have sterile morphologically distinct soldiers and workers, caste changes are simpler and less flexible (see figure at end of box).

Several primer pheromones, acting specifically on the corpora allata (endocrine glands) to affect juvenile hormone (JH) levels and other endocrine glands of receivers, mediate caste changes in the colony (see reviews by Henderson 1998, and Kaib 1999). This is best understood in one of the lower termites, the African drywood termite *Kalotermes*. As in Hymenoptera, development of new queens is inhibited by the presence of the queen. In termite colonies there is also a king who has a similar effect on the development of new kings. In the absence of the queen, the king produces pheromones that encourage the development of replacement queens.

The pheromones inhibiting differentiation of the pseudogates come from the anus of the queen, are taken up by proctodeal trophallaxis (anal excretion feeding) and passed around the colony from worker to worker by oral trophallaxis (from mouth to mouth) or mutual grooming. The pheromones are not yet identified but growing evidence suggests that one may be juvenile hormone, produced by the queen's corpora allata (Henderson 1998).

In some species, soldiers inhibit the production of more soldiers so homeostasis of numbers is maintained. If soldiers are lost in colony defence, inhibition is relaxed and more pseudogates differentiate into pre-soldiers, then into soldiers, to bring up the numbers. At some times of the year, a higher proportion of soldiers is produced, for example when the winged reproductives are about to leave the colony, which is a reminder that many other factors including nutrition, season and hormones interact with pheromones in controlling these equilibria (Kaib 1999). Just as for Hymenoptera, it is possible to explain pheromones influencing caste changes as signals rather than as control (Keller & Nonacs 1993; Shellman-Reeve 1997). For example, high levels of soldier-produced pheromone could signal to pseudogates that the colony had no need of more and that they could make greater contributions to colony survival as workers. It will benefit the colony to have an optimum number of soldiers, with between-colony selection for a local optimum based on local predator pressure.

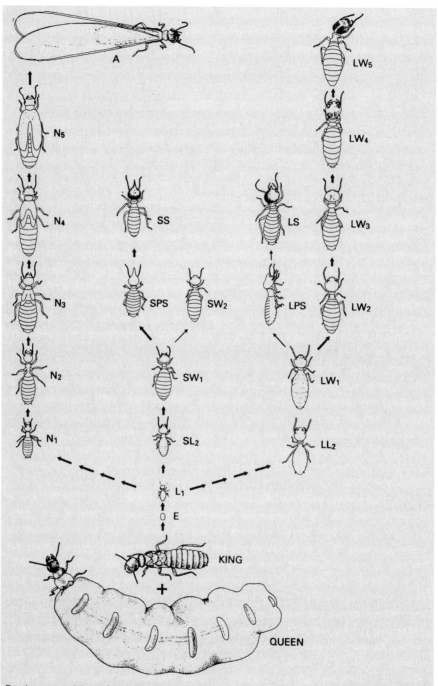

Development pathways of the higher termite *Nasutitermes exitiosus*. Heavy arrows indicate the main lines of development, light arrows the minor lines. A, alate; E, egg; L, larva; LL, large larva; LPS, large presoldier; LS, large soldier; LW, large worker; N, nymph; SL, small larva; SPS, small pre-soldier; SS, small soldier; SW, small worker. The numbers indicate the stages. Figure from Gullan & Cranston (1994). Artist Karina McInnes.

dramatically demonstrated by removing the queen and seeing the rapid changes in worker behaviour and physiology; these can start in as little 30 minutes in honeybees (*Apis mellifera*) (Box 6.4). Without the queen pheromone, workers start to rear new queens and the workers' ovaries start to develop.

Keller & Nonacs (1993) (following Seeley 1985, and others) argue that the pheromone effects are not the consequence of pheromone control by the queen but instead that workers are using the queen pheromone as an honest signal that the queen is there and that the workers' response to the pheromone increases their inclusive fitness as much as that of the queen.

The queen could perhaps control the colony by deception, fooling workers to act in her interest rather than theirs, but Keller & Nonacs (1993) argue that dishonest signalling in the colony is unlikely to be evolutionarily stable

---

Box 6.4 | Queen mandibular pheromone of the honeybee

Of queen pheromones only that of the honeybee (queen mandibular pheromone, QMP), has been chemically identified so far: it consists of five principal components (Winston & Slessor 1992). If the queen is removed, the orphaned worker bees respond rapidly to the sudden loss of QMP. Within 24 hours the workers start to rear new queens, enlarging cells and feeding some of the youngest larvae, previously destined to be workers, with royal jelly. The workers must act quickly as developmental paths for larvae are fixed at 6 days after egg laying and without a new queen the colony will die.

The relatively involatile pheromone is spread from the queen around the colony by messenger bees. The queen is always surrounded by a retinue of eight to 10 workers (see photograph in Box 6.3), which constantly change as new workers approach and lick or touch her with their antennae. After picking up the pheromone from the queen, these workers groom themselves and then act as messengers by running through the rest of the colony for about 30 minutes, making frequent reciprocal antennal contacts with other workers, and passing on the QMP by contact, as if playing chemical tag (Seeley 1979). Tracking of radio-labelled components has demonstrated that the pheromone is carried from bee to bee as a unit (Naumann et al. 1992).

The queen produces about 300 μg of QMP (the quantity in her mandibular gland, termed one queen equivalent, Q_{eq}) during each 24-hour period. The workers are extremely sensitive to QMP, down to a concentration of about 10^{-7} Q_{eq}. However, a high percentage of the QMP produced is ingested by the queen and by each bee that takes up the message, causing the pheromone to disappear as it is passed on. Constant production and rapid removal of the pheromone message means that if the queen dies and production of QMP ceases, the extinction of the message is almost immediate. Another inherent feature of pheromone message removal as it passes from worker to worker is the lessening of its effects as colony numbers increase, perhaps signalling to the workers that the size of the colony is reaching the optimum for fission (Keller & Nonacs 1993), which explains why workers start building new queen cells as the colony size moves towards swarming size, despite the presence of an active queen secreting QMP.

Although QMP has a major effect on worker physiology, pheromones produced by the brood have an even stronger inhibiting effect on ovary development in workers (Arnold et al. 1994).

and conclude that queen-produced pheromones are honest messages of queen activity or presence.

There are possible scenarios for the evolution of control by pheromones (Bourke & Franks 1995, p. 239) but, nonetheless, if it was not in their interest to respond, workers or subordinate queens would evolve to ignore queen pheromones. The genetic variation in worker sensitivity to queen pheromones, on which selection could act, has been demonstrated in honeybees (Slessor *et al.* 1998). Similarly, there will be selection for queen behaviour that avoids costly queen–worker conflict that reduces colony productivity (Keller & Nonacs 1993). Evolutionary solutions in the eusocial nest may be most stable where benefits are shared between workers and queen, pulling in the same direction (Seeley 1995, p. 11).

If queen pheromone is a signal to say that 'I am laying eggs' then one should expect the time course of pheromone production to match egg laying and to correlate with fecundity, which it does in honeybees and fire ants (Winston & Slessor 1992; Vargo 1998, 1999). A honeybee queen's queen mandibular pheromone blend changes with age (Box 6.4) (Winston & Slessor 1992). Newly emerged, she has no pheromone but she produces the decanoic acids by the time she mates. The full blend, including the aromatics, is only produced after mating and when she begins to lay eggs.

The reason that physical fights do not occur is not because the pheromone controls the workers but because their interests often match those of the queen. The strong morphological specialisation of the queen as an egg and pheromone factory and of the workers for their many colony-sustaining roles, means that an individual worker gains more by helping to rear the queen's eggs than by laying its own. With specialised morphological castes, the queen pheromone may be a relatively low-cost cooperative signal (Keller & Reeve 1999). In a cooperative signal, with benefits to both sides, evolutionarily stable signals do not have to be differentially costly to signallers with high values of the signalled attribute (which is the central assumption of the Zahavi handicap model for signals) (Chapter 1) (Keller & Reeve 1999).

A further pheromone-mediated effect is the way that any eggs that *are* produced by workers are destroyed by other workers, termed 'worker policing'. Worker honeybees have ovaries and although they cannot mate, they can lay unfertilised eggs which become males. Workers destroy the eggs laid by other workers because if the queen is multiply mated, workers are on average more related to the sons of the queen than to the sons of other workers (Ratnieks 1993). Worker policing is made possible because the queen's eggs can be recognised by a pheromone mark from her Dufour's gland (Ratnieks 1995). In nests with a queen, almost all the eggs produced by workers are destroyed (Ratnieks & Visscher 1989). Workers also attack workers with well-developed ovaries (Visscher & Dukas 1995). A genetic basis for worker policing has been found in honeybees (Montague & Oldroyd 1998). Worker policing is evolutionarily stable because it benefits the queen *and* the average worker (Bourke 1997) (for queen pheromone marking of the eggs in ants see Vander Meer & Morel (1995), and in bumblebees see Ayasse *et al.* (1999)).

Even in once-mated single queen colonies, in which workers would be more related to their own sons than to the queen's sons (Bourke 1997), workers might not reproduce if it reduces the efficiency of the colony.

6.3.1.1 Queen pheromones in fire ants

Queen pheromones in the fire ant (*Solenopsis invicta*) inhibit virgin daughter queens in the nest from losing their wings (dealating) and starting to lay male eggs (Vargo 1998) (Fig. 9.12). Dealated virgin females reintroduced into the mother nest are killed by workers. Keller & Nonacs (1993) wonder if rather than being under pheromone control, workers use the queen signal as an indication that the queen is alive (in which case they benefit by killing the virgin reproductives) or dead (in which case they let the virgin queens live as the colony needs reproductives). Virgin queens might use absence of the pheromone as an indication that they will be able to reproduce without being killed by the workers.

In social insects with many queens in each colony, there may be an inverse relationship between queen number and individual fecundity. In the fire ant this effect is caused by a mutual inhibition mediated by pheromones (Vargo 1992). However, in some other species such as *Leptothorax acervorum* this effect is not shown (Bourke 1993).

6.3.2 Primer pheromones and reproduction in social mammals

It is in social mammals, those living in groups on shared territories and especially those breeding cooperatively, that mammal primer pheromone interactions have reached their greatest complexity and subtlety. Pheromone stimuli in social mammals can induce hormonal changes, affect the success of pregnancy, alter the course of puberty, modulate female cyclicity and ovulation, and modulate reproductive behaviour and aggression. These physiological effects include the Bruce and Whitten effects in mice (the hormonal basis of these and other primer pheromones is discussed in Chapter 9).

Cooperative breeding, with alloparental care in which members of the social group assist in rearing young that are not their own, is common in some mammalian taxa, in particular rodents and canids. For example, cooperative or communal nesting and care of young have been reported for 35 species and from nine of 30 rodent families (Solomon & Getz 1997). Cooperative breeding covers a wide range of behaviour depending on reproductive skew in the species, from plural breeders with all females reproducing through to singular breeders, social groups in which only one female breeds together with 'helpers-at-the-nest' (Solomon & French 1997) (Fig. 6.1).

Most of the pioneering work on mammal primer pheromones was on social rodents such as house mice which are plural cooperative breeders (all females breed, although not all males). Female house mice suckle each other's young and cooperatively defend the nest. Characteristic of these societies is an interplay of dominance (in particular between males), sex, and population density.

In plural breeders, the effects of females on each other are mutual, but in singular cooperative breeding species, such as beavers, prairie voles or the common marmoset, the dominant female suppresses reproduction by the subordinate females. The parallels between social organization in these species and social insects are explored in Section 6.3.4. It is worth noting that some of these social effects, such as influences on puberty timing, are also seen in solitary rodent species under some conditions.

6.3.3 Reproduction in singular cooperatively breeding mammals with high reproductive skew

Reproductive suppression is common in singular cooperatively breeding mammals in which typically only one dominant female breeds (Fig. 6.1). As in many social insects, the subordinate females are often her daughters, and in mammals, as in social insects, signals affecting the reproduction of subordinates range, in different species, from physical dominance to pheromones. Most mammal social groups do not use pheromones for this. For example, in the most eusocial mammals, naked mole rats (*Heterocephalus glaber*), with colonies of up to 300 non-breeding workers, the suppression of worker fertility by the queen is not pheromonal (Fig. 6.2) (Faulkes & Abbott 1993). Instead, the queen, which is larger than other colony members, exerts her reproductive suppression on the non-breeding workers by physical dominance, 'shoving' and pushing subordinates down the tunnels (Bennett *et al.* 1999). In singular breeding canids the mechanism has only been identified in the grey wolf (*Canis lupus*): subordinates could reproduce but do not because their mating attempts are interrupted by their parents (Asa 1997).

It is in some of the singular cooperatively breeding rodents and the New World primates that there are strong pheromone parallels with advanced social insects (Carter & Roberts 1997; Solomon & Getz 1997; Abbott *et al.* 1998). We know most about the reproductive biology of prairie voles (*Microtus ochrogaster*) and the common marmoset (*Callithrix jacchus*). In both species, many of the effects are mediated by odours for recognition, signal or primer pheromones.

In prairie voles, monogamous pairs and their offspring form the core of a communal breeding group. Continued breeding by the original pair and

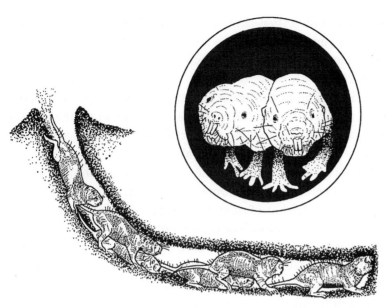

Fig. 6.2. Naked mole rats are eusocial mammals with a worker caste, here shown digging a tunnel together. Although only one female reproduces, pheromones do not seem to be important in causing the workers not to reproduce (Faulkes & Abbott 1993). From a photograph by David Curl. Drawing by Priscilla Barrett in Manning & Dawkins (1998).

concurrent inhibition of reproduction of other members of the group is promoted by reproductive suppression of offspring, incest avoidance, social preferences for the familiar sexual partner and active defence of territory and mate (Carter & Roberts 1997). Almost two-thirds of prairie voles young remain in their parents' nest (philopatry). These non-breeding subordinates engage in all parental behaviour except nursing (Solomon & Getz 1997).

The young of the common marmoset also stay within their natal group into adulthood and do not breed. All group members, of both sexes, contribute to infant care, and may groom, tend (babysit) and transport young, but in addition may help with post-weaning feeding of infants (French 1997; Tardif 1997). The evolution of cooperative breeding may be a two-step process (Lacey & Sherman 1997). The first step is the presence of ecological conditions that encourage natal philopatry: staying on the parental territory rather than trying to breed on one's own. This could be because of high costs, or low success, of independent breeding or dispersal, and would lead to groups containing two or more generations of related adults. The second step is the evolution of alloparental care, depending on the benefits to kin and ultimately on inclusive fitness. Long-term studies of the costs and benefits of helping in mammals and birds give widespread confirmation that helpers frequently do gain large indirect genetic benefits by helping to rear collateral kin (Emlen 1997).

For prairie voles, indirect benefits from alloparenting could include better survival of sibling pups, faster pup development and reduced workload for parents, thus allowing the parents to produce more litters (Solomon 1991; Wang & Novak 1994).

For common marmosets, the initial benefits of alloparenting might originally have been increased survival of young, but once set on the path of helping, it has become a requirement as the energy costs of breeding are so high that a lone pair is effectively incapable of reproducing successfully (French 1997). Cooperation between animals may also be needed for the successful founding of new marmoset groups (Abbott *et al.* 1998).

The importance of ecological factors for the fine balance of benefits and costs to helpers is shown by the patchy distribution of singular and plural cooperative breeding across related genera. Even in the same genus there may be species that are singular breeders and others that are plural breeders, for example prairie voles, and common voles respectively. Populations of the same species in different places, for example prairie voles (Roberts *et al.* 1998), may show more or less alloparental care according to local ecological conditions.

6.3.3.1 Inhibition or suppression of subordinate reproduction

The size of social groups is not the deciding factor for the transition to pheromonal control in mammals, as species that use pheromones in reproductive suppression tend to have small family groups. More species of cooperatively breeding mammals may turn out to use pheromones than is currently realised.

Two effects keep subordinate female prairie voles *Microtus ochrogaster* prepubescent (see review by Carter & Roberts 1997). First, they delay puberty as long as they are exposed to only familiar males (father or male sibs) recognised by odour. Second, the stimulatory effect of urine from an unfamiliar male is overruled by inhibitory pheromones in the urine of their mother and sisters. Subordinate females thus remain functionally pre-pubescent and provide support to the communal family (Chapter 9). The suppression of subordinate males is likely to be behavioural as they still produce sperm but do not mate.

Suppression of ovulation in subordinate common marmoset females is by a combination of olfactory, visual and behavioural cues but once reproductively suppressed, this can be extended by odour alone: if a subordinate female is taken from the group, she will start her ovarian cycle but disinhibition is delayed by about 20 more days if she is exposed to the scent marks of the dominant female (Chapter 9).

A feature of reproductive suppression in mammals is the variety of mechanisms controlling singular cooperative breeding, even in closely related species. For example, unlike the case of the prairie vole, pheromone cues are not sufficient to suppress oestrus in the pine vole (*Microtus pinetorum*) (Brant *et al.* 1998). Similarly, in the golden lion tamarin (*Leontopithecus rosalia*), a member of the same family as the marmosets, subordinate females ovulate and are physiologically capable of mating but do not do so (French 1997).

6.3.4 Parallels between social mammals and social insects

The response of subordinates in marmosets or prairie voles may be an adaptive response to signals from the principal female, analogous to the worker responses to signalling by social insect queen pheromones (Section 6.3.1) (Keller & Nonacs 1993). Subordinates in marmosets, prairie voles and social insects may have evolved specific, adaptive responses to signs of subordinate status that lead them to respond with alloparental and other behaviour that increases their inclusive fitness by helping the society or family group (Abbott *et al.* 1998). Like workers in social insect colonies, subordinate female marmosets show many behavioural, neuroendocrinological and physiological differences from dominant females (Abbott *et al.* 1998). The differences include both the alloparental tasks undertaken by subordinate marmosets and also their physiological responses to pheromones and other cues from the dominant female (Chapter 9). Abbott *et al.* suggest that the behaviour and physiology of subordinates seem to be a stable alternative to dominant status, not a state of generalised stress imposed by the dominant female and endured by the subordinates to their physiological detriment (there is no elevation in the circulating hormones, cortisol or prolactin, associated with stress).

A further parallel comes from developmental pathways. The spontaneous alloparenting behaviour and high likelihood of remaining in the parental nest (philopatry) of subordinate prairie voles are influenced by their prenatal hormonal environment in the uterus (Roberts *et al.* 1996). I wonder how different this is from developmental influences on social insect larvae as they are directed to worker or queen roles?

6.3.5 Proximate and ultimate routes for effects

The roles of pheromones in influencing who reproduces in social groups of both insects and mammals are clearly complex. The interplay between pheromones and hormones, and the way that closely related species achieve similar ends by either a pheromone or behavioural dominance route, should make us reconsider rigid categories. What pheromones and behavioural dominance share in their mechanism of action is, ultimately, an effect on hormone release from the hypothalamus in mammals, and from the corpora allata in insects (Chapter 9). Could one argue that pheromones and behavioural dominance are equivalent at the ultimate physiological level?

6.4 | Conclusion

The individuals in animal societies interact via a complex web of semio-chemical signals. Odour recognition is also important for kin recognition for mate choice and other behaviour in less social species. A common feature of recognition is the learning of odour cues at a certain life stage. Social insects, such as ants, bees, wasps and termites use hydrocarbon odour cues on their cuticles. Colony odours are usually the combination of secretions under genetic control and odours gained from the environment. The mixture of odours may serve to mask the underlying genetic heterogeneity of even a bees' nest. Social mammals use mixtures of odours that are just as complex. In both insects and mammals, the odour signatures are the passport to acceptance by the group.

Eusocial species of social insects and social mammals are characterised by reproductive division of labour. In some species, group members fight to establish which animals will reproduce. Other species use pheromones that act as signals rather than as coercion. The mechanisms used in social insects and in mammals have many similarities.

6.5 | Further reading

Hölldobler & Wilson (1990) give a masterly and detailed overview of the ants. Bourke & Franks (1995) give a detailed and helpful account of the hypotheses about reproduction in ants and other social insects. Chapters in Vander Meer et al. (1998) cover pheromones more widely in social insects. Seeley (1995) gives an excellent overview of honeybee adaptations including a review of intra-colony conflict and resolution. Chapters in Solomon & French (1997) cover mammalian cooperative breeding. Dixson (1998) provides an in-depth review of primate reproductive biology. Krebs & Davies (1997) and Alcock (1998) give introductions to kin selection theory.

Chapter 7

Pheromones and recruitment communication

7.1 | Introduction

The ability to recruit group members to new sources of food, to defend the territory, or to protect the group against enemies are crucial to the success of social species across the animal kingdom. This ability is one of the most important factors behind the extraordinary ecological dominance of social insects in so many habitats. Recruitment brings nestmates to the place and task required.

Recruitment signals are commonly pheromones, in part because the taxa that show the most development of recruitment behaviours are those most reliant on pheromones (Table 7.1), but also because pheromones enable mass recruitment to tasks. 'Call to arms' pheromones for collective defence are also common in social insects (Chapter 8). New nest site finding and colony moving are also often pheromone mediated (Fig. 7.1). Even elaborate nest building may be organised this way (Section 7.3.2).

The coordinated activity of social insect colonies puts them at a pinnacle of biological complexity, but although hundreds of thousands, even millions, of individuals may be involved, activity is not commanded from the centre: perhaps the complexity is possible precisely because of this. Instead, the social integration and assembly of colony-level patterns come from simple interactions between individuals and individual responses to local conditions, largely mediated by pheromones (Section 7.3) (Traniello & Robson 1995; Bonabeau *et al.* 1997).

The range of recruitment tasks mediated by pheromones in social insects is illustrated by African weaver ants (*Oecophylla longinoda*), which have the most complex set of signals yet identified in ants (Hölldobler & Wilson 1978). These ants have five different recruitment systems, which use combinations of different tactile signals and pheromones from two glands (Table 7.2) (for the names of ant glands see Fig. 1.6).

Table 7.1. *Some examples of trail pheromones in insects*

Insect	Principal compounds	Example	Origin
ISOPTERA **Termitidae** *Coptotermes*	Aliphatic hydrocarbons	*Z,Z,E*-3,6,8-dodecatrienol	Ventral abdominal gland[a]
LEPIDOPTERA **Lasiocampidae** *Malacosoma* (larva)	Steroids	5β-Cholestane-3,24-dione	Ventral abdominal gland
HYMENOPTERA **Formicidae** Ponennae *Megaponera*	Heterocyclic nitrogen compounds	N,N-dimethyluracil	Poison gland
Myrmicinae *Atta*	Heterocyclic nitrogen compounds	Methyl-4-methylpyrrole-2-carboxylate	Poison gland
Solenopsis	Terpenoids	*Z,E*-α-farnesene	Dufour's gland
Dolichoderinae *Iridomyrmex*	Aliphatic hydrocarbons	*Z*-9-hexadecenal	Sternal gland
Formicinae *Lasius*	Aliphatic hydrocarbons	Hexanoic acid	Hindgut

[a]Ventral abdominal gland = sternal gland.
From Chapman (1998).

Fig. 7.1. Honeybee workers use Nasonov pheromone to recruit other workers outside the hive when they swarm during colony division (Left) and in particular to mark a good new nesting cavity and guide the swarm inside (Right) (Schmidt 1998). Arrows point to exposed Nasonov glands. Photographs by J. Schmidt.

Table 7.2. *Basic properties of the five recruitment systems of the African weaver ant Oecophylla longinoda*

The recruitment systems are directed respectively to food, potential new territory, new nest sites, organisation of emigration, and responses to enemies at short range and at long range. The weaver ants are strongly territorial and defend their trees against conspecifics and other ant species (Box 5.1).

System	Chemical signals	Tactile signals	Pattern of movement	Apparent function
Recruitment to food	Odour trail from rectal gland; regurgitation of liquid crop contents	Antennation; head waving; mandible opening associated with food offering	Occasional signpost marking with looping trails laid around food source; main trail directly to nest	Recruitment of major workers to immobile food source, especially sugary materials
Recruitment to new terrain	Odour trail from rectal gland	Antennation; occasional body jerking	Main trail directly to nest; broad looping movements resembling signpost marking, but only after foragers physically contact terrain; increase in frequency of anal spotting	Recruitment of major workers to new terrain
Emigration	Odour trail from rectal gland	Antennation; physical transport of nestmates and tactile invitation of signals leading to transport	Main trail directly to nest site; no signpost marking; predictable sequence of categories of nestmates carried	Emigration of colony to new nest site
Short-range recruitment to enemies	Short looping trails from sternal gland and exposure of gland surface with abdomen lifted in air	None	Trails short, looping and limited to vicinity of contact with enemy	Attraction and arrest of movement of nestmates; inducement of clumping and quicker capture of invaders and prey
Long-range recruitment to enemies	Odour trail from rectal gland	Antennation; at higher intensities, body jerking	Main trail directly to nest; no signposts laid	Attraction of major workers to vicinity of invaders and prey operation in conjunction with short-range recruitment; especially intense during territorial wars with conspecifics

From Hölldobler & Wilson (1978).

7.2 | Foraging ecology and evolution of recruitment communication

Eusocial insects (ants, wasps, bees and termites) and the naked mole rat are central-place foragers, going out from the nest in search of food. While all termites show foraging recruitment behaviour, the development of such behaviours in ants, bees and wasps varies greatly between species and only a minority of social caterpillar species show recruitment (Box 7.1). In ants,

Box 7.1 | Tent caterpillars

Caterpillars of some moth and butterfly species live in large groups and use pheromone trails. Social caterpillars living in silk tents have enhanced predator defence, thermoregulation and foraging efficiency (Fitzgerald 1993). The most sophisticated trail communication has developed in species such as the eastern tent caterpillar (*Malacosoma americanum*), which forages from a fixed base on a patchily distributed resource. *M. americanum* lives colonially, with some 50 to 300 sibling individuals from a single egg mass in a silk tent in the branches of cherry or apple trees (see figure below). Hungry caterpillars forage in groups, leaving the tent in search of food every 6 hours or so. They travel to fresh areas, ignoring previously visited, now defoliated, sites. The nest serves as an information centre: animals that were not successful in finding a good feeding site of their own will find productive areas if they follow the trails of successful conspecifics.

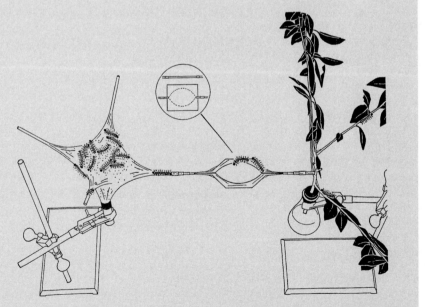

The eastern tent caterpillar (*Malacosoma americanum*) lives socially in a silk nest (left) and caterpillars travel to the feed on leaves (right). Caterpillars returning from good feeding sites mark the trail with a pheromone, identified using this Y-maze bioassay. Figure from Fitzgerald & Gallagher (1976).

different recruitment mechanisms include **tandem running** in which the scout ant leads one nestmate to the resource, **group recruitment** which recruits tens of nestmates, and **mass communication** which uses pheromones alone to recruit large numbers of nestmates (see Box 7.2).

Which recruitment mechanism is used depends as much on the ecology of the species as its taxonomic position. Comparing closely related species with different foraging strategies reveals that key factors that select for recruitment behaviour are clumped, patchy food resources (Traniello & Robson 1995). For example, the ant *Pachycondyla obscuricornis* hunts small arthropods, and it has no need for foraging recruitment as each prey item can be carried back to the nest by one forager, the finder. However, for nest moving this species does use tandem running (see Box 7.2), facilitated by pygidial gland pheromones. The congener *P. (Termitopone) laevigata* is a specialist predator of termites: when a scout ant finds a single termite, she returns to her nest, laying a pygidial gland trail (Box 7.3). The pheromone acts as a mass recruitment signal stimulating nestmates to leave the nest and follow the trail

Box 7.2 | Mechanisms of recruitment

Ants show the whole range of recruitment behaviour from none through to mass communication by pheromones alone (see reviews by Hölldobler & Wilson 1990, Vander Meer & Alonso 1998).

One of the more primitive forms of ant recruitment is **tandem running**: in ants such as *Leptothorax acervorum* a single nestmate is recruited by the returning scout and led to the new nest or food source. The follower touches the leader's abdomen with her antennae as they run. This behaviour may have evolved from tandem running in ant courtship. The scout's recruitment signals often include motor displays (which are ritualised, stereotyped movements such as tugging) and in some species are accompanied by pheromone release to attract workers. Recruitment in tandem running species is relatively slow as each scout only brings back one nestmate with her.

More rapid recruitment is accomplished by species that use **group recruitment** to excite tens of nestmates at a time and then lead them out to the site. In the ant *Camponotus socius*, scouts leave chemical signposts round new food sources and lay a hindgut contents trail back to the nest. More than the trail is needed to recruit the nestmates: the recruiting ant performs a waggle display when facing nestmates head-on. Workers then follow the scout along the trail. Stingless bees have a similar behavioural display that excites nestmates to follow the pheromone trail. Naked mole rats may use group recruitment.

Mass communication, with recruitment by chemical signals alone, is the only method of termite and tent caterpillar recruitment communication and is characteristic of many species in the majority of ant subfamilies. Recruitment by pheromones allows rapid recruitment of large numbers of nestmates as no physical contact between individuals is required, for example in the fire ant (*Solenopsis invicta*), so that the signal can travel quickly throughout the colony.

Box 7.3 | Trails in and out

Trails will be familiar to anyone who has watched a line of ants running across a wall or path, appearing to follow a strong invisible 'thread'. In most ant species, the trail, which can be erased by rubbing a finger across it, is a pheromone laid down by the workers as they travel back from the food source. Similarly, tent caterpillars only mark on return trips if they have encountered a good food resource (Fitzgerald & Peterson 1983). Termites, ants and tent caterpillars show similar behaviour when laying trails, touching the ground with the sternal or other trail gland (see figure below). Conspecifics follow the trails to the food.

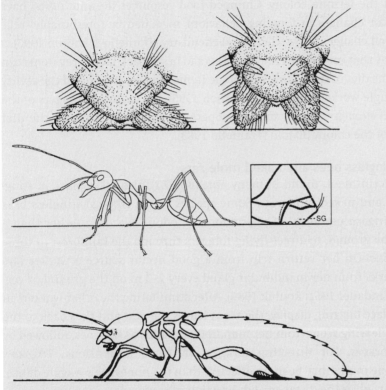

Three different taxa show very similar behaviour using glands for trail marking. (Top) The tent caterpillar (*Malacosoma americanum*): the position of the tip of the abdomen showing marking by the sternal gland (left) when dragged on the substrate, compared with the position when not marking (right). Figures from Fitzgerald & Edgerly (1982). (Middle) An African weaver ant (*Oecophylla longinoda*) worker depositing sternal gland (SG) pheromone onto the ground (by rotating the terminal abdominal segment, other glands can be exposed for other signals). Figures from Hölldobler & Wilson (1978). (Bottom) A nymph of the termite *Zootermopsis nevadensis* laying an alarm trail. Figure from Stuart (1969).

Most ant species rely on celestial or visual landmarks for return to the nest and do not lay trails outbound (Wehner 1997). However, some social insects lay trails on both the outward trip, to guide them back to the nest, and on return trips (if they have found food). These include tent caterpillars, most termites (which are blind) and blind ants such as *Eciton* army ants (Section 7.3.1).

The great sensitivity of worker ants to trail pheromone, with a threshold of 0.08 pg/cm (3.5×10^8 molecules/cm) found for the leaf-cutter ant *Atta texana*, means that only tiny quantities are needed (for trail following behaviour, see Chapter 10) (Tumlinson *et al.* 1971): at any one time, each worker might contain as little as 0.3 ng of the main component (methyl 4-methylpyrrole-2-carboxylate). The total amount of trail pheromone in one nest of *Atta texana* (about 1 mg) could theoretically lay a trail leading a column of ants round the world three times.

to attack the termite colony. Clumped food resources (termite nests) have selected for changes in foraging behaviour mediated by (presumably) relatively small changes in pygidial gland chemistry (Traniello & Robson 1995).

Species that can recruit many workers to bring back large prey items can expand the diet of the colony to include items that greatly exceed the ability of any single worker (Traniello & Robson 1995). For example, foragers of the ant *Lasius neoniger* are all small. Cooperative foraging increases the diet breadth of the colony 30-fold (Traniello 1983).

7.2.1 Stingless bees and naked mole rats
Group recruitment, found in many ants (Box 7.2), also takes place in stingless bees and naked mole rats. Some species of the tropical stingless bees such as *Trigona* use 'dots' of pheromone as 'odour beacons', rather than a line on the ground, to direct flying foragers through the rainforest in three dimensions. On her return trip from a good nectar source a worker lays down marks from her mandibular gland every 2–3 m on the ground or vegetation (Lindauer 1961; Roubik 1989). After stimulating the other workers in a face-to-face buzzing display, the scout leaves the nest and flies back to the source, releasing scent from her mandibular glands as she goes, followed by other workers. For directing foragers in three dimensions, *Trigona*'s pheromone marks may be more effective than the honeybee's waggle dance, which only gives the distance and direction of resources.

Naked mole rats (*Heterocephalus glaber*) are eusocial, burrowing rodents that inhabit arid regions of north-eastern Africa (Fig. 6.2) (Sherman *et al.* 1991). They forage from a central nest to find concentrated food sources, such as plant tubers, that are patchily distributed – bigger colonies can dig more tunnels, increasing the chance of coming across tubers. Judd & Sherman (1996) studied captive colonies of naked mole rats to see if a successful forager could recruit its nestmates. They allowed a scout to search a ring maze of Plexiglas tunnels (with sweet potato in only one of many boxes). When it found food, the scout went directly back to the nest. On the way back it made 'chirp' sounds (not given by unsuccessful scouts or later recruits) and, when it got to the nest, it waved the new food round. The experimenters then filled other food boxes in the maze before allowing nestmates into the maze: the nestmates tended to go to the site found by the scout more often than by chance, even though they might have had to pass some other sites (that had been filled with food meanwhile) on the way. Nestmates were following the odour trail (perhaps urine?) left by the particular mole rat that had found a new food source. They preferred to follow this route over that travelled by another nestmate carrying the same food type. Like ants meeting on a trail, if the scout met with nestmates on its way back to the nest they would often briefly interact and then the nestmates would follow the scout's trail to the food.

What about recruitment in non-eusocial group-living mammals? While formal recruitment may not exist, later foragers may still gain information from earlier foragers. For example, Norway rats (*Rattus norvegicus*) moving between foraging and harborage sites tend to follow trails made by fellow group members. Further, laboratory tests suggest that Norway rats lay trails back from good food sources. Using a T-maze with a cleanable floor, Galef & Buckley (1996) showed that a second rat would choose the return path of a previous rat that had found food. It was not simply a matter of following any trail: outward trails or trails of rats returning unfed were not followed.

7.2.2 Foraging patterns

Just as the kind of food drives the evolution of different recruitment communication mechanisms, how long the food supply lasts affects the longevity of pheromones used for trails and the form of foraging. Ants such as *Solenopsis* fire ants, with insect prey that is rapidly depleted or moves, lay transient trails that fade in minutes. The short life of the pheromone ensures that the trail remains relevant: if not renewed, it fades (Chapter 10). Species with longer-lasting food supplies such as trees, for example *Atta* leaf-cutter ants that collect leaves for their fungus garden (Chapter 11), have long-lasting trunk trails. Ants that collect a renewable resource, honeydew from aphids, have trails between trees where they tend the aphids and the nest that follow the same routes for years (and individual ants may follow particular trails for more than one season) (Quinet & Pasteels 1996). Some termite species have trail pheromones still active after 4 years (Traniello & Robson 1995).

Termites feed mostly on dead vegetation or wood so foraging is more like slowly mining a coal face (see reviews by Pasteels & Bordereau 1998, and Kaib 1999). It is quite rare to see termites on the surface as they are usually hidden in their covered walkways, roofed over with 'carton' made of earth or wood mixed with faeces and saliva (Fig. 7.2). Rather than sending individual scouts foraging far from the nest, soldiers go a short distance from the end of the covered trails and recruit workers to any food they find (Fig. 7.3). Sternal gland secretions lead workers of the termite *Schedorhinotermes*

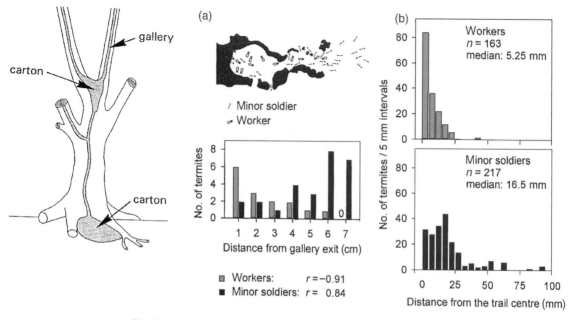

(a)

Minor soldier
Worker

No. of termites

8
6
4
2
0
    1 2 3 4 5 6 7
Distance from gallery exit (cm)

Workers: $r = -0.91$
Minor soldiers: $r = 0.84$

(b)

No. of termites / 5 mm intervals

Workers
$n = 163$
median: 5.25 mm

Minor soldiers
$n = 217$
median: 16.5 mm

0   25   50   75   100
Distance from the trail centre (mm)

Fig. 7.2. (Left) Nest of the termite *Schedorhinotermes lamanianus*. Brood and reproductives are found in the carton structures. Galleries lead from these to foraging sites. Figure from Kaib *et al.* (1996). (Right) Termites are in constant risk of ant attack. The vulnerable workers are always protected by a wall of soldiers, which travel further from the exit of foraging galleries and the edges of pheromone trails. Figure from Kaib (1999).

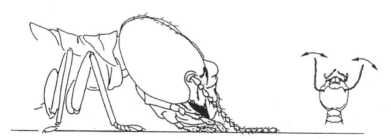

Fig. 7.3. Food collection by workers of the termite *Schedorhinotermes lamanianus* is organised by pheromones. Gnawing is stimulated by long-lasting labial gland pheromone laid down by workers. Other workers detect this with contact receptors on their antennae, which they sweep from side to side. The workers do not gnaw if they detect sternal gland trail pheromone, so trails are kept clear of gnawers. The result is a flexible division of the 'mining zone' into feeding and trail areas. Figure from Reinhard & Kaib (1995).

lamanianus to areas of food (Kaib 1999). Once there, workers gnawing food release a persistent, non-volatile gnawing pheromone from their labial glands that causes other workers to gnaw there (Fig. 7.3).

Matching foraging effort to the value of food resources is also a key feature of ant recruitment behaviour. Aspects of self-organisation are discussed in Section 7.3 but here I touch on ways in which pheromone concentration

Nest   Foraging area   Food

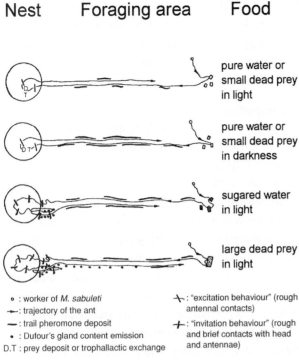

pure water or
small dead prey
in light

pure water or
small dead prey
in darkness

sugared water
in light

large dead prey
in light

∘ : worker of *M. sabuleti*

⟶: trajectory of the ant

�—: trail pheromone deposit

• : Dufour's gland content emission

D.T : prey deposit or trophallactic exchange

⊥ : "excitation behaviour" (rough antennal contacts)

⟙ : "invitation behaviour" (rough and brief contacts with head and antennae)

Fig. 7.4. Signals in the ant *Myrmica sabuleti* vary according to the prey they encounter. In the light they appeared to lay less pheromone. In the dark, the ants lay poison-gland pheromone trails. If a worker finds sugar water or carryable insect prey (a fruit fly for example) she then lays a trail with Dufour's gland secretions just outside the nest entrance. If the prey is a large insect, scouts deposit Dufour's gland secretions all along trail. Figure from Cammaerts & Cammaerts (1980).

or other communication is graded according to food value. Foragers of the ant *Lasius niger* exploiting a 1 M sugar source laid 43% more trail marks than those exploiting a 0.05 or a 0.1 M source (Beckers *et al.* 1993). The greater trail laying is enough to account for choice of the richer source both in experiments and in models of trail recruitment (Section 7.3.1). Signals in the ant *Myrmica sabuleti* vary according to the food encountered (Cammaerts & Cammaerts 1980) (Fig. 7.4). The proximate cue that leads workers of the ant *Pheidole pallidula* to mass recruit other workers rapidly to bring home large prey items species appears to be the 'tractive resistance' of the prey. If the scout cannot move the prey it gives up quickly and goes back to the nest, laying a more continuous pheromone trail than if it finds a small prey item that it can carry by itself (Detrain & Deneubourg 1997).

Patterns of trails can adjust searching to match areas of high productivity, adjust movement of workers to avoid conflict with neighbouring nests, and direct foragers to sites needing recruitment (Traniello 1989; Hölldobler & Wilson 1990; Traniello & Robson 1995). For example, colonies of the Mediterranean harvester ant (*Messor barbarus*) cut smooth trunk trails cleared of vegetation, that lead from the nest to foraging areas with seeds (Lopez *et al.* 1994). Where there is a low density of seeds the trails are long and have few branches. In areas richer in seeds, they have shorter trunk trails leading from many nest entrances of the same colony (thus reducing foraging transport costs).

7.2.3 Recruitment and competition

Recruitment is also important to secure resources against intra- and inter-specific competition with other ants, particularly where species do not have exclusive territories so ownership of any prey item could be disputed (Chapter 5) (Hölldobler & Wilson 1990; Traniello & Robson 1995). When a scout of the large desert ant *Aphaenogaster* (= *Novomessor*) finds prey that is too big to be carried unaided, such as a dead insect, it recruits colony mates from up to 2 m away by releasing poison-gland pheromone into the air and stridulating. As soon as enough colony mates arrive, they gang-carry the prey back to the nest (Hölldobler *et al.* 1978). Speed is of the essence because they must move the prey to safety before formidable but slower moving ants, including fire ants and *Iridomyrmex pruinosum*, can claim the prize. Small ant species such as the tiny North American *Monomorium minimum*, which are unable to carry large food items, can monopolise these by rapid mass recruitment and chemical repellents to deter other species (Adams & Traniello 1981).

Some species of stingless bee, including some *Trigona* spp., use pheromones to recruit massively to rich nectar or resin sources and aggressively displace other foraging bees of the same and other species (Roubik 1989, p. 105).

7.2.4 Trail specificity

Trail communication systems are very ancient in termites and may have evolved only once (Pasteels & Bordereau 1998). Trail pheromones in all termites are secreted from the sternal gland. The major component is one of just two compounds: either (Z,Z,E)-dodecatrienol (by *Reticulitermies flavipes* for example) or (E,E,E)-neocembrene-A (E-6-cembrene A) (by *Nasutitermes exitiosus* for example) (Pasteels & Bordereau 1998) (Table 7.1). These same compounds are used by reproductive castes as sex pheromones, but in quantities about 1000 times greater than those in workers. Termite species sharing the same major component can distinguish their own trail so the pheromones may have many minor components (Kaib *et al.* 1982; Pasteels & Bordereau 1998).

In contrast, among the ants, trail communication has evolved many times: some ten different glands are sources of trail pheromone and there is great diversity in the glands and compounds used in different ant genera (Table 7.1) (Hölldobler & Wilson 1990). At the species level, trail pheromones do not differ chemically as much as sex pheromones. However, although different species in a genus may share the main components of their trail pheromones, as for example in sympatric fire ant *Solenopsis* spp., workers will only follow those of their own species. Similarly, the recruitment pheromone seems to be the same in all species of the harvester ant *Pogonomyrmex* but the Dufour's gland secretion, used to mark persistent trunk routes, has a hydrocarbon blend that varies between species (and other markers may even differ between colonies, and, in some species, between individuals) (see Chapters 1 and 6) (Hölldobler & Wilson 1990).

In some species with mass communication, for example in *Solenopsis* fire ants, individual pheromone components may have different roles, some being involved in recruitment (alerting) and others in orientation (the tendency to follow a trail once alerted) (Chapter 2) (Vander Meer & Alonso 1998).

7.3 | Social insects as self-organising systems

Self-organisation models offer promising ways of understanding how colonies organise foraging effort so that it matches food profitability and how colonies construct nests. Complex, characteristic foraging or nest-building patterns, for example, can be generated from simple interactions between many simple units (individual workers) each with simple pheromone trail-laying and trail-following behaviour (see reviews by Traniello & Robson 1995, Bonabeau *et al.* 1997).

7.3.1 Ant foraging

The swarm raids of army ants, one of the wonders of the natural world, provide one of the best examples of self-organised, decentralised control (Franks *et al.* 1991). The hundreds of thousands of ants forming an *Eciton burchelli* swarm raid are virtually blind and the complex structure of the raid system is generated by the blind leading the blind (Franks *et al.* 1991).

Deneubourg *et al.* (1989) used a computer simulation model to investigate the way different food resource patterns may help explain the raiding patterns of different army ant species (Fig. 7.5 top). Their model has virtual 'ants' moving across a two-dimensional landscape: at each point the 'ant' can move ahead to the left or right (Fig. 7.5, inset bottom left). Like a real foraging army ant moving forwards, the model ant adds pheromone at the point chosen. The pheromone left by each ant that passes a point increases the probability a later ant will follow the same path so, although the movement starts as a random walk, a positively-reinforcing 'trail' rapidly forms. Like real pheromone, the virtual 'pheromone' evaporates over time. Ants capturing food then return from the swarm front, using similar rules (but laying more pheromone). Keeping the model rules the same, but varying the food densities at the swarm front, was enough to produce different

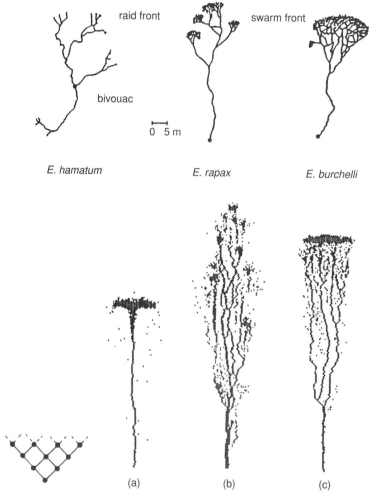

Fig. 7.5. (Top) Foraging patterns of three army ant species: *Eciton hamatum*, *E. rapax*, *E. burchelli*. (Bottom) The foraging of army ants can be modelled as ants moving across a 2-D landscape; the inset on the left represents the network of points in 2-D space for a virtual 'ant'. Three distinct foraging patterns develop in simulations using the same model with three different food distributions (Deneubourg *et al.* 1989). (a) no food; (b) each point has a 1/100 probability of containing 400 food items (as for an army ant such as *E. rapax* specialising on social insect nests); (c) each point has a 0.5 probability of containing one food item (as might be the case for a generalist predator such as *E. burchelli*). Figures after Deneubourg *et al.* (1989).

colony-level foraging patterns (Fig. 7.5, bottom). When each point has a fixed probability of containing one non-renewable food item, transportable by one ant, trails from that are roughly equivalent to those of *E. burchelli* (which feeds mostly on scattered single arthropods). With rare concentrated food densities, trails similar to those of *E. hamatum* form (this species feeds on concentrated prey – social wasp nests). In field experiments, Franks *et al.* (1991) were able to change the foraging behaviour of *E. burchelli* colonies by placing large bags of dead crickets (mimicking concentrated wasp nests) in the path of the swarm: the foraging trails became similar to those of *E. hamatum*. With intermediate densities and probabilities, trails similar to those of *E. rapax* formed.

The finding of new sources (and shorter routes to get there) may in part be stochastic – by the mistakes ants make in following trails. Pasteels *et al.* (1987) included this stochasticity in their models. They studied the finding and exploitation of new food sources by the mass-recruiting ant *Tetramorium caespitum*. When presented simultaneously with two non-depleting food sources, many mass-recruiting ant species initially recruit in equal numbers but then effort switches to one source only. The switch of foraging to one source can be explained by the responses of individuals responding to the 'trails': if there is a stochastic, chance, initial difference in pheromone concentration between the trails, the one with more will become amplified by positive feedback as more individuals will follow it and lay down pheromone, whereas the other will diminish. Similarly, if the resources are unequal in value, more ants will go to the richer one as more workers will lay pheromone coming back from it (Section 7.2.2) (Beckers *et al.* 1993; Bonabeau *et al.* 1997; Nicolis & Deneubourg 1999).

However, it is worth noting that no particular self-organised system is necessarily the optimum solution for a given problem: there can still be natural selection *between* alternative forms of self-organisation (Reeve & Sherman 1993; Traniello & Robson 1995).

7.3.2 Building a termite nest

Termites are among the most impressive builders in nature, constructing elaborate nest structures up to 30 m in diameter, on a scale some 10^4 to 10^5 times bigger in size than the individual termite workers (Fig. 7.6) (Bonabeau *et al.* 1997, 1998). However, there is no termite architect with a plan directing the workers, who must in any case work literally and figuratively in the dark. Instead, the structure emerges from simple responses of individual termites to local cues. The termite *Macrotermes* uses soil pellets impregnated with pheromone to build pillars. In the first phase, pellets are deposited at random. If enough pellets collect, the termites respond by building a pillar or strip (a feedback effect called 'stigmergy' by Grassé 1959). The process can be explained by mathematical models of pheromone concentrations over time and space that incorporate features such as the attraction of the cement pheromone to workers to lay more pellets, how the pheromones diffuse and fade, and thresholds for laying pellets at different trail pheromone concentrations. The models suggest that the same simple responses by individual workers are enough to create the different complex elements of the termite mound. The pheromone cues are influenced by, for example,

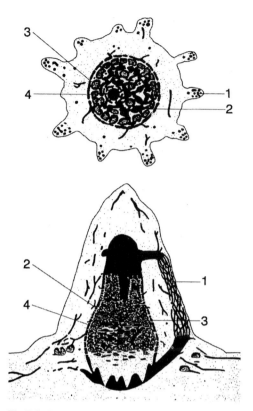

Fig. 7.6. The giant air-conditioned mounds of *Macrotermes* termites have a complex internal structure of air spaces, revealed here in horizontal and vertical cross section. The tunnels include an intricate system of air ventilation ducts in the walls for gas exchange with the outside air (1), brood chambers for the young termites and fungus gardens (2), a base plate (3), which is probably involved in cooling the air, and a royal chamber for the queen and king (4). The mound can be 3 to 5 m high. Figure from Bonabeau *et al.* (1998) after Lüscher (1961).

responses to the structures already built, the air currents that these create and additional sources of pheromone such as the queen (Fig. 7.7).

7.4 | Conclusion

The ability to recruit group members is one of the most important factors behind the extraordinary ecological dominance of social insects in so many habitats. Recruitment signals are commonly pheromones. Eusocial insects (ants, wasps, bees and termites) and the naked mole rat are central-place foragers, going out from the nest in search of food. While all termites show foraging recruitment behaviour, the development of such behaviours in ants, bees and wasps varies greatly between species and only a minority of social caterpillar species show recruitment. In ants, different recruitment mechanisms include **tandem running** in which the scout ant leads one

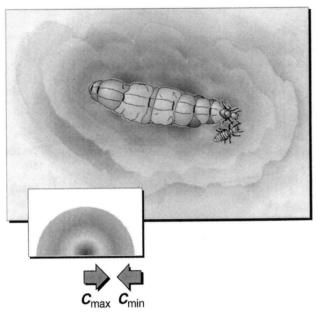

Fig. 7.7. The termite queen produces a pheromone which diffuses from her (different pheromone concentrations are represented by different levels of grey). Workers respond to this 'chemical template' and construct the walls of the royal chamber about 1.5 cm from her, where pheromone concentrations are between particular thresholds, $C_{min} - C_{max}$, represented in the inset, where the queen is at the darker centre. Figure from Bonabeau *et al.* (1998).

nestmate to the resource, **group recruitment** which recruits tens of nestmates, and **mass communication** which uses pheromones alone to recruit large numbers of nestmates. Key factors that select for recruitment behaviour are clumped, patchy food resources.

Despite the overall size and complexity of the colony, the social integration and assembly of colony-level patterns come from self-organised behaviours, emerging from simple interactions between individuals and individual responses to local conditions. Most of these interactions and responses are mediated by pheromones.

7.5 | Further reading

Traniello & Robson (1995) provide a stimulating overview of recruitment behaviour in ants and termites. Hölldobler & Wilson (1990) cover trail recruitment in ants comprehensively. Vander Meer & Alonso (1998) discuss ant pheromones. Pasteels & Bordereau (1998) and Kaib (1999) review recruitment behaviour in termites. Roubik (1989) discusses tropical stingless bees. Chapters in Detrain *et al.* (1999) discuss recruitment and self-organisation in social insects.

Chapter 8

Fight or flight:
alarm pheromones

8.1 | Introduction

Alarm signals offer perhaps one of the greatest challenges to evolutionary explanations of behaviour. Why should animals make alarm signals, at potential risk to themselves, to warn conspecifics of danger? Evolution of such apparently altruistic signals should be difficult if it makes signallers more conspicuous and at greater risk of predation than silent neighbours (see Chapter 11 for predators attracted to alarm pheromones). Benefits to kin probably account for the evolution of alarm signals in most species (Section 8.2) but alarm signals in groups of unrelated individuals are more difficult to explain.

Alarm pheromones may be more widespread than any other type of alarm signal. Their importance is underlined by the observation that in social insects alarm pheromones are the most commonly produced class of chemical signal, after sex pheromones, and have evolved independently within all major taxa (Blum 1985).

Like other pheromones, alarm pheromones are likely to have evolved as secondary or modified uses of existing compounds (Chapter 1). Two common evolutionary routes seem to be from either chemicals used in defence or those released by injury, both events being linked to predation.

Responses to alarm pheromones may be more adaptive than a simple response to the presence of predators as many prey species must live alongside their predators. Only when the predators are hunting (as indicated by release of alarm pheromone from prey) are they actually a danger.

8.2 | Evolution of alarm signals between related individuals

When already in the predator's jaws, giving an alarm signal might not save the victim, but the victim's signal could evolve by kin selection if it can save the lives of relatives such as offspring, siblings, or clones, by alerting them to danger.

8.2.1 Parental care
In the most straightforward cases, alarm pheromones have evolved in species with family groups.

8.2.1.1 Subsocial insects
The female of the subsocial aubergine lace bug (*Gargaphia solani*) stays with her eggs and nymphs. She runs toward any predator, fanning her wings, an effective defence against predators such as coccinelid beetles or small spiders (Tallamy & Denno 1982). If caught by a predator, the larva emits an alarm pheromone, geraniol, from its dorsal glands (Aldrich *et al.* 1991). Nearby nymphs stop feeding and run.

8.2.1.2 Alert signals in deer
Many mammals release odours when alarmed or pursued, but these 'alert' alarm signals are more like alarm calls in birds, not given on death but given when predator is detected. Alert signals by prey may be directed at the predator itself, an 'I've seen you' signal (Hasson 1991). The signaller may benefit if the predator is less likely to attack once it has lost the element of surprise. Alternatively, alert signals may be directed at the signaller's kin.

Alert signals in family groups of black-tailed deer (*Odocoileus hemionus columbianus*) may have evolved in response to both selection pressures. When disturbed, black-tailed deer expose the metatarsal gland on the hindleg, releasing a strong garlic-like odour (Müller-Schwarze *et al.* 1984). Females exposed to the odour from either male or female glands were more likely to show increased alert behaviour, lifting the head and scanning the surroundings, and were more likely to leave, compared with animals exposed to control odours. The recipients of the signal are likely to be kin, as mothers, their female yearlings and fawns tend to associate within larger aggregations.

In black-tailed deer, odours are only part of the alert signal: visual signals include an erect tail, raising of anal hair, cocking of the ears, and a characteristic stiff-legged gait. The signaller also hisses and stamps its feet. Similar combinations of visual and odour signals are found in pronghorn

(a)   (b)

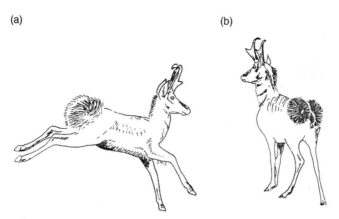

Fig. 8.1. On detecting a predator, pronghorn antelopes (*Antilocapra americana*) raise the long white hairs of the rump patch with a jerk (a, b). The hairs surround a musk gland and fold into it. When they are raised, musk odour is released and can be smelt downwind, even by people (Müller-Schwarze & Müller-Schwarze 1972). This combined olfactory and visual alert signal is directed at conspecifics and could also be directed at predators. Figure from Stoddart (1980).

antelopes (Fig. 8.1). The redundancy of signal, with simultaneous visual, sound or odour signals, and the subtlety of receiver responses, helps to explain why mammal alarm pheromones are so difficult to study.

8.2.2 Kin selection

Along the continuum of relatedness, clones (with all individuals genetically identical) ought to offer the most extreme cases of altruism, and altruism should be greatest where clonal individuals stay close to each other (Hamilton 1987), as in the sea anemones and aphids discussed in Sections 8.2.2.1, 8.2.2.2 and 8.2.2.3. We might also expect altruism, including alarm signals, in eusocial insects with a high degree of relatedness within the colony.

8.2.2.1 Clonal sea anemones

The sea anemone *Anthopleura elegantissima* forms clonal groups on rocky shores along the Pacific coast of North America. The anemone shows a characteristic rapid response to the wounding of nearby individuals: first, it gives a rapid shake of the tentacles, second, it withdraws the vulnerable tentacles and oral disc and, third, the whole anemone contracts, all within 3 seconds. The response is to a pheromone, anthopleurine, released by damaged individuals (Table 8.1) (Howe & Sheik 1975). As near neighbours are genetically identical individuals produced by asexual division (binary fission), the likely receivers of the pheromone would be fellow clone members (although it would benefit the receiver to respond whether or not it was related to the signaller).

The behaviour of one of the anemone's specialist predators, the nudibranch sea-slug (*Aeolidia papillosa*), suggests that anthopleurine evolved

Table 8.1. *Alarm pheromones in different animal species.*

In species such as the honeybee, the alarm pheromones are complex mixtures.

Animal	Principal compounds	Example	Origin
Coelenterata Sea anemone		 Anthopleurine	Tissue
Insecta **Isoptera** Termitidae	Terpenes	 Limonene	Cephalic gland
Thysanoptera Thripidae	Hydrocarbons	 Decyl acetate	Unknown
Hemiptera Aphididae	Terpenes	 E-β-farnesene	Cornicles
Tingidae	Hydrocarbons + Monoterpenes	 E-2-hexenal Geraniol	Dorsal abdominal gland
Coreidae	Hydrocarbons	 Hexanal	Metathoracic gland
Hymenoptera Apidae	Hydrocarbons	 2-Heptanone	Mandibular glands + poison gland

Table 8.1. (*cont.*)

Animal	Principal compounds	Example	Origin
Hymenoptera (*cont.*)			
Formicidae Ponerinae	Heterocyclic nitrogen compounds	 2, 5-Dimethyl-3-isopentylpyrazine	Mandibular glands
Myrmicinae	Branched hydrocarbons	 4-Methyl-3 heptanone	Mandibular glands
Vertebrata Fish (minnows)		 Hypoxanthine-3-*N*-oxide	Club cells in the skin

Largely from Chapman (1998).

originally for defence: the predator prefers to feed on the parts of the anemone with the least anthopleurine. Incidentally, enough anthopleurine remains in the tissue of the predator, days after eating anemones, for sufficient to diffuse out to alarm anemones ahead of it (Howe & Harris 1978).

8.2.2.2 Aphids

Aphids are another clonal organism with separate individuals. For much of the year parthenogenetic (asexually reproducing) mother aphids are surrounded by their clonal offspring. As expected, aphids show a well-developed alarm pheromone system (Hardie *et al.* 1999). If an aphid is attacked by a predator or parasitoid, alarm pheromone is released in cornicle secretions (Fig. 8.2), which can themselves glue predator mouthparts together. In response to the alarm pheromone, nearby aphids stop feeding abruptly, pulling their mouthparts from the leaf, back away, run away or even jump or fall off the leaf to escape. The result is an aphid-free halo around the predator. In addition, aphids once alarmed are less easily

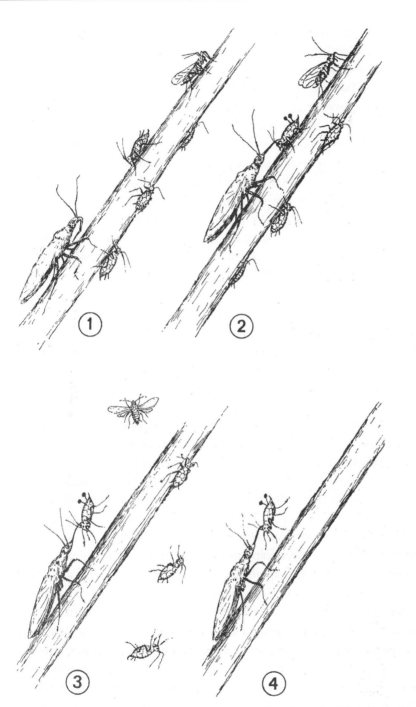

Fig. 8.2. Drawing shows step-by-step occurrences as a predatory insect (nabid) attacks a colony of aphids feeding on a plant stem. (Top) The nabid has seized its prey and the aphid has released the alarm pheromone from its cornicles. (Bottom) Other aphids detect the alarm pheromone and scatter by flying or dropping off the plant. Although the nabid continues to devour its original captive, other aphids in the vicinity have departed. Figure from Nault (1973).

caught by ladybird predators (Montgomery & Nault 1978). Signalling aphids benefit by kin selection as the majority of receivers will be clonemates.

The main component of the alarm pheromone of many aphids is E-β-farnesene (Table 8.1) (Hardie *et al.* 1999). The response to alarm pheromone can depend on context. For example, aphid species attended by ants respond less to alarm pheromone, particularly if ants are present (Chapter 11). Some conditional factors may have evolved to reduce the chance of false alarms (especially as plants also produce some of the compounds). For example, in the turnip aphid (*Lipaphis* (*Hyadaphis*) *erysimi*), individuals do not respond unless E-β-farnesene is combined with odours produced by aphids feeding nearby on the plant. In some species, aphids respond more to alarm pheromone after the plant has been vibrated in the same way that a predator would during its hunt (Clegg & Barlow 1982).

Responding to alarm pheromone has its costs. For example, the costs of dropping off the plant include potential lost feeding time, death on the ground and the chance of not finding the host plant again. The costs will vary according to the value of the feeding site the aphid left: aphids feeding on nutritionally richer artificial diets are less likely to move in response to alarm pheromone than those on poorer diets (Dill *et al.* 1990). Costs also vary by habitat. In hot, dry habitats there is higher mortality of aphids on the ground due to desiccation: pea aphid (*Acyrthosiphon pisum*) clones from hot dry inland sites in Canada, where soil temperatures reach up to 42°C, were less likely to drop in response to alarm pheromone than those from moister, cooler, coastal sites (Dill *et al.* 1990). Within each clone, some individual aphids respond to alarm pheromone by dropping off the leaf, while others do not: the difference in numbers dropping between clones from inland and coastal sites reflected the proportion of each. The proportion 'dropping' in a clone can be quickly changed by selection experiments (Andrade & Roitberg 1995).

Parasitism may also affect responses to alarm pheromone. In the hot inland populations of *A. pisum*, aphids parasitised by a wasp are more likely to drop off the plant than unparasitized individuals if approached by a predator or stimulated by alarm pheromone (McAllister & Roitberg 1987). This could be an adaptive host-suicide response by the parasitised aphids – they are more likely to die on the ground than staying on the leaf with the predator (but in that case why don't aphids simply drop as soon as they are parasitised?).

Aphid alarm pheromones have been exploited deceitfully by plants to repel boarders (Chapter 11).

8.2.2.3 Social aphids

Altruism is taken one step further in social aphid species with soldiers; this was first investigated by Aoki (1977). Some 50 species of aphids in two families, the Pemphigidae and Hormaphididae, have sterile soldiers (see review by Stern & Foster 1996). The soldiers, often first instar nymphs, attack intruders. In many species they have enlarged mouthparts, legs or horns, which they use to stab invertebrate enemies (Fig. 8.3). The aphid soldiers may die defending their clone. In the Japanese social aphid *Ceratovacuna lanigera*, although alarm pheromone is secreted by all stages, it elicits attack in soldiers but escape in adults and other non-soldiers (Arakaki 1989).

8.2.2.4 Other social insects

Social insects are not clones, but in the case of the social Hymenoptera at least, individuals are more closely related than usual diploid animals because of their unusual haplo-diploid reproductive system. Like social aphids,

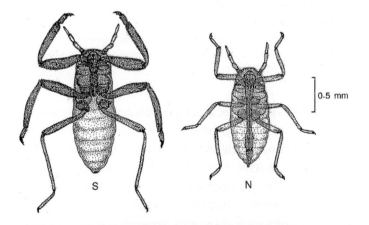

Fig. 8.3. (Top) The first instar aphid soldier (S) of *Colophina monstrifica* is much larger than the normal larva (N) and has a shorter beak and thickened cuticle with enlarged horns and mouthparts. Figure from Stern & Foster (1996).
(Bottom) First instar soldiers of the social aphid *Ceratovacuna lanigera* attacking a predatory syrphid larva which they have marked with alarm pheromone (Arakaki 1990). Photograph by N. Arakaki.

Fig. 8.4. In a suicidal attack the worker honeybee leaves its autotomised sting while simultaneously releasing alarm pheromone and injecting venom into the victim. The alarm pheromone alerts other workers to danger, reduces the threshold sensitivity to attack and directs other bee attacks where there are already stings (Schmidt 1998). Photograph by J. O. Schmidt.

these animals show 'suicidal' responses in defence of the nest (Fig. 8.4). Indeed, eusocial wasps, bees, ants and termites (Chapter 6) show some of the most highly developed alarm pheromone systems of any animal (Blum 1985). In ants, alarm systems have been found in every species tested (Hölldobler & Wilson 1990). The pattern and range of compounds used as alarm pheromones, and the great variety of different glands they are secreted from even within a group such as the ants, strongly suggests multiple independent evolution across the social insects (Chapter 1) (Blum 1985). Most alarm pheromones are multi-component, with two or more compounds (which can come from different glands). The compounds may simultaneously alert, attract and evoke aggression (Chapter 10) (Hölldobler & Wilson 1990, p. 261).

For eusocial species, alarm pheromones are a key evolutionary development that enables the colony to recruit its collective force to defend the nest against attack (Blum 1985). Defence is needed because colonies represent rich concentrated resources of brood and in honeybees, wax, pollen and stored honey (Winston 1987). Predators of honeybee nests include humans and other vertebrates such as bears, honey badgers, anteaters and birds. Invertebrate threats include other social Hymenoptera and even conspecifics from other nests (see Chapter 11).

The effectiveness of colony defence is illustrated by the behaviour of the Japanese honeybee (*Apis cerana japonica*) when threatened by the giant hornets *Vespa mandarinia japonica* (Ono *et al.* 1995). If the giant hornets get into the nest, they will destroy it, taking the brood to feed their own larvae. An attack starts when a hornet scout finds the nest and marks the site with its sternal gland. Soon, nestmate hornets flying in the same area congregate and attack the bee nest together (Matsuura cited in Ross & Mathews 1991).

Japanese honeybees have evolved a coordinated response triggered by the hornet's marking pheromone. When a giant hornet approaches a Japanese honeybee nest, guard bees signal alarm by releasing a pheromone and those in the entrance retreat into the nest. Meanwhile, more than 1000 workers leave the comb and wait just inside the entrance. If a foraging hornet tries to enter, a buzzing ball of more than 500 bees forms around it (Fig. 8.5). Pheromones in the bee venom and mandibular secretions attract yet more workers and focus attack on the hornet, which is now chemically marked like a beacon. Remarkably, the hornet is roasted to death by the heat generated by the swarm, which can reach about 47°C at the centre, but the bees survive as they have a higher upper thermal limit than the hornet. The aim appears to be to kill the hornet before it can mark the nest further. Using pheromones to gather assistance, the Japanese honeybees can collectively overcome a predator bigger and more powerful than any individual bee.

The cascade and escalation of response through recruitment is characteristic of eusocial bees, wasps, ants and termites. The wasps *Vespula vulgaris* and *V. germanica* spray an attacker with fine droplets of venom from the sting (Machwitz 1964 cited in Blum 1985). Similarly, *Nasutitermes* termite soldiers spray more alarm pheromone and defensive secretions from their enormously enlarged head nozzles at intruders (Ernst 1959, cited in Blum 1985) (Pasteels & Bordereau 1998).

However, not all social insects have the *aggressive* alarm responses shown by the examples above (Hölldobler & Wilson 1990). Some species of ant, for example *Lasius alienus,* show a **panic & escape** response to its own alarm pheromone: the colony as a whole flees from the stimulus or dashes around in erratic patterns (Regnier & Wilson 1969). If disturbed enough, individuals may even evacuate the nest. This may be an adaptive response for species of small ants that form small colonies with little possibility of realistic defence.

Responses within a species of social insect differ not only between castes and age classes (with major, or older, workers and soldiers more active in attack of intruders) but also according to context. Some usually aggressive species may give hair-trigger responses, including panic, to dangerous species, termed **enemy specification** by Hölldobler & Wilson (1990, p. 261). For example, a single fire ant (*Solenopsis invicta*) approaching the nest is enough to cause minor workers of the ant *Pheidole dentata* to recruit majors to start nest defence.

Fig. 8.5.

Other factors may influence the type of response. In many social insects, individuals that would react very aggressively to alarm pheromone near the nest, respond away from the nest by dispersing (Maschwitz 1964, cited in Blum 1985). Recent experience also affects the sensitivity of response in honeybees and wasps: workers are more alert and more likely to pursue potential intruders in the hours after an attack. The powerful effects of alarm pheromones on social insect colonies can be exploited by other species to devastating effect (Chapter 11).

8.2.2.5 Tadpoles, kin, and alarm pheromones

Toad tadpoles form large, conspicuous shoals in ponds. When a crushed tadpole is brought near the shoal there is a dramatic fright response: the shoal breaks up and those closest to the victim swim into deeper water (Kulzer 1954, cited in Pfeiffer 1974). The recipients of the alarm pheromone are likely to be kin as toad tadpoles associate preferentially with their siblings, using learnt odour cues (Box 6.1) (O'Hara & Blaustein 1982; Waldman 1982, 1985). The conspicuous black colour of toad tadpoles may be aposematic, as tadpoles of many toad species are distasteful to predators (see Waldman 1982). In experiments, North American western toad (*Bufo borealis*) tadpoles exposed to a conspecific alarm signal were more successful at escaping predation by an invertebrate predator, a dragonfly nymph (*Aeshena umbrosa*) (Hews 1988).

8.3 | Evolution of alarm signals in unrelated individuals

If victims are unlikely to be surrounded by kin, the evolution of alarm pheromones released when the jaws of a predator damage the skin is especially perplexing as there would seem to be little benefit to the signaller. It could be that most of the responses we see are adaptive to the *responder* (odours released by injured conspecifics indicate danger nearby) but that these odours are not evolved as signals on the part of the *signaller* (Sections 8.3.2.1, 8.3.2.2 and 8.3.2.3).

Fig. 8.5. (*cont.*) The Japanese honeybee *Apis cerana japonica* has a mass defence of the nest, coordinated by responses to alarm pheromone, when (Top) scouts of the predatory giant hornet *Vespa mandarinia japonica* are detected.
(Middle) Hundreds of tightly aggregated bees surround the giant hornet and their combined heat, up to 47°C, kills the hornet.
(Bottom) The defensive ball of bees can be held on the hand with no danger of stings (Ono *et al.* 1995).
Photographs: Top © M. Ono & HSRC (Honeybee Science Research Center, Tamagawa University);
Middle © M. Ono; Bottom © M. Ono & Nature.

8.3.1 Marine invertebrates

Many marine invertebrates show alarm responses to chemicals released by damage to conspecifics. *Diadema antillarum*, a coral reef sea urchin protected by needle-sharp spines 30 cm long looks invincible but nonetheless many predators, such as the trigger fish, can get round the spines. Neighbouring *D. antillarum* down current of a crushed conspecific race away for a metre or two, 'running' on their ventral spines, before settling down again (Snyder & Snyder 1970). There is no evidence of a special alarm pheromone.

8.3.2 Fish

Karl von Frisch found the 'fright response' in the European minnow (*Phoxinus phoxinus*) almost by accident while observing the behaviour of fish individually marked with nicks in their skin. When he put a fish back into the water just after marking it, he was surprised to find that the other fish in the shoal swam rapidly away, apparently in alarm (von Frisch 1938, cited in Pfeiffer 1974). The skin damage had released alarm pheromone, characteristic of many species of freshwater fish in the superorder Ostariophysi, which elicits alarm responses in conspecifics. These include anti-predator behaviours such as increased shelter use, shoaling, or 'freezing'. The fright reaction in different fish species can be related to their ecology (see review by Smith 1999).

Minnow skin, and that of other members of the superorder Ostariophysi, has large numbers of specialised 'club cells' that contain the alarm pheromone. These cells have no outlet to the surface so the pheromone is released only when the skin is damaged. Laboratory and field tests suggest that the alarm pheromone is hypoxanthine 3-*N*-oxide (Table 8.1) or similar molecules with a nitrogen-oxide functional group (Brown *et al.* 2000). The alarm pheromone system in fish has evolved independently from the toad tadpole system (Section 8.2.2.5): there is no cross reaction of fish to toad tadpole alarm pheromones or *vice versa* (Pfeiffer 1974).

Each club cell of the North American fathead minnow (*Primiphales promelas*, a member of the same cyprinid (carp) family as the European minnow), contains enough alarm pheromone to create an active space of about 80 litres (Lawrence & Smith 1989). One square centimetre of skin would thus provide an active signal in about 58 000 litres, equivalent to a cube almost 4 m on each side. In flowing water, only fish downstream would be affected by the pheromone (although eddies carry signals upstream too, Dahl *et al.* 1998). However, the fright response of fish that have sensed the pheromone is transmitted visually to other fish in any direction (demonstrated in experiments with fish in separate tanks but in view of each other). Other prey fish species, which share the same predators in a shared habitat, respond to each other's alarm pheromones (either directly or by observing the visual signs, Krause 1993b).

The fright reaction to the alarm pheromone itself seems to be innate, but fish learn to associate predator odour and visual signs with danger when they sense these with alarm pheromone. The predator stimuli can then elicit the fright reaction by themselves (Chapter 12). This learning is more rapid in populations exposed to predation (Levesley & Magurran 1988). Learning gives flexibility because fish can learn to respond to the predators found in their local habitat. Field populations of predator-naïve fathead minnows learnt to respond to first the odour and then sight of predators within days after the natural predatory fish (the pike *Esox lucius*) were added to a pond (Brown *et al.* 1997).

Recent experiments on the European minnow have cast some doubt on the role of alarm pheromone (Magurran *et al.* 1996; Henderson *et al.* 1997; Irving & Magurran 1997). The more natural the conditions in the experiments, the lower the intensity of the fright reaction. With the same population of minnows, a strong response in laboratory aquaria gave way to a weak one in a fluvarium (a glass-sided semi-natural stream bed) and was lost in the wild. The experiments in a river used underwater video to observe the effects of remotely released alarm pheromone vs. a control treatment. They found no change in the behaviour of the fish around the source for the 30 minutes before and after alarm pheromone release, as measured by the number of fish in view and their behaviour – the rapid dispersal, predicted by laboratory experiments, did not occur. Nor were pike attracted in the field experiments (see Section 8.3.2.1).

Despite these results, the bulk of evidence currently supports the conclusion that fright responses to alarm pheromone are adaptive (Smith 1997). These include the original field observations made by von Frisch (above) and three other field observations of fish alarm responses, in addition to numerous field trapping studies, which all show aversive responses to alarm pheromone. Nonetheless, more underwater video of field responses would be helpful.

What factors could select for responses from receivers and the presence of alarm pheromone in all of these fish?

8.3.2.1 Benefits to receivers of alarm pheromone signals

One laboratory study with pike suggests that fathead minnows, tested in small groups, survived 40% longer if exposed to alarm pheromone. The effect was largely due to significant increases in shoaling and use of shelter (Fig. 8.6) (Mathis & Smith 1993).

Individuals responding to alarm pheromone by shoaling could benefit from risk dilution or 'selfish herd' effects. According to Hamilton's 'selfish herd' theory, per capita predation risk is lower in the centre than the edge

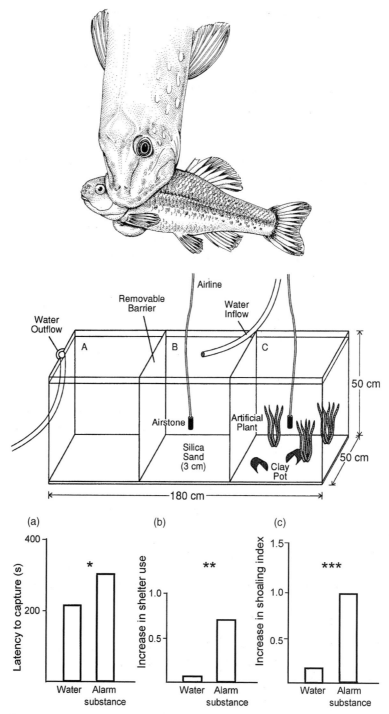

Fig. 8.6.

of a group (Hamilton 1971; Krause 1994). Krause (1993a) observed the behaviour of a single European minnow in a shoal of dace (another freshwater fish). Before the experiment, the dace had been habituated to alarm pheromone, so they would not respond to further alarm pheromone. On exposure to alarm pheromone in the experiment, the minnow moved towards the centre of the shoal – the safest place, furthest from the exposed outside of the shoal.

Alarm pheromone builds up in the body and faeces of pike that feed on minnows containing alarm pheromone. These chemical cues leak from the pike and its faeces – minnows and other prey fish can use these cues to learn the high-risk areas best avoided because a minnow-eating pike is resident there (Brown *et al.* 1995). As minnows can learn to avoid areas on the basis of alarm pheromone in predator faeces, pike fed on minnows defecate away from their hunting grounds (whereas pike fed on mice defecated anywhere in the fish tank) (Brown *et al.* 1996).

8.3.2.2 A benefit to the signaller from kin selection in shoals?

For fish alarm pheromones to evolve by kin selection, shoals would need to be composed, on average, of related individuals more often than not. Naish *et al.* (1993) used DNA fingerprinting to investigate European minnow (*Phoxinus phoxinus*) shoals in the field. They found no evidence of greater relatedness within than between shoals. However, there remains the hypothetical possibility of a finer structure of sibling groups within shoals, aided by an ability to recognise familiar shoalmates and siblings, by odour cues (Box 6.1) (Brown & Brown 1995; Olsen *et al.* 1998).

8.3.2.3 A benefit to the signaller: the 'Attraction of a second predator' hypothesis

The club cells of ostariophysian fish may have evolved with a primary function such as control of disease on the skin. Only later might other advantages have emerged. Could the club cell contents benefit the victim in the moment of danger?

Fig. 8.6. (*cont.*) Fish alarm pheromone can reduce the success of the predatory pike *Esox lucius*, by increasing the time taken to catch a fathead minnow, *Primiphales promelas*.
(Top) Pike with minnow in mouth. Drawing from Alcock (1998).
(Middle) At the beginning of each trial, compartment A contained one pike and C four fathead minnows. Opaque barriers were removed after introducing either a conspecific (minnow) skin extract (alarm pheromone) or distilled water control into B.
(Bottom) Results: (a) the mean time to capture of the first minnow by the pike is increased by exposure to alarm pheromones. The reasons seem to be the behaviours elicited by the alarm pheromone, which include increased shelter use by minnows (b) and a greater increase in shoaling (c). (* $p < 0.05$, ** $p < 0.025$, *** $p < 0.01$). Middle and bottom figures from Mathis & Smith (1993).

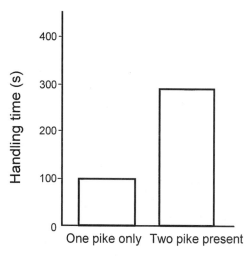

Fig. 8.7. When the minnow *Primiphales promelas* is bitten by the predatory pike *Esox lucius*, club cells in its skin are broken, releasing 'alarm pheromone' compounds. Other predators are attracted by the odour and may interfere with the feeding of the first, increasing the time it takes a pike to orient its prey for swallowing. In addition, in the confusion, the arrival of the second predator allowed the minnow to escape in five out of 13 experiments, which never happened when the pike was undisturbed. Figure redrawn from Chivers *et al.* (1996).

Surprisingly, minnows can be injured enough to release alarm pheromone but survive. For example, in one location 16% of fathead minnows had injury marks that would have released alarm pheromone. When brought into the laboratory the fish survived 1 year (Smith & Lemly 1986). So there could be advantages for the signaller *if* alarm pheromone could increase its chance of escape even after capture. One way this could happen is if fish alarm pheromone acts as a 'distress signal', which enables the victim to escape by attracting conspecifics or secondary predators that interfere with the primary predator (Smith 1992). While there is no evidence of mobbing as seen in birds, northern pike (*Esox lucius*) are attracted by minnow alarm pheromone in aquarium tests and predatory beetles are attracted to field traps baited with alarm pheromone (Mathis *et al.* 1995). In aquaria, when a second pike was released once a first pike had a minnow in its mouth, the second pike did interfere, increasing the handling time for the first pike and often allowing the minnow to escape (Fig. 8.7) (Chivers *et al.* 1996).

8.4 | Conclusion

Alarm signals offer perhaps one of the greatest challenges to evolutionary explanations of behaviour. Benefits to kin probably account for the evolution of alarm signals in most species. These include alarm pheromones released between siblings, offspring and parents in subsocial insects and in family groups of deer. Alarm pheromones are found in clonal animals such as aphids and sea anemones, in particular in social aphid species with soldiers. In social insects alarm pheromones are the most commonly produced class of chemical signal, after sex pheromones, and have evolved independently within all major taxa. In ants, bees, wasps and termites alarm pheromones coordinate attack on enemies threatening the nest. Some species show a **panic & escape** response to situations where defence is not realistic. Kin selection may explain toad tadpole alarm responses and aposomatic colouration.

Where victims are unlikely to be surrounded by kin, the alarm responses may be adaptive to the *responder* rather than an evolved signal from the sender. The principal example is the alarm response of ostariophysian fish, which release alarm pheromone when their skin is broken. One suggestion is that there could perhaps be advantages for the signaller if the pheromone attracts other predators, which interfere with the first.

8.5 | Further reading

Diamond (1997) gives a brief introduction to alarm pheromones and Smith (1999) gives a good review of alarm signals in general and in fish in particular. Alarm pheromones in arthropods are reviewed by Blum (1985). Hölldobler & Wilson (1990) comprehensively cover the ants. Hölldobler (1995) discusses the integration of alarm signals in the behaviour of ants.

Chapter 9

Perception and action of pheromones:
from receptor molecules
to brains and behaviour

9.1 | Introduction

All animals detect and react to chemicals in the external environment, including pheromones and the chemical cues that indicate food, shelter or predators. The key sense used to detect these chemical cues, in air or water, is smell (olfaction) rather than taste (Box 9.1). In this chapter we examine how the molecular basis of smell and the organisation of neural connections for olfaction help to explain behavioural and physiological responses to pheromones – and the evolution of pheromones as enormously varied and powerful signals.

Box 9.1 | Taste and smell

The sense of taste is largely found in vertebrates and is always used in feeding (Finger 1997). The difference between taste (gustation) and smell (olfaction) is not the medium, air or water, or even the distance between signaller and receiver. Instead it is the kinds of sensory cells that are stimulated. Whereas olfactory cells have axons that go into the brain, taste cells are secondary sensory cells that do not themselves extend into the brain but instead pass on their signal to a sensory nerve fibre. Taste cells, which are grouped in taste buds, can be anywhere on the surface of the animal, but regardless of location, taste buds always connect to nerves leading to a different area of the brain from the olfactory system, and in a much more simple manner, without the integration of information and coding found in olfactory systems. The number of sensory cells, ranges of substances that stimulate them, and qualities or categories of stimuli that can be discriminated, are all smaller for taste than for olfaction (Hildebrand 1995). The enormous range of flavours that we can distinguish in food, for example, is almost entirely by olfaction – and depends on having an unblocked nose. When considering non-vertebrates it may be better to use 'chemoreception' as a more inclusive term.

Olfactory pathways in the brain

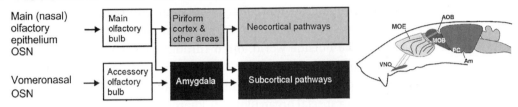

(a) **Rodents & other vertebrates**

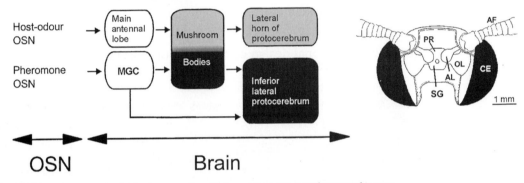

(b) **Moths & other insects**

OSN　　　Brain

Fig. 9.1. There are many similarities between the olfactory pathways in vertebrates and insects.
(a) In rodents and other vertebrates, the main olfactory bulb (MOB) receives input from olfactory
sensory neurons (OSNs) in the main olfactory epithelium (MOE) and sends projections to higher areas
such as the piriform cortex (PC). Rodents and many other terrestrial vertebrates also have an
accessory olfactory bulb (AOB) which receives inputs from sensory neurons in the vomeronasal organ
(VNO) and sends projections to areas such as the vomeronasal amygdala (Am). The VNO plays a major
role in the perception of social chemosensory stimuli in many species but it has not been demonstrated
in adult humans (Chapter 13).
(b) In moths, the antennal lobe (AL) receives input from OSNs along the length of the antenna (AF). In
many insects, the AL in males is divided into two subsystems, one for processing information about
general odorants (Main AL), the other devoted to species-specific information about female sexual
pheromones (the macroglomerular complex, MGC). Many AL projections converge on the mushroom
body in the protocerebrum (PR), whereas some projections from the MGC bypass the mushroom body
completely. Other abreviations: OL, optic lobe; CE, compound eye; SG, suboesophageal ganglion.
Caption and figure adapted from Christensen & White (2000).

I first describe common features of olfactory systems across the animal
kingdom, then discuss some well-studied and contrasting systems in invertebrates and vertebrates (including the dual olfactory systems of many
vertebrates, the main olfactory system and the vomeronasal system)
(Fig. 9.1), some factors affecting responses to pheromones, the effects of
primer pheromones on reproduction, olfactory cues and recognition learning and, finally, developmental paths prompted by pheromones.

9.2 | Chemical cues: perception and interpretation by the brain

Across the animal kingdom, olfactory systems are remarkably similar. All have olfactory sensory neurons (OSNs) which are nerve cells with one end exposed to the outside world, often through an 'olfactory window' in an otherwise impermeable skin or cuticle (the other end of the nerve cell extends directly into the brain). In many animals the sensory neurons are gathered into a special organ (Box 9.2). Pheromone or other odour molecules are converted into signals to the brain by first binding to the olfactory receptor protein (OR) in the cell membrane of the olfactory sensory neuron (Fig. 9.2). This activates a so-called G-protein, also in the cell membrane,

Box 9.2 | Functional design of olfactory organs: flow and scale

In most animals, olfactory sense cells are concentrated in organs such as antennae or nasal cavities, specialised for detecting chemical signals. Some animals also have chemosensory organs on their feet or ovipositors. Some get greater spatial separation between the sensory cells by having long antennae or widened heads like hammer headed sharks (other animals move their heads from side to side to achieve the same effect, Chapter 10).

The designs of olfactory organs are, like other biological structures, an evolved compromise between conflicting selective forces. Olfactory organs must, first, maximise exposure of the olfactory sensory neurons to the outside environment while protecting these relatively delicate cells and, second, maximise the effective capture of odour molecules within the biomechanical constraints of size, flow and fluid dynamics.

Insect sensory neurons are protected inside cuticular 'hairs' and bathed in sensillum lymph (Fig. 9.2). Terrestrial vertebrates enclose their delicate olfactory tissue in the nasal cavity, protected by a continuously secreted layer of mucus (Pelosi 1996).

Scramble competition to be the first male to detect and locate an upwind female releasing pheromone has been a powerful selection pressure on male moths (Chapters 3 and 10). The result of selection is antennae exquisitely sensitive to the specific pheromone with an estimated behavioural threshold of just 4.5×10^{-18} molecules/l (for silk moth males) (Kaissling 1998a, b). A stimulus pulse delivering as little as one molecule per pore tubule is a very strong stimulus for a pheromone olfactory sensory neuron.

The high sensitivity of male insects comes from the specificity of the receptor protein and the pheromone binding protein (present in large quantities) combined with the design of the antennae, which carry thousands of the specialised olfactory sensory neurons. For example the male silk moth (*Bombyx mori*) has some 17 000 hairs on each antenna, each containing the ends of two olfactory sensory neurons, one tuned to the sex pheromone bombykol (Kaissling 1998a) (see figure). The combined area of the antennae is more than 170 mm^2. Radioactive-label studies suggest that most pheromone molecules from the airstream passing the cross-sectional area of the antennae are adsorbed, and that 80% of these are caught by the long hairs of the sensilla hairs (Kanaujia & Kaissling 1985). However, scale is very important: structures this small are affected by the treacle-like viscous characteristics of airflow at these low Reynolds numbers (Re, a term used to describe characteristics of flow speed, viscosity, size and shape, Vogel 1994) (see Chapter 10). Direct measures of the giant silk moth (*Actias luna*) showed that only 8–18% of the air directly upstream passed *through* the antenna, while the rest went around (Vogel 1983) (in contrast to air, some 38% of a light beam passed through). More, finer, hairs increase the probability of pheromone molecules hitting the antenna but lead to less air flowing through. The Re of the fine hairs themselves is around 0.2. At these Re-values, rows of hairs tend to function as paddles rather than as sieves, which helps to explain why so much of the air passes around the antenna (Koehl 1995).

Close to the antennal surface there is an almost stationary boundary layer (Fig. 10.6); the lower the Reynolds number the thicker it is (Vogel 1994). The shape of the antenna will affect the thickness of the boundary layer, in turn affecting the perceived signal, including the relative importance of diffusion effects on pheromone encounters of the sensilla hairs (Schneider *et al.* 1998) (see Box 9.4 for sampling by flicks of the antenna).

Vertebrates, such as dogs, which rely on their sense of smell for prey detection, increase their sensitivity to odours by having multiple copies of generalist olfactory sensory cells, in a large area of olfactory epithelium ($\sim$30 000 mm^2 compared with the olfactory area in humans of 400 mm^2). The large olfactory area is reflected in the very rough estimate of dog olfactory

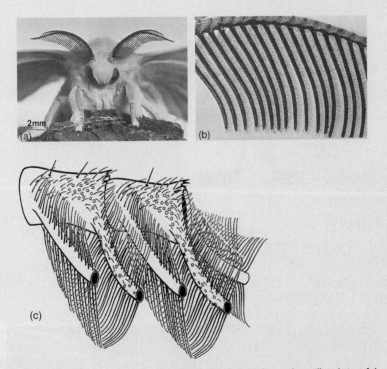

(a) Male silk moth (*Bombyx mori*) showing the numerous side branches and sensillary hairs of the antennae which form an effective olfactory sieve for molecules of the pheromone, bombykol.
(b) Scanning electronmicrograph close-up of the antenna.
(c) Two segments of the male antenna of the silk moth *Antheraea polyphemus*. The long sensillary hairs are each innervated by 1–3 pheromone sensitive cells and are regularly arranged to form an 'odour sieve'. From Kaissling (1987) in Steinbrecht (1999). Photo/micrographs (a), (b) by R.A. Steinbrecht.

thresholds for some compounds being 1000 to 10 000 times lower than for humans (Marshall *et al.* 1981) although superiority in distinguishing complex odours may be more important than sensitivity by itself. However, measuring odour thresholds is notoriously difficult: much depends on the relevance of the olfactory discrimination task to the life of the animal (dog or human). The required sensitivity of olfactory systems for pheromones may depend on whether signals are perceived at close range.

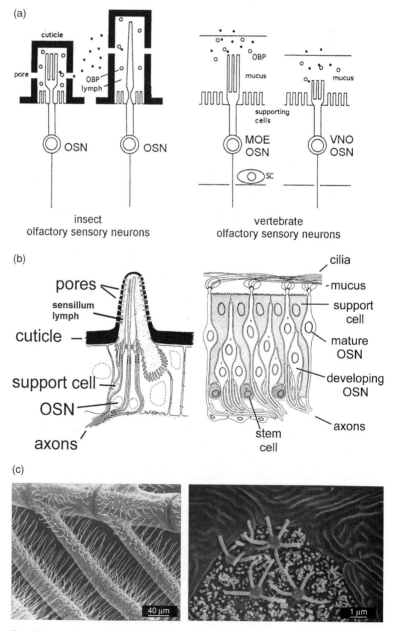

Fig. 9.2.

triggering a cascade of reactions leading to electrical impulses being sent down the axon of the olfactory sensory neuron to the brain. Olfactory sensory neurons are called 'receptors' by some authors but to avoid confusion, only the olfactory receptor proteins (OR) themselves are called receptors in this book.

The olfactory systems of most animal species have the ability to detect, discriminate and distinguish innumerable different molecules as different odours, including ones they have never smelt before. The ability comes from the combination of the range of receptor types (far outnumbered by the number of possible odorants, but large compared with the number of receptor types in vision, for example), and the characteristic broad but overlapping specificities of the receptor proteins. This means that a huge 'olfactory space' can be covered, and the signals can be distinguished by comparing combinatorial responses across receptor types, a comparison made possible by the brain design features described below.

9.2.1 Olfactory receptor proteins on olfactory sensory neurons

The olfactory receptor proteins (ORs) of vertebrates are encoded by a large multigene family, in rats and mice estimated to be about 1000 genes or 1% of the genome, by far the largest family of genes in the genome of any species (Buck 2000; Zhang & Firestein 2002). While the putative ORs of vertebrates have been known since 1991, ORs were only identified in insects in 1999, in the fruit fly *Drosophila melanogaster* (for humans, see Box 9.3) (Buck & Axel 1991; Clyne *et al.* 1999; Vosshall *et al.* 1999).

Fig. 9.2. (*cont.*) Olfactory sensory neurons (OSNs) are similar across the animal kingdom, from insects (left) to vertebrates (right).

(a) In these schematic diagrams, odour molecules (black dots) diffuse through pores in the cuticle of arthropod sensory hairs (Left) or through the mucus protecting vertebrate olfactory tissue (Right). The odours may be carried to the receptors by binding proteins. The binding of the odour molecule to a receptor protein initiates a signal sent down the nerve axon to the brain. Figure after Hildebrand & Shepherd (1997).

Abbreviations: OBP, odorant binding protein; MOE, main olfactory epithelium; SC, stem cell; VNO, vomeronasal organ.

(b) Schematic diagrams of: (Left) an insect olfactory sensillum with two OSNs, based on electron micrographs of moth sensilla trichodea. Figure from Kaissling (1998a); (Right) vertebrate OSNs in olfactory epithelium. Unlike other neurons, OSNs are short lived and continually replaced from stem cells in the adult, every 30 days or so. Figure from Farbman (1992).

(c) Scanning electron micrographs (SEMs) of olfactory sensory organs and neurons. (Left) Silk moth (*Bombyx mori*) male antenna, showing long trichoideal hairs projecting from the antennules. About 1% of the total antenna is shown. Figure from Steinbrecht (1999). Photo/micrograph by R. A. Steinbrecht. (Right) Vertebrate olfactory epithelium. In terrestrial vertebrates the two kinds of olfactory sensory cell (ciliated and microvillous sensory neurons) occur in different organs, in the MOE and VNO, respectively. However, in many fish and salamanders, both cell types are found mixed in the same epithelium. This is illustrated here by the SEM of the olfactory epithelium of the catfish (*Plotsus lineatus*). Figure from Theisen *et al.* (1991). Photo/micrograph by B. Theisen.

Box 9.3 | Human olfactory receptors

Humans have between 500 and 750 OR-like genes (Mombaerts 1999). Unlike most other gene families which are concentrated in a few locations, OR genes in humans and other vertebrates are distributed over almost all the chromosomes (Rouquier et al. 1998a). Human ORs cover a similar 'receptor space' to mouse ORs, suggesting that the human olfactory system has retained the ability to recognise a broad spectrum of chemicals even though the majority of the genes in humans are non-functional pseudogenes (Zhang & Firestein 2002). It is possible that a reduced reliance on smell reduced selection on human ORs so that mutations have not been selected against. A new family of OR genes was identified in humans, chimpanzees and gorillas (Rouquier et al. 1998b). The gene is functional in chimpanzees, has a nonsense codon in humans and is mutant in gorillas. Human VNO genes are discussed in Chapter 13.

In vertebrate main olfactory epithelium, each OR gene is expressed in about 1/1000 olfactory sensory neurons (OSN), suggesting that each olfactory sensory neuron expresses just one OR type. However, each odour receptor is broadly tuned and can be stimulated by a range of particular different odour molecules – and each kind of odour molecule stimulates a range of different olfactory receptors (Duchamp-Viret et al. 1999; Malnic et al. 1999). These possibilities are illustrated by an imaginary molecule with both a pointed *and* a round end. This imaginary molecule can stimulate both the round-end-accepting receptors and the pointed-end-accepting receptors (a sophisticated version of the young child's toy with different shaped pegs and holes) (Fig. 9.3). A given odour molecule's characteristics, which include its functional groups, charge and size as well as shape, stimulate a characteristic unique selection of receptors and thus a unique combination of glomeruli (Section 9.2.2). Some receptor proteins in the vomeronasal olfactory (VNO) epithelium may be more specific (Section 9.3.1) (Leinders-Zufall et al. 2000).

9.2.2 Glomeruli and combinatorial coding

Each OSN sends a single axon to the brain. The axons of all the olfactory sensory neurons expressing a particular receptor protein converge on the same few glomeruli (singular: glomerulus); these are modules in the brain (Fig. 9.4) (Hildebrand & Shepherd 1997; Mori et al. 1999; Buck 2000; Dulac 2000; Firestein 2001). The number of glomeruli is thus related to the number of types of olfactory sensory neurons in that organism; the number of glomeruli ranges from about 50 in *Drosophila melanogaster* up to 100 in zebra fish, 160 in the honeybee and about 1800 in rats (Pilpel & Lancet 1999) (Fig. 9.5). We do not know how many glomeruli humans have (Mombaerts 1999). A glomerulus is a relatively large spherical collection of nerve cells

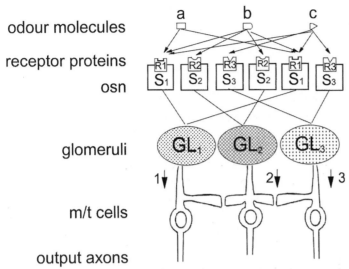

odour molecules

receptor proteins

osn

glomeruli

m/t cells

output axons

Fig. 9.3. Schematic representation of a hypothesis for coding the molecular features (odotypes) of odours in the olfactory system (Christensen & White 2000). The odour molecules **a**, **b**, and **c** contain some common features, as well as different ones, represented here in their geometric shapes. Any given odour molecule will interact with particular receptors, depending on its shape and the receptor active sites. In this hypothetical example, odour a only stimulates olfactory sensory neurons (OSNs, marked osn in the diagram) S_1, carrying receptor R1 whereas odour b stimulates OSNs S_1 and S_2, carrying both R1 and R2, with different parts of its shape. The OSNs with the same receptor Rn are scattered in the olfactory tissue but converge on the same glomerulus, GL_n, so for example all the S_1 OSNs converge on glomerulus **1** and all the S_2 OSNs on glomerulus **2**. Activity in an individual glomerulus then leads to output to other parts of the brain via the m/t (mitral/tufted) cells, modified by interactions with periglomerular and granule cells (not shown). The brain recognises an odour molecule by the particular pattern of glomeruli stimulated by it, so in this example odour a stimulates GL_1 whereas odour **b** stimulates both GL_1 and GL_2. Figure after Lancet (1991) and Christensen & White (2000).

(neuropil) 100–200 μm in diameter, where the axons of one type of OSN converge and transmit the signals to other nerve cells (called mitral and tufted cells in vertebrates, and interneurons and output or projection neurons in insects) (Fig. 9.4). These other cells have numerous links with each other and to and from other parts of the brain. It is likely that they have a key role in processing information before transmitting it to other parts of the brain, leading to the pheromone responses described in the rest of this book, including behavioural responses such as movement, olfactory learning and hormone release (Section 9.7).

The glomeruli, along with the neurons linking them, together form part of the olfactory bulb in vertebrates or the antennal lobe in insects (Fig. 9.1). Each kind of odour molecule will stimulate a unique combination of glomeruli, giving a combinatorial code. Because combinations of different types of receptors respond to each type of odour molecule this is called 'across-fibre patterning'.

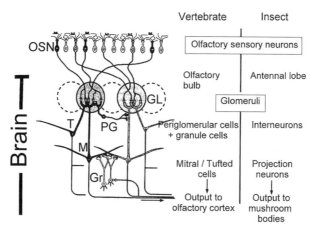

Fig. 9.4. The synaptic organisation of the mammalian olfactory bulb (illustrated) and the insect antennal lobe. On the left, the diagram shows two glomerular modules representing two different types of odorant receptors in vertebrates (mammals). Mitral cells (M) and tufted cells (T) are output neurons, and granule cells (Gr) and periglomerular cells (PG) are local interneurons. OSN, olfactory sensory neuron; GL, glomerulus. On the right, the equivalent terms are given for the analogous cells and structures in insects. Diagram on left from Mori *et al.* (1999).

Organism	Cell types	Receptor types	Receptors per OSN	Odorant universe coverage	Sensory-cell discrimination	Integrated chemosensory power	Glomeruli
Nematode	15	600		+++	±	+	0
Fruit fly	50	100		++	++	++	~50
Mammal	1000	1000		+++	+++	+++	~1800

Fig. 9.5. Evolution of olfactory cells and receptor proteins. Nematodes and mammals have similar numbers of chemosensory receptor types so they should detect the same (large) number of odorants with similar sensitivities. However, because in the nematode sensory cells each bear many different types of receptor, individual cells may show poorer discrimination between odours. In mammals each olfactory sensory neuron (OSN) seems to have only one type of receptor, giving the best possible discrimination at the level of single neurons. The fruit fly may represent an intermediate evolutionary step; the number of cell types has increased modestly but the sensory neurons have gone most of the way towards clonal exclusion. Figure and caption from Pilpel & Lancet (1999).

Molecular, electrophysiological, and now imaging investigations confirm that glomeruli are functional units as well as anatomical units for the initial processing of odour information (Kauer & White 2001). Work on zebra fish, rats and honeybees (Fig. 9.6) uses fluorescing dyes to show the pattern of glomeruli in the brain 'glowing' as the animal is stimulated by different odours. Particular odours elicited essentially the same patterns in each of the honeybees tested and, as predicted, each glomerulus has a broad but different response spectrum (also shown in zebra fish and rats). It is worth

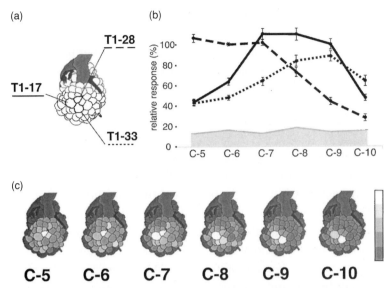

Fig. 9.6. Fluorescing dyes reveal which glomeruli in a honeybee's antennal lobes (see Fig. 9.1) are stimulated by different odours and the response shows the combinatorial olfactory response to each odour.
(a) Schematic view of the antennal lobe with three glomeruli marked (T1-17, T1-28, T1-33).
(b) Responses of these three glomeruli to a series of alcohols varying in carbon chain length from C5 (1-pentalol) to C10 (1-decanol). Glomerus T1-28 responds most strongly to the short chain alcohols, whereas T1-33 responds most to longer chain alcohols. Each point represents the average of 14–21 individuals. Responses are shown as a percentage of the response of T1-28 to hexanol.
(c) Spatial activity patterns elicited by the same alcohol series as in (b). Note the continuous shift in activity (the lighter areas) between the three focal glomeruli and that other glomeruli are also stimulated. Figure from Galizia & Menzel (2000).

noting that in the honey bee, responses to pheromones such as isoamyl acetate, an alarm pheromone component, stimulate as broad a range of glomeruli as floral odours.

While the glomeruli are clearly important, they are only part of the story. In the rat, if the glomeruli that 'light up' in response to certain odorants are removed, the animal can still detect and discriminate these odours (Lu & Slotnick 1998). Even rats with relatively small remnants of one olfactory bulb can perform a variety of odour detection and discrimination tasks as well or nearly as well as controls. These counter-intuitive outcomes suggest that odours may be coded by a highly distributed pattern of inputs.

One puzzle about vertebrates is how their long memories for odours can be matched with the observation that individual olfactory sensory neurons are short lived and continuously replaced. The constancy of the olfactory glomeruli map provides an explanation: perhaps the new OSN axons are guided to the correct glomerulus by molecular tags (Buck 2000)?

In both mice and *Drosophila*, there is a stereotypical pattern of links from the glomeruli up to the higher levels of the brain (Jefferis *et al.* 2001;

Zou *et al.* 2001). Via its glomerulus, each type of olfactory sensory neuron sends its signal to particular parts of the olfactory cortex (in mammals), thus transferring the olfactory 'map' in the glomeruli to higher parts of the brain. Inputs to the olfactory cortex from different OSNs overlap spatially, and could allow integration of signals as part of the combinatorial code.

9.2.3 Temporal coding in brains

The activation of unique patterns of glomeruli with different odours has provided strong evidence for the spatial arrangement of olfaction processing in the olfactory lobe. However, different odours are coded not just by which output neurons fire but also the pattern of firing of those neurons in relation to others over millisecond periods during and after the stimulus (Laurent *et al.* 2001). Oscillations of neural activity (at a frequency of 30–50 Hz) have been recorded from olfactory systems in many vertebrates including bony fish, frogs, turtles, rodents, rabbits, cats and humans. Similar oscillations to these have been found in insects and molluscs, and in insects, and perhaps other organisms, the synchronisation of the firing seems to be an important feature. The ubiquity of fast oscillations in olfactory systems suggests that they are the basis of higher neuronal processing of odour stimuli. The start of such oscillations in vertebrates is often prompted by a 'sniff' (Box 9.4).

Box 9.4 | Sniff and flick: the behaviour of smelling

For many animals, olfaction involves active sampling of the environment: 'In the realm of odors, scents, perfumes, effluvia, stenches, and vaporous aphrodisiacs, mammals sniff, lobsters, insects, and snakes flick' (Dethier 1987). Sniffs, flicks and pulses may be essential for animals to sense the world: with a continuous stimulus, sensory cells adapt and stop signalling, the central nervous system habituates, both responses being ways of reducing useless or irrelevant information (Dethier 1987). Interrupted sampling allows the sense cells to sense. Increasing the speed of the medium, air or water, during the sniff will also reduce the thickness of the boundary layer over the olfactory sensory cells (Chapter 10), speeding access of chemical signals.

The boundary layer is especially important for aquatic animals. The lobster olfactory sensillae form a dense 'hairbrush' on the antennule (Atema 1995). At low flow rates, <5 cm/s, a thick boundary layer prevents the access of odour molecules. To detect odours, lobsters must flick their antennules, effectively taking a 'sniff'. Flicking drives water through the hairs at high velocity (>12 cm/s). Excited lobsters flick at up to four times a second (4 Hz). This flicking rate matches the flicker fusion frequency of lobster chemoreceptor cells, which can integrate stimuli over periods of 200 ms. Another marine crustacean, a stomatopod, alters its flicking characteristics through life to match its tenfold growth in size (Mead *et al.* 1999).

In humans, a sniff (which lasts about half a second) produces turbulence inside the nose, creating eddy currents that bring odours to the olfactory epithelium hidden high in the nasal cavity (Dethier 1987). Using functional magnetic resonance imaging (fMRI) to visualise brain

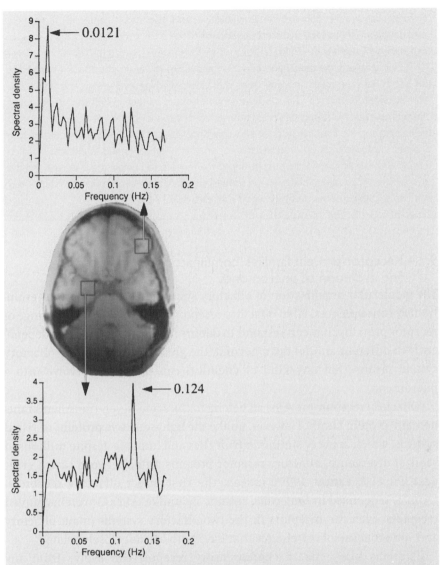

The separation between sniff-induced and odour-induced brain activity in humans can be seen within the same subject at the same time (during the same fMRI scan lasting 320 s). The subject performed a smell task (sniffing once every 8 s (0.125 Hz)); odorant presence alternated with odorant absence every 40 s (0.0121 Hz). A Fourier transform of activity in different brain areas shows the piriform cortex is activated by sniffing (centre and bottom) and the lateral orbito-frontal gyri are activated by smelling (odour present) (centre and top). In this image the brain is seen performing different tasks at two different frequencies simultaneously. Figure from Sobel et al. (1998).

activity (see figure), Sobel et al. (1998) found that the act of sniffing and the stimulus from odour molecules activate different parts of the brain (see figure). Sniffing, whether odorant was present or not, induced activation of the olfactory cortex and primed the oscillations in the part of the brain that will receive the odorant information (Section 9.2.3). Odour molecules, regardless of sniffing, induced activation mainly in the frontal lobe.

Incidentally, women do much better in olfactory tests than men (Chapter 13). It may be a coincidence but fMRI shows that much larger parts of the brain respond to odours in women than in men (Yousem et al. 1999).

Flow rate affects the perception of odours and although we do not notice it happening, one nostril takes in much more air than the other, switching over every few hours. The turbinate swelling which partially blocks off one nostril can be seen in MRI scans (Sobel et al. 1999a). The strength of signal perceived from an odour depends on its rate of absorption into the mucosa and this interacts with the flow rate. Sobel et al. (1999b) tested an equal mixture of odours with different absorption rates (octane and L-carvone) and found that if sniffed through the low-airflow nostril, people judged the mixture to contain more octane and the reverse when they used the high-airflow nostril. The different airflow through the nostrils may give two snapshots of the olfactory world with each sniff.

9.2.4 Receptor-protein families, combinatorial processing and evolution of pheromones

The glomerular organisation of olfactory systems has two important evolutionary consequences. First, olfactory receptors can evolve from any range of receptor proteins that can respond to odours (and thus may arise independently in different animal taxa). Second, the great flexibility of the olfactory system means that any kind of chemical can potentially evolve into a pheromone.

Olfactory receptor proteins all belong to the same 'seven-transmembrane-domain' protein family. However, unlike the light-sensitive proteins of visual systems, which are very similar in fruit flies and humans despite millions of years of divergence, olfactory receptor proteins differ between taxa a great deal (Pilpel & Lancet 1999). Indeed, the first insect olfactory receptors, recently sequenced in *Drosophila*, seem to be unlike other G-protein-coupled receptors. Even the receptors in the two olfactory systems (main olfactory and vomeronasal) of vertebrates that have both are not closely related.

It seems likely that a chemosensory receptor can evolve from any G-protein-coupled-receptor protein whether originally functioning as a receptor for hormones or neurotransmitters, so long as it is in the cell membrane and can interact with external odorants (Dryer & Berghard 1999).

A wide range of relatively non-specific receptor proteins, each expressed on different olfactory sensory neurons linked to glomeruli, gives evolutionary flexibility to signals. The system can respond to the widest possible range of odours, even as entirely new compounds are created (in evolving biological systems). Although a given odour may be unlikely to fit any one receptor perfectly, it is likely to stimulate some. This offers a mechanism for the observed evolution of pheromone signals from odours that are present and that turn out to have signal value to receivers (and senders), for example sex hormones 'leaking out' of females and detected by male goldfish

(Chapter 1). Natural selection can work on improving selectivity and sensitivity in the receiver, and greater production in the sender, if these lead to greater reproductive success or survival. The result is the wonderfully diverse range of compounds – which can seem weird and haphazard at first sight – used as pheromones by organisms across the animal kingdom (Chapter 1).

9.2.5 Sensitivity and perireceptor events

Some animals have evolved enormously sensitive olfactory systems by increasing the number of detector units (generalist olfactory sensory cells), carried on bigger sense organs, for example, the large olfactory areas inside dogs' noses (Box 9.2).

Two further important factors in increasing the sensitivity of olfactory sensory cells are proteins (pheromone- and odorant-binding proteins), and clean-up enzymes.

9.2.5.1 Pheromone- and odorant-binding proteins

Many pheromones used by terrestrial animals are hydrophobic molecules yet they have to reach the OSN membrane through a protective layer of watery mucus (vertebrates) or a watery sensilla lymph (in insects) (Fig. 9.2 a, b). Pheromone- and odorant-binding proteins (PBPs and OBPs, respectively) bind to odorant molecules and ferry them to the OSN membrane (Pelosi 1996). Insect pheromone-binding proteins bind specifically to the pheromones of that species, providing a first level of selectivity (Steinbrecht 1996).

9.2.5.2 Clean-up enzymes

As the only part of the central nervous system exposed directly to the outside world, olfactory tissue has a particular need for ways to deactivate the wide range of molecules, biologically important or not, that contact it. Enzymes found in both insects and mammals break down odorant molecules on sensory epithelium, degrading this chemical flotsam (see review by Thornton-Manning & Dahl 1997). In pigs and humans, the enzymes are active against putative steroid pheromones as well as other molecules.

A chemical signal which remained bound to the receptor, stimulating it for long periods, would be no use to a male moth needing to respond to the contact and loss of a pheromone plume over milliseconds (Chapter 10). As in other signalling systems, an 'OFF' is needed to recognise the 'ON'. There are large quantities of a specific pheromone-degrading enzyme in the sensillum lymph (Fig. 9.2 a, b) (Vogt et al. 1985) but conformational changes in the PBP when bound to the pheromone may also be involved in the 'OFF' process (Kaissling 2001).

9.2.6 Behavioural consequences of pheromone perception

Just as neurons in other parts of the brain process the information coming from the olfactory bulb, still other neurons integrate these processed signals with inputs from different senses (Stein & Meredith 1993). For example, neurons which integrate olfactory and visual inputs have been identified in the moth brain (Olberg & Willis 1990) (see also Chapter 10).

The integrated inputs will be further modulated by the animal's internal state and experience, for example the hormonal state of male hamsters affects their responses to female odours (see Box 9.5). An organism in a natural setting interweaves signals from its external and internal environment to yield an experience richer than the sum of the individual inputs (Stein & Meredith 1993).

The final outcome of pheromone stimulation comes from the integrated signals from the brain prompting immediate behaviours, or hormonal changes over the medium term, or a complete developmental switch affecting the rest of the animal's life as in phase change in locusts (Chapter 4).

9.3 | Vertebrate dual olfactory system

Most amphibians, reptiles and mammals have a dual olfactory system: the main olfactory epithelium (MOE), sending its signals to the olfactory bulb in the brain, and the vomeronasal organ (VNO, also known as Jacobson's organ), sending its signals to the accessory olfactory bulb (Fig. 9.1) (Eisthen 1997). Depending on the species, pheromones and olfactory recognition cues may be detected by the MOE or the VNO, or both.

In mammals, the VNOs where present are a pair of elongated tubes, lying along the base of the nasal septum, opening at the anterior end only, via a narrow duct (see reviews by Døving & Trotier 1998, Keverne 1999, Meredith 1999). In rodents the VNO duct opens into the floor of the nasal cavity. In carnivores, ungulates and Old World primates, the VNO ducts open into the nasopalatine canal (the nasoincisor canal) which joins the mouth and nose, so that stimuli can enter the VNO via the mouth in these species, following licking of the stimulus source. For example, male antelopes, other ungulates, and felines sniff and lick the urine of oestrous females, often showing a characteristic lip-curling 'flehmen' behaviour (Fig. 9.7). In mammals, odour stimuli reach the VNO sensory epithelium in a mucus stream which contains lipocalin odorant-binding proteins (OBPs) (Pelosi 1998). Entry into the VNO may be facilitated by pumping, which occurs in response to novel stimuli (Meredith 1994). In reptiles, such as snakes, the VNO openings are in the roof of the mouth and stimuli are delivered by the tips of the tongue (Chapter 10).

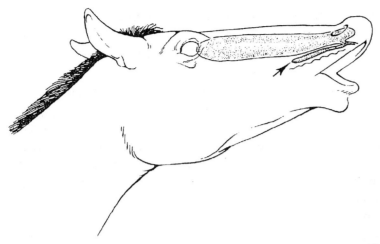

Fig. 9.7. The flehmen response or lip curl in the horse in response to oestrous urine. The location of the vomeronasal organ is indicated by the arrow. Figure from Houpt (1998).

9.3.1 Molecular architecture of the vomeronasal olfactory system

The VNO system shows at least four major intriguing differences from the main olfactory epithelium (MOE): in its receptor proteins, in the specificity of these receptors, in the role of proteins as stimuli, and lastly in the patterns of connections between the sensory neurons and their glomeruli (see reviews by Buck 2000, Dulac 2000, Keverne 1999).

9.3.1.1 Receptor proteins

In mammals, the VNO has two distinct families of olfactory receptors, vomero-receptors 1 and 2 (VR1 and VR2), which are only very distantly related to those of the main olfactory epithelium, suggesting that the VNO receptor system has evolved separately from the MOE (Fig. 9.5). The two families of VNO receptors project to different parts of the accessory olfactory bulb.

9.3.1.2 Specificity of receptors

Whereas the olfactory receptor proteins are broadly tuned to ligands (odour molecules) (Section 9.2.1), and more and more different receptors are activated as ligand concentration is increased, the receptor proteins in the mouse VNO are tightly tuned to, and very sensitive to, the individual small molecule pheromone components tested (see Table 9.1) (Leinders-Zufall *et al.* 2000). The detection threshold for some of these chemicals is remarkably low, near 10^{-11}M, placing these neurons among the most sensitive chemodetectors in mammals. Each of the pheromones activates a unique, non-overlapping subset of vomeronasal sensory neurons. However, vomeronasal neurons also respond to other, non-pheromone, odorants (Sam *et al.* 2001).

9.3.1.3 Proteins
Odours detected by the VNO have long been thought to be associated with high molecular weight fractions, such as the mouse urinary proteins (MUPs) (Section 9.6) which characteristically bind the small signal molecules. Studies seem to show that the proteins and the small molecules can stimulate the VNO but act on receptors in different areas.

9.3.1.4 Mapping of vomeronasal organ sensory neurons to glomeruli
The pattern of connection of VNO sensory olfactory neurons to their glomeruli also differs from the pattern in the main olfactory system. The VNO olfactory sensory neurons expressing a particular receptor-protein connect to 10–30 distinct glomeruli, albeit in a particular zone of the accessory olfactory bulb, and each glomerulus seems to receive input from more than one type of sensory neuron. It is not yet clear what the implications of these, and other differences in neuronal organisation in the accessory olfactory bulb (AOB), are for the processing of stimuli.

The recent work on VNO receptors is very exciting. However, it is simply too early to say that because a receptor, odorant-binding protein, or G-protein is expressed in VNO sensory neurons, that it is necessarily involved in perception of pheromones rather than more general odours, as many authors have done. The next section explores the overlapping roles of the MOE and the VNO. The VNO is likely to have some receptors with broad response specificities as well as putative pheromone receptors (Johnston 1998).

9.3.2 The overlapping roles of vomeronasal and main olfactory systems
The importance of the vomeronasal organ (VNO) in many odour-mediated behaviours in reproduction and the exciting molecular evidence for separate families of receptors in the VNO and main olfactory epithelium (MOE) has led some biologists to propose functional differences between the MOE and the VNO. It has been suggested that the MOE is for smelling general odorants and the VNO is for 'pheromones' and other odours associated with sexual reproduction. However, behavioural and endocrine evidence suggests that the story is more complicated and interesting: not every substance that stimulates the VNO is necessarily a pheromone, and, conversely, not everything that is a pheromone need act via the VNO. Depending on species, signal and previous experience, either the VNO or the MOE

may be used and, for some behaviours, the sensory inputs work together (see reviews by Meredith 1998, 1999 and Johnston 1998, 2000). Neither system has an exclusive role for pheromone perception. Either olfactory system can be involved in chemical recognition of individuals (Section 9.7). No single group, such as rodents or ungulates, can be a model for all mammals. Before describing some examples in more detail, I would like to discuss contrasting cases which show the varied roles of the VNO and MOE (Johnston 1998, 2000).

9.3.2.1 The vomeronasal organ mediates responses to some pheromone signals

In a wide variety of mammal species, important behavioural and physiological effects in both sexes are mediated by odours perceived through the VNO (Wysocki & Lepri 1991; Meredith 1999). Both priming and signal (releaser) effects are found. There are many priming pheromone responses in females that are dependent on an intact vomeronasal system. These include modulation of oestrus in mice by urine signals from both males and females, modulation of oestrus in rats, the acceleration of puberty in female mice and voles, and the block to pregnancy in mice produced by 'strange' males (see Section 9.6) (Meredith 1999). There is a similar need for an intact VNO for the induction of oestrus in prairie voles (but not meadow voles), and in *Monodelphis* opossums.

Male hamsters, mice and prairie voles are particularly dependent on chemosensory input for normal reproductive behaviour. In naïve hamster males, the VNO plays a key role in stimulating copulation in response to vaginal fluid, but *after* sexual experience, learned olfactory cues, perceived through the main olfactory epithelium, can provide some necessary cues (see Box 9.5).

9.3.2.2 Some pheromone signals are mediated by the main olfactory system, not the vomeronasal organ

There are numerous species in which behavioural evidence suggests that the main olfactory system, *not* the vomeronasal system, may be the sensory system that mediates responses to particular pheromone signals (Johnston 1998). Two of the best understood examples are the pheromone which induces nipple search and attachment by young rabbits and, in pigs, the attraction and 'standing' by oestrous sows induced by androstenone in the boar's saliva.

To reduce the risk of predators finding the nest, the young of the European rabbit (*Oryctolagus cuniculus*) are only nursed once a day, for just 3 to 4 minutes (Hudson & Distel 1995). The pups respond to pheromone from

Box 9.5 | Pheromones, hormones, experience, and sex in the hamster

The responses of male hamsters to female vaginal fluid pheromones have given a useful system for investigating the different roles of the main olfactory and vomeronasal systems and their interdependence (Johnston 1998; Meredith 1998). Vaginal fluid marks left by the female around her territory are detected by the male via his main olfactory system (MOE). The MOE also seems to be important for distinguishing a familiar mate from other females. Once he is with the female, aphrodisin, the large molecular weight pheromone component (or smaller molecules bound to it) of vaginal secretion is perceived by the male VNO, stimulating investigation and mounting. However, sexually experienced males no longer need the VNO input to stimulate copulation as they have learned other odour cues associated with females, detected by the MOE.

Many species and both sexes show a prompt surge in luteinising hormone blood levels on contact with the opposite sex, mediated by the VNO (Johnston 1998; Meredith 1999). The hamster male hormone surge is mediated by VNO stimulus with vaginal odours. However, sexually experienced males show the hormone surge even if they have their VNO removed, so long as they are brought in to contact with females (such males show no hormone surge to the odours alone). A separation between mating behaviour and the hormone surge is also suggested by the observation that sexually experienced males with neither main olfactory nor VNO inputs will show the hormone surge on contact with females but they will not mate, as olfactory stimuli are needed for copulation.

The VNO is thus vital for the male hamster's first sexual experiences to be successful, even if this experience allows later substitution by learned odours detected through the MOE.

Brain connections

Links between the outputs of the VNO and the MOE have been demonstrated at the neuroanatomical level. In the hamster, some outputs of the main olfactory and VNO systems both converge onto single neurons in the amygdala (Licht & Meredith 1987).

An antibody technique allows mapping of brain areas activated at a high level, indicated by expression of the c-*fos* gene. Such mapping has explained how sexually experienced animals can use inputs from the MOE, in place of the VNO, and where these link to the outputs from the vomeronasal system (Meredith 1998, 1999). Inexperienced, intact hamster males show low levels of c-*fos* activation in the medial preoptic area (MPOA) after they have investigated vaginal fluid, and there is no activation in inexperienced males with vomeronasal lesions. Sexually experienced intact males show a high level of activation after an equivalent exposure to vaginal fluid, and vomeronasal lesions made after experience do not reduce it. Experience seems to sensitise the MPOA to chemosensory input and to re-route main olfactory lobe input so that it can substitute for vomeronasal input in driving the MPOA (see Fig. 9.1 and figure below).

Hormones

Sexual behaviour in the male hamster requires the female chemosensory cues *and* enough testosterone in the brain (Wood & Swann 2000). Castrated hamster males or males with no olfactory bulb show no mating interest in females. Males need both testosterone from the testis and chemosensory input from the olfactory bulb. Mating is restored in castrated, but olfactory intact, males if they are injected with testosterone (see figure below). In most mammalian species, steroid hormones produced in the testis are an essential signal to the brain for expression of copulatory behaviour. Gonadal steroids reflect social status, nutritional status, state of maturity, and stress levels, so they represent an integrated signal from the internal milieu to the central nervous system.

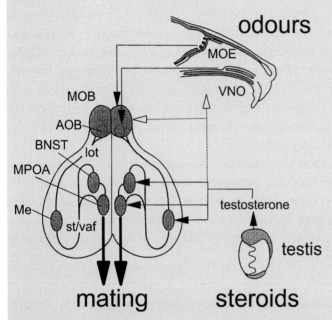

Schematic diagram of the neural circuitry for sexual behaviour in the male Syrian hamster (*Mesocricetus auratus*) illustrating the pathways through which odours and hormones facilitate mating. An intact olfactory bulb and adequate testosterone on the bed nucleus stria terminalis (BNST) and medial preoptic area (MPOA) are essential for sexual behaviour. Other abbreviations: AOB, accessory olfactory bulb; lot, lateral olfactory tract; Me, medial amygdaloid nucleus; MOB, main olfactory bulb; MOE, main olfactory epithelium; st, stria terminalis; vaf, ventral amygdalofugal pathway vomeronasal organ. Figure labelled from Wood (1998).

their mother's nipple region, which elicits stereotyped searching, usually successful in just 6 seconds. If their main olfactory system is disabled, the pups are unable to suckle from the mother. However, VNO removal does not disrupt nipple finding. Responsiveness to the pheromone may be acquired before birth and postnatally pups can learn to associate suckling with novel odours. (Newly born humans also use olfactory cues to find the nipple, Chapter 13.)

A number of studies of reproductive behaviour in ungulates support the suggestion that the main olfactory system is important for reproductive functions (Dorries *et al.* 1997). For example, in sheep the odour of a male (from his wool) elicits a luteinising hormone (LH) surge in ewes (the 'ram effect') but the LH surge in female sheep is not affected by lesioning the VNO or cutting the vomeronasal nerve (Cohen-Tannoudji *et al.* 1989).

In another ungulate, the domestic pig (*Sus scrofa*), the steroid 5α-androstenone is a pheromone in the boar's saliva that attracts the oestrous female to the male and elicits a receptive mating stance (called 'standing' or 'lordosis') (Chapter 12). Blocking access to the female's vomeronasal duct with surgical cement had no effect on attraction to the odour of androstenone, the threshold sensitivity, or the proportion of females 'standing' in response to it (Dorries *et al.* 1997). It is just possible that the VNO is needed for expression of androstenone-mediated behaviour in sexually inexperienced pigs, and Dorries *et al.* point out that their sows had been housed with boars before the experiments and were likely to be sexually experienced. Nonetheless the specific pheromone, androstenone, is being perceived by the olfactory system.

These examples show that responses to some pheromones in mammals are mediated by the main olfactory system and not the vomeronasal system. In other vertebrates, such as snakes, the VNO has a wide range of olfactory functions including prey discrimination and finding (Chapter 13) but is also used to perceive social signals. A similar diversity of functions may exist in mammals (Johnston 1998).

9.3.2.3 The vomeronasal system and main olfactory system often interact in mediating communication

Although some functions may be mediated by the main olfactory system *or* the vomeronasal system, the most common case may be that the two systems interact to mediate responses to odour signals (Johnston 1998). Responses to pheromones in the hamster offer an excellent demonstration of the different roles of the main olfactory and VNO systems *and* their interdependence (Box 9.5). The behavioural responses of hamsters also demonstrate how factors such as social status can influence behaviour, via hormone levels that affect neurons in the brain.

9.3.2.4 Hormone release is not a consequence limited to stimulation of the vomeronasal organ

The basis of the well-studied mouse pheromone effects is stimulation of the VNO, sending signals to the AOB (accessory olfactory bulb), which in turn sends signals to ('projects to') the hypothalamus, resulting in hormone release with behavioural or physiological effects (Sections 9.6 and 9.7.2). However, two important points argue against a unique role for the VNO in social and sexual communication outside the rodents. First, as mentioned above, in some mammals, notably sheep, it is possible to have hormone surges released by odour stimulation of the olfactory system (i.e. hormone responses are *not* a unique consequence of VNO stimulation). Second, in rodents and other mammals the outputs of the VNO *do* link to other higher parts of the brain, even if they do this indirectly. The work on hamsters shows how the main olfactory and VNO outputs interact in the amygdala (Box 9.5).

9.3.2.5 Conscious and unconscious responses to odours

Some authors have argued that the VNO–AOB projections do not directly link to higher brain areas and thus that VNO perception is 'unconscious', whereas the main olfactory system, through the olfactory bulb, does link to cognitive areas and could be said to be 'conscious'. Putting aside debates about animal consciousness (see Dawkins 1993 for a stimulating discussion), the case that odours perceived unconsciously are more likely to be pheromones is not proven (Dorries 1997). What are we to make of the many pheromone effects leading to hormone release (in sheep, for example) that are perceived by the olfactory system – do such animals consciously perceive these odours but not other pheromone odours? Or do garter snakes, which use the VNO both for finding prey and social pheromones, consciously perceive the chemicals they use to follow prey but not those they use to find and aggregate with conspecifics?

Are humans affected by odours without consciously perceiving them? The question is relevant to discussion of human odour perception and putative human pheromones (Chapter 13). For example, in a study that demonstrated that axillary (armpit) extracts can affect the timing of the human menstrual cycle, women found all extracts odourless (Stern & McClintock 1998). Do undetected odours have measurable effects on brain activity? Lorig *et al.* (1991) found that concentrations of odours (in this case galaxolide, a synthetic musk) so low that people did not notice them, nonetheless had effects on electroencephalogram (EEG) readings. Functional magnetic resonance imaging (fMRI) allows more sophisticated localisation of brain activity: fMRI scans (Sobel *et al.* 1999a) report brain activity stimulated by undetectable levels of one of the putative human pheromones proposed by Monti-Bloch *et al.* (1998) (see Chapter 13). One set of controls that seems to be lacking are those of undetectable concentrations of other odorants not thought to be pheromones. In the context of putative pheromones, some research seems to be influenced with assumptions extrapolated from the mouse work suggesting that all mammalian pheromones *should* be detected by a VNO and are thus unconscious. However, for the same reasons as in the discussion of sheep, pigs and garter snakes (Section 9.3.2.2), potential human pheromones need not be perceived by a VNO, nor need they be odourless.

9.4 | Moths and sex pheromones

Male moths fly upwind to find a conspecific female releasing a plume of sex pheromone. The specialisation of male moth antennae for catching pheromone molecules (Box 9.2) is matched by specialised pheromone-sensitive sensillae on the antennae and glomerular structures in the brain. Most moth sex pheromones are multi-component and each component stimulates a different glomerulus in a special structure, the macroglomerular complex (MGC), found only in males (Fig. 9.1) (Hansson 1997). When the glomeruli for the different pheromone components are stimulated at the same time, it indicates that the male's antenna has hit a wisp of pheromone-laden air that has come from a female upwind. He responds with brief upwind surge of flight (Chapter 10).

However, males will not fly to calling females of related, sympatric species even though many of the pheromone components in a plume are the same (Chapter 3). If the pheromone components characteristic of such sympatric moth species are released in the plume, the males will turn back (this is adaptive because there is little point in finding the female if she is of the wrong species). The sensory mechanism for the turn-back behaviour has been explained recently (Fig. 9.8).

9.4.1 Evolution of specialised, tuned receptors and macroglomerular structures for sex pheromone detection in male moths

The physiological and structural features of the male moth antennae and brain offer an exquisite example of functional anatomy and sensory response. The antennal lobe of the male moth was an inspired choice as a model experimental system: it offers known pheromones as stimuli, specialised glomerular structures, the ability to use the female moth brain as an electrophysiological and anatomical comparison to highlight features of the male system, and a range of species for evolutionary comparisons.

The characteristic specialisation of the male olfactory system of the moth has evolved in response to intense selection pressure to be the first to detect and locate a female releasing sex pheromone (scramble competition, Chapter 3). Selection is thus for greater sensitivity through large numbers of highly tuned olfactory sensory neurons. Unlike the relatively unspecific receptor proteins of most olfactory sensory neurons, the receptor proteins and pheromone-binding proteins in these specialised olfactory sensory neurons are themselves very specific in their binding and are present at a high concentration. Such is the specificity and sensitivity of the specialised pheromone sensory neurons that at the low air concentrations of pheromone critical for males to catch the first trace of a female (and track her upwind), they are the only sensory cells on the antennae to respond.

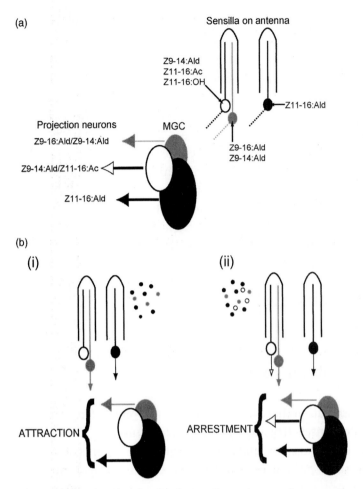

Fig. 9.8. Male moths make split-second orientation responses to the temporal structure of a pheromone plume and the chemical content of individual filaments in the plume (Chapter 10).
(a) The corn ear worm moth (*Helicoverpa zea*) uses a multi-component pheromone.
(b) If the antennal cells for its own two pheromone components (shown as black and grey) are both stimulated then the outputs from their respective glomeruli in the macroglomerular complex (MGC) (Fig. 9.1) cause the male moth to fly forward (i). However, *H. zea* shares the major component (black) with sympatric species which live in the same region of North America. *H. zea* has specialised receptor cells sensitive to the minor, characteristic antagonistic pheromone component of sympatric species (white). If these cells (white) are stimulated then *H. zea* males will turn back (ii). See Appendices A1 and A2 for the naming of pheromones. Figure from Vickers *et al.* (1998).

At low concentrations, only the specialist olfactory sensory cells will be stimulated – giving an effect sometimes called 'labelled lines'. I think that these may not be different from ordinary olfactory responses except in the difference between the specificity of the receptors and the sensitivity of those olfactory sensory neurons in relation to others in the olfactory system. If the concentration of pheromone is raised high enough, many less specific olfactory sensory cells respond (which explains the importance of using biologically relevant concentrations in bioassays of putative pheromones, in order to avoid ambiguous results (Chapter 2)).

The functional beauty of the male moth macroglomerular complex (MGC) may have obscured more general observations on pheromone processing in other kinds of animal. Males of many other insects with scramble competition for mates do have MGCs for the perception of female sex pheromones, for example, male honeybees and male American cockroaches. However, other organisms that use pheromones for communication do not necessarily have a specialised anatomical structure in the brain to deal with them. For example, worker (female) honeybees produce and respond to alarm pheromone and Nasonov gland pheromone (see Chapters 6 and 8), but they do not have an MGC. Nonetheless they give sensitive and predictable responses to these pheromones, which are processed in the glomeruli of the antennal lobe in the same way as other odours (Galizia *et al.* 1999). Similarly, pheromones in *Drosophila* are processed by ordinary combinatorial approaches (Carlson 1996).

9.5 | Factors affecting behavioural and physiological responses to pheromones

It is a common observation that different animals respond differently to the same stimulus. Responses vary according to context and to many variables including the receiver's age, sex, and hormonal state, dominance status and experience. The same animal may also respond differently on different occasions.

Responses to pheromone signals may depend on the context. For example, in a parasitoid wasp, rejection of marked hosts depends on the frequency of available unparasitised hosts (Hubbard *et al.* 1999) (Chapter 4). Honeybee alarm pheromone causes bees to attack the intruder if the pheromone is released near the nest but to flee if detected when the bee is far from the nest (Chapter 8). Ant-associated membracid bugs do not respond strongly to their own alarm pheromone if they are being guarded by ants (Chapters 8 and 11) (Wood 1977).

Environmental conditions can greatly affect the pheromone responses of 'cold-blooded' (poikilothermic) animals. For example, both temperature and relative humidity affect the responses of male moths to sex pheromones (McNeil 1991).

In some species and stimuli, the changes in responsiveness occur in the central nervous system processing, in others changes occur in the peripheral sensory system.

9.5.1 Peripheral and/or central nervous system effects on pheromone response

Hormones help to orchestrate appropriate responses to pheromone signals, according to social status, reproductive maturity, sex and environmental variables such as day length or temperature. Hormones may influence either the olfactory sensory cells (peripheral effects) or the neurons in the central nervous system involved in processing olfactory information.

Hormones affect peripheral olfactory responses in males of the southeast Asian cyprinid fish *Puntius* spp.: higher blood levels of androgen increase the electroolfactogram (EOG, Chapter 2) response to a putative female sex pheromone (15-keto-prostaglandin-F-2α), giving a ten times increase in sensitivity specifically to this compound (Cardwell *et al.* 1995). In contrast, hormone effects on parts of the central nervous system seem to explain sex differences in behavioural responses in goldfish (*Carassius auratus*). Male goldfish show a strong physiological and behavioural response to female sex pheromone (whereas females respond little). However, EOGs show similar signals in both males and females are sent to the brain in response to detection of female pheromone (Sorensen *et al.* 1987).

In male hamsters, brain circuits involved in responses to female pheromones are not active unless testosterone levels in the blood are above a threshold value. This mechanism provides an internal monitor of readiness to mate as only sexually mature, well-fed males produce significant testosterone levels. Thus the male is primed to respond if female odour is encountered, only when he is sufficiently mature and dominant to be fertile (see Box 9.5).

Hormones also mediate changes in the central nervous system leading to changes in responsiveness in insects (see reviews in Cardé & Minks 1997). For example, one of the hormones acting on the moth brain is octopamine, which is responsible for controlling the responsiveness of male moths such as the cabbage moth (*Trichoplusia ni*) to pheromones (Linn 1997). A daily circadian-cycle secretion of the hormone, influenced by photoperiod and environmental factors such as temperature, ensures that the response to pheromone is greatest during the evening. This matches the time when the female moth of this species signals with pheromone, her timing being controlled by her hormones and neurohormones.

As well as daily cycles, hormones also control longer-term changes in responsiveness. When male black cutworm moths (*Agrotis ipsilon*) first emerge as adults they do not fly to female sex pheromone. However, electrophysiological measurements (Chapter 2) show that the pheromone sensory cells and the rest of the peripheral nervous system are working and sensitive from emergence. Older males do respond, as a result of activation of the antennal lobes in the brain (Section 9.3) by juvenile hormone (JH) produced by the corpora allata (CA) secretory glands (Gadenne *et al.* 1993). In the rest of their adult life, continued male response to female sex pheromone depends on juvenile hormone released by the CA: mature males with the CA removed do not respond to female pheromone, but their response is regained if JH is injected (reminiscent of the responses in the male hamster brain dependent on testosterone, Box 9.5).

The responses of honeybee workers to pheromones also depend on age, although the mechanisms are not known. Newly emerged workers show little behavioural or electrophysiological response to queen-produced pheromones and alarm pheromones, but strong responses are apparent by the time workers are 5–10 days old (Chapters 6 and 8) (Winston 1987, p. 136).

9.5.2 Sex and caste differences in peripheral odour reception and brain circuits

Sexual differentiation of olfactory sensory epithelium and the brain during development leads to profoundly different behavioural and physiological responses to pheromones. These have been well studied in both mammals and insects (Dorries 1992; Hildebrand 1995). The morphological specialisation of the antennal lobe into a macroglomerular complex (MGC) in male moths specialised for reception of sex pheromone is a good example (see Section 9.4). Sexual dimorphism also characterises the expression of pheromone receptor proteins and odorant binding proteins in the olfactory sensory neurons forming the sensilla on the male antenna. Analogous changes occur in mammalian olfactory systems. Although the morphological differences in brains between the sexes in mammals are not as marked as in moths, there are differences in the sensitivity of males and females to pheromones. For example, female pigs have a threshold for the male pheromone, androstenone, that is fivefold lower than males (Dorries *et al.* 1995). Studies in a carnivore, the ferret (*Mustela putorius furo*), show sex differences in the processing of pheromone odours in the main olfactory bulb and the hypothalamus (Kelliher *et al.* 1998). Sex differences in the vomeronasal system are explored by Guillamon & Segovia (1997).

Some of the most remarkable developmental effects in pheromone perception and production are in social insects. Early developmental switches caused by rearing an individual as a worker or as a reproductive lead to lifelong differences in response to pheromones in social insects of different castes (pheromone production is similarly affected, see Chapter 1, Box 6.2).

9.5.3 Experience

For some pheromone systems, the responses of animals change with experience. For example, after she has laid an egg, the female apple maggot fly

(*Rhagolitis pomonella*) marks the fruit with a pheromone which deters further egg laying in the fruit by herself or other females (Chapter 4). However, females require experience of the marking pheromone before they can discriminate between marked and unmarked fruit (Roitberg & Prokopy 1981). Contact with her own marking pheromone is sufficient, so normally in the wild she would have this experience while criss-crossing her trail as she marks her first fruit.

The responses of both male and female mammals to pheromones are facilitated by experience. Male mammals such as rats and mice (but not, apparently, lemmings) may need sexual experience before they can distinguish oestrous from dioestrous female odours, which is surprising as the ability to distinguish the two could be just as important for their first mating attempts, as for later ones (Vandenbergh 1994). The importance of sexual experience for responses to female pheromones by the male hamster is discussed in Box 9.5. Maternal behaviour elicited by pheromones in a number of mammals, including sheep and mice, has elements of learning. More specific cues, usually olfactory, are needed by animals giving birth for the first time, whereas more experienced (multiparous) females appear to have learned a wider range of associations (see Section 9.7.1).

There are learned elements in the response of freshwater fish to their alarm pheromone (Chapters 8 and 12). In addition, predator odours associated a first time with alarm pheromone later cause a conditioned alarm response by themselves. Learning gives flexibility of response because fish then respond appropriately to the particular predators found in their habitat.

Intriguing results suggest that there can be effects of experience (exposure) on the sensory perception at the peripheral level as well as on the central nervous system. Nearly half the adult human population does not perceive an odour when sniffing 5α-androstenone (and this is in part genetically determined) (Chapter 13). However, about half of initially anosmic subjects who sniffed 5α-androstenone three times daily for 6 weeks could perceive it as a smell by the end of the 6 weeks (Wysocki *et al.* 1989). Sensitivity to control odours was unaffected. Recent experiments suggest that for women, but only those of reproductive age, the response is not limited to specific anosmias or to 5α-androstenone: a wide variety of odours could induce the effect (Dalton *et al.* 2002). Men did not show this response. In mice, repeated exposure to odorants increases peripheral olfactory sensitivity. For two unrelated odours, androstenone and isovaleric acid, induction of olfactory sensitivity was odorant specific and occurred only in inbred strains that initially had low sensitivity to the exposure odorant (Wang *et al.* 1993).

9.6 | Primer pheromones and reproduction

Primer pheromones are important for coordinating reproduction in many types of organism (Chapters 1 and 3). The role of primer pheromones has been particularly well studied in mammals. For species which breed only at certain times of year, there may be suppression of oestrus for most of the year, often controlled by day length, broken only by contact with male odours in the breeding season (for example, the 'ram effect' in sheep, Section 9.3.2.2). Similarly, rams may be stimulated to produce sperm by the odours of females. Among vertebrates, it is in social mammals, living in groups on shared territories, and especially those breeding cooperatively (see Chapter 6), that primer pheromone interactions have reached their greatest complexity and subtlety (see reviews by Vandenbergh 1994, 1999a and Keverne 1998).

Most of the pioneering work on mammal primer pheromones has been on social rodents such as house mice, which are plural cooperative breeders (all females breed, although not all males) (Chapter 6). Female house mice suckle each other's young and cooperatively defend the nest. Characteristic of these societies are an interplay of dominance (in particular between males), sex and population density.

In plural breeders, the effects of females on each other are mutual, but in singular cooperative breeding species, such as beavers, prairie voles and the common marmoset (a New World primate), the dominant female suppresses reproduction by the subordinate females (see Section 9.6.8).

9.6.1 Male pheromones induce oestrus and accelerate puberty in females (Whitten and Vandenbergh effects)

In most mammals, hormone cycles lead females to be sexually receptive during discrete intervals known as oestrus, which are signalled by changes in behaviour, smells and visual signs. In many mammal species, contact with an adult male or his pheromones induces oestrus. In mice the oestrus-inducing effects of males on adult females (the Whitten effect) and accelerating puberty in young females (the Vandenbergh effect) are caused by the same male pheromone signals (Table 9.1) (Ma et al. 1999; Novotny et al. 1999a). In mice and voles these primer pheromones act via the vomeronasal organ and the accessory olfactory bulb (Keverne 1998). In social species such as mice, usually only the urine of the dominant male is effective. Female urine has the opposite effects.

The effects of urine pheromone signals are not limited to laboratory populations of rodents. Populations of house mice in large outdoor enclosures and exposed to additional male urine reached the highest density, those exposed to additional female urine reached the lowest density (Drickamer & Mikesic 1990).

The suppression of oestrus cycles occurs in adult female mice kept together in groups without males (see Section 9.6.2, the Lee–Boot effect). If exposed to a male or male urine, more than 50% of these anoestrous females come into oestrus three nights later (Marsden & Bronson 1964). Similar effects are also found in the Norway rat, golden and Djungerian hamsters, prairie voles (see later), goats, sheep and cattle (see Chapter 12 for applied uses) (Vandenbergh 1999b). In most cases, tactile or other social stimuli have an additive effect with the male odour.

Puberty in other rodents, pigs, cattle, sheep, goats and non-human primates (e.g. the tamarin *Saguinus fuscicollis*) can be advanced by 1–4 weeks by exposure to adult males or their odours (Vandenbergh 1999c).

9.6.1.1 Identity of the male pheromones in mice

Dominant male mice produce urine with many activities – first, as a primer pheromone with the related effects of inducing oestrus, accelerating puberty in females and causing the Bruce effect (Section 9.6.3); second, as a signal pheromone, attracting sexually active females but repelling immature females and subordinate males, and eliciting aggression (Chapter 5). Higher levels of testosterone in the dominant male cause the two principal differences between the urine of dominant and subordinate males: first, the urine of the dominant male has high concentrations of some small non–polar volatile molecules and, second, it has much higher levels of major urinary protein (MUP), up to 20 mg/ml. The MUPs are compact proteins with a large hydrophobic cavity that can reversibly bind these smaller, non-polar volatile molecules (Beynon *et al.* 1999).

The non-polar primer pheromone molecules in male urine are dehydro-*exo*-brevicomin (abbreviated brevicomin or DHB), 2-(*sec*-butyl)-dihydrothiazole (thiazole or SBT) and 6-hydroxy-6-methyl-3-heptanone (from bladder urine), and *E,E*-α-farnesene and *E*-β-farnesene (secreted into urine from the preputial gland) (Table 9.1). All four are active in puberty acceleration and oestrus induction: Ma *et al.* (1999) and Novotny *et al.* (1999a, b) report that these small molecules are active in puberty acceleration when presented in water at appropriate concentrations, in the absence of protein. Some researchers have suggested that the primer effect is due to the major urinary proteins themselves (Mucignat-Caretta *et al.* 1995, 1998).

MUPs reversibly bind odour molecules and release them slowly – the small molecules themselves are very volatile and rapidly disappear, but, when bound to the MUP, yield a long-lasting slow-release signal active for days (Hurst *et al.* 1998; Beynon *et al.* 1999). The MUPs are highly variable between individuals (Pes *et al.* 1999) and indeed may allow individual recognition (Hurst *et al.* 2001). The volatile odour molecules attract attention to the marks, even attracting the mice into well-lit areas they would normally

Table 9.1. *The structure and function of mouse pheromones produced by the dominant male and pheromone produced by crowded females*

Some of these small molecules are active on their own (but are usually associated with major urinary proteins). Others, such as thiazole and brevicomin require urinary proteins (MUPs) to be present for activity as releaser signals. Thiazole and brevicomin, and, separately, the farnesenes, are the pheromones that accelerate puberty and induce oestrus in females (Ma et al. 1999; Novotny et al. 1999a, b).

Name	Chemical structure	Origin	Possible chemosignalling function in female mice	Detection threshold EVG response[a]
2,5-Demethylpyrazine		Female urine	Puberty delay[b]	$10^{-8} - 10^{-7}$ M
2-sec-Butyl-4, 5-dihydrothiazole		Male bladder urine	Oestrus synchronisation Puberty acceleration	$10^{-10} - 10^{-9}$ M
2,3-Dehydro-exo-brevi comin		Male bladder urine	Oestrus synchronisation Puberty acceleration	$10^{-10} - 10^{-9}$ M
α- and-β- Farnesenes		Male preputial gland	Puberty acceleration	$10^{-11} - 10^{-10}$ M
2-Heptanone		Female or male urine	Oestrus extension	$10^{-11} - 10^{-10}$ M
6-Hydroxy-6-methyl-3-heptanone		Male bladder urine	Puberty acceleration	$10^{-8} - 10^{-7}$ M

[a]EVG is electrovomerogram. Each detection threshold range is based on field potential measurements from at least three different mice at multiple locations on the VNO (vomeronasal organ) luminal surface.

[b]Lee−Boot effect

From Leinders-Zufall et al. (2000).

avoid (Mucignat-Caretta & Caretta 1999). The MUPs in the male urine stimulate competitive countermarking (see Chapter 5) (Humphries *et al.* 1999; Hurst *et al.* 2001).

Two of the small molecules, brevicomin and thiazole, act synergistically as a releaser pheromone eliciting aggression and challenge from other males. However, the compounds are only active when added to the urine of castrated animals (inactive in itself), and not simply when presented in water (Novotny *et al.* 1985). MUPs in the urine might be the cause of this effect and also that of the greater activity of an elephant pheromone when presented in elephant urine rather than in water (Chapter 1) (Rasmussen *et al.* 1997).

The hormone changes causing the implantation failure in the Bruce effect are likely to be triggered by the small molecules in male urine. However, the blocking of this effect by the individual odours of the particular male the female mated with will require more complex odours, perhaps related to the major histocompatibility complex (MHC). Males that differ only at one MHC locus (Chapter 3) can cause pregnancy failure (Box 3.1) (Yamazaki *et al.* 1983). MUPs can carry many different kinds of odour molecules so they may have a role in carrying the odour molecules of individual difference even if they do not create them.

9.6.2 Female pheromones inhibit oestrus in adult females and delay puberty in juvenile females (Lee–Boot effect)

Female mice kept crowded together in groups, without males, have suppressed oestrus cycles instead of the normal cycle every 4–5 days (the Lee–Boot effect) and odours from their urine produce this effect in other females. Similarly, urine odours from grouped females delay puberty in younger females (Vandenbergh 1999a). A number of odour molecules are absent from the urine of subordinate crowded females if their adrenal glands are removed and one of these, 2,5-dimethylpyrazine, seems to be the key pheromone responsible for the oestrus suppression and puberty delay (Ma *et al.* 1998). This effect tends to overrule the female puberty acceleration pheromones produced by males. While adult female odours usually act to accelerate the development of juvenile male mice, urine odours from crowded females delay it.

9.6.3 Pregnancy block by male pheromones (Bruce effect)

In female mice who have recently mated, urinary pheromones of most other males (except the one she mated with – or his identical twin) stop the newly fertilised egg from implanting and the female returns to oestrus; the more genetically dissimilar the males, the more likely she will return to oestrus (see Section 9.7.2). The Bruce effect has been found in a number of other rodents including voles, lemmings and prairie deer mice. However, not all male-induced pregnancy block is a pheromonal Bruce effect. For the vole *Microtus pennsylvanicus*, with blocks possible up to 17 days, contact with the male is required and reflex ovulation stimulated by mating, rather than pheromones, may be the signal to start a new pregnancy.

9.6.4 Female pheromones and ovarian synchrony

In many mammalian species, females living in groups synchronise their oestrus, often by using pheromones. For example, odours produced by female rats bring about synchronous oestrus (McClintock 1983). It seems that there are two pheromones, one that advances ovulation and one that delays it. It is possible that humans have a similar system (Chapter 13). The effects of synchronised oestrus will be a changed operational sex ratio, closer to 1:1, thus reducing opportunities for the selection of multiple mates by males (Chapter 3). Synchronous oestrus would also allow communal nesting (Chapter 6).

9.6.5 Male pheromone effects on males

While lengthening days trigger reproduction, social interactions and pheromones determine which individuals reproduce in the lesser mouse lemur (*Microcebus murinus*), a nocturnal prosimian primate in Madagascar (Perret 1992). This lemur lives in population nuclei whose individuals have overlapping home ranges and animals often share the same sleeping area. Urine marks from the dominant male suppress sexual activity in subordinates within his territorial area by depressing their testosterone levels.

A similar interaction occurs between males of the marsupial sugar glider (*Petaurus breviceps*). Most scent marking is made by the dominant male, who is the heaviest in the group and has the highest blood concentration of testosterone, and the scent marks may affect subordinate males (Stoddart *et al.* 1994).

9.6.6 Female pheromone effects on males

Females may produce primer pheromones affecting the physiology and behaviour of males. The female urinary pheromones in the lesser mouse lemur attract males and volatiles from her urine marks stimulate spermatogenesis and testosterone secretion in all males (this stimulus is essential for the dominant male to produce urine active against subordinates) (see Section 9.3.2 for more examples) (Perret 1995).

9.6.7 Pheromones and sex ratios

In several primate species a male-biased sex ratio is found when there is competition between females for locally limited resources. To reduce intrasexual competition with daughters it would be an advantage to females to produce more sons – the local resource competition model of Clark (1978) cited in Perret (1996). The density of female pheromone marks could give an indicator of female numbers: in the lesser mouse lemur female urinary pheromones have been shown to bias offspring sex ratios. Exposure to high concentrations of female pheromone during a sensitive period leads to production of male-biased litters, about 70% male, whereas in the absence of female urinary pheromones the ratio was about 70% female (Perret 1996).

9.6.8 Suppression of reproduction in subordinate females

In cooperatively breeding species in which only the dominant female breeds, such as the common marmoset (a New World primate) and prairie voles, the dominant female suppresses reproduction by subordinate females (whether this should be thought of as 'signalling' rather than 'suppression' is discussed in Chapter 6). In both these examples, most or all of the subordinate females are daughters of the dominant female.

Suppression of ovulation in subordinate females of the common marmoset (*Callithrix jacchus*) is by a combination of olfactory, visual and behavioural cues. Once reproductively suppressed, this can be extended by odour alone: if a subordinate female is taken from the group, she will start her ovarian cycle but disinhibition is delayed by about 20 more days if she is exposed to the scent marks of the dominant female (Barrett *et al.* 1990). Intriguingly, the odour of unfamiliar dominant females was not effective (Abbott *et al.* 1998). Visual cues could also delay disinhibition in anosmic subordinates (those without a sense of smell) isolated from the group. Olfaction may be important for initiating ovarian suppression as well as for maintaining it: females made anosmic before joining a newly formed social group did not stop ovulation if they became subordinates. Reproductive suppression in another member of the marmoset family, *Saguinus fuscicollis*, may be similarly mediated by odours from the dominant female (Epple & Katz 1984). Pheromone-mediated effects on reproduction are likely to be widespread in marmosets and other New World primates and prosimian primates, as they have many scent glands and a fully functioning VNO (vomeronasal organ). In contrast, adult Old World monkeys, apes and humans do seem not to have a VNO (Chapter 13) (Dixson 1998, p. 185).

Subordinate female prairie voles (*Microtus ochrogaster*) tend to remain prepubescent so long as they are in their natal family group (see reviews by Carter *et al.* 1995, Carter & Roberts 1997). This is due to two effects: first, subordinate females delay puberty as long as they are exposed to only familiar males (father or male sibs), recognised by odour; second, the stimulatory effect of urine from an unfamiliar male is overruled by inhibitory pheromones in the urine of her mother and sisters. Subordinate females thus remain functionally pre-pubescent and provide support to the communal family. The

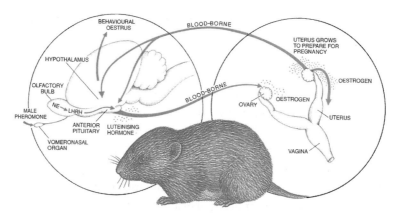

Fig. 9.9. Hormonal cascade that triggers oestrus in the female prairie vole (*Microtus ochrogaster*) begins when she sniffs an unfamiliar male (she does not sniff her father or sibs). The vomeronasal organ picks up pheromones, stimulating the olfactory bulb. Norepinephrine (NE, noradrenaline) and luteinising hormone-releasing hormone (LHRH) are secreted and start the production of luteinising hormone. Luteinising hormone reaches the ovaries via the blood stream and stimulates them to produce oestrogen. Oestrogen is then carried to the hypothalamus, where it induces oestrus. Figure from Carter & Getz (1993). Artist Patricia Wynne.

proximate reason that familiar males do not stimulate puberty (the Vandenburgh effect) is due to recognition-dependent behaviour: because familiar males do not evoke mutual nasogenital grooming, the urinary pheromone does not get into the VNO of the female. (However, if urine is placed on the vole's nose by the experimenter, it is effective even from a sib male. The hormonal processes involved in triggering oestrus in the prairie vole are illustrated in Fig. 9.9.)

9.7 | Olfactory cues and recognition learning

Many discriminatory behaviours (for example mate choice, feeding a juvenile, a guard bee letting another bee into the nest) are based on learnt olfactory cues (see Recognition mechanisms, Box 6.1). In vertebrates such as mammals, these may be forms of imprinting, biologically relevant learning during a sensitive period defined by a particular developmental stage or physiological state (Hudson 1993). Learning of important odour cues can happen early in life or when adult, as appropriate. The response that is prompted by recognition could vary from a long-lasting hormonal change to an immediate behaviour, such as initiation of courtship.

Early imprinting on species-specific odours can be important in recognition of a mate of the correct species when adult (Chapter 3) (Owens *et al.* 1999). Normally, as the parents are of its own species, this leads to appropriate courtship choices, but cross-fostering experiments can demonstrate that these choices are learned, at least in some species such as pygmy mice and house mice (Vandenbergh 1994). Young male black-tailed deer will imprint on the odour of pronghorn antelope if cross-fostered (Müller-Schwarze & Müller-Schwarze 1971).

Subtler is the imprinting of odours used to recognise kin and colony members or mates (Chapter 6). Thus far the learning processes used by social insects to recognise kin have not been examined but it is likely that the antennal lobe and mushroom bodies and lateral protocerebrum, parts of the brain to which the antennal lobes project, are involved (Menzel *et al.* 1996).

Two mammalian odour recognition systems described in Chapter 6 have provided detailed model systems for investigating the mechanisms of olfactory memory in vertebrates. The first is the way a mother sheep (ewe) learns the odour of her lamb, via the main olfactory system; the second, via the vomeronasal system, is the female mouse's learning of her mate's odour, which is the basis of the Bruce effect of pregnancy block. Both model systems require simultaneous appropriate odour stimulation *and* noradrenaline release in the olfactory bulbs by nerves stimulated by birth or mating, respectively (Keverne & Brennan 1996). One exposure to the odour is enough for learning. In both systems, the sensitive period is characterised by changes in the animal's hormonal state that facilitate the learning (Hudson 1993). In the ewe, learning depends on priming by ovarian steroids (such as progesterone and oestradiol) in the blood through the pregnancy. In the mouse, the presence of oestrogen is a prerequisite for mating and for the formation of olfactory memory.

9.7.1 Maternal behaviour and kin recognition of offspring: sheep as a model

In sheep an enduring bond between a mother and her lambs is established very rapidly, usually within 2 hours of giving birth (parturition) (see reviews by Lévy *et al.* 1996, Kendrick *et al.* 1997). Olfactory cues perceived by the main olfactory system are key to the process. Stretching of the vaginocervical area while giving birth sends nerve signals to the brain that trigger a cascade of neurobiological and hormonal mechanisms resulting in three changes in behaviour (Fig. 9.10): first, release of oxytocin, a peptide hormone, mainly in the paraventricular nucleus of the hypothalamus, leads to maternal behaviour towards lambs in general; second, oxytocin released in the olfactory bulb changes the ewe's response to amniotic fluid making it very attractive (whereas before it was repulsive); and, third, oxytocin and other neurotransmitters released in the olfactory bulb cause her to learn the individual odour of her lamb while sniffing and licking the amniotic fluid. Afterwards she will only suckle her own lambs, recognised by smell (see also Chapter 6). The sensitive period for learning lasts for between 4 and 12 hours after giving birth. In this early period, but not later, an orphan lamb will be accepted

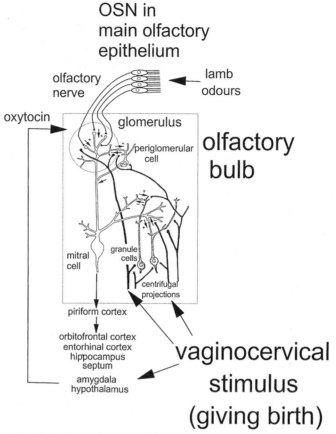

Fig. 9.10. Ewes learn the odour of their newly born lambs. The changes in brain neurochemistry stimulated by giving birth that facilitate the mother's changes in behaviour and odour learning have been mapped. The + and − signs in the diagram represent excitatory and inhibitory influences; OSN is olfactory sensory neuron. The olfactory memory underlying selective recognition of lamb odours involves plastic changes occurring within the olfactory bulb as well as in its secondary (piriform cortex) and tertiary projection sites (orbitofrontal cortex, entorhinal cortex and hippocampus) (see Fig 9.1). Within the olfactory bulb, memory formation is associated with a strengthening of the mitral to granule and periglomerular cell synapses to bias the network to respond to that lamb's odours. Figure after Kendrick *et al.* (1997).

and adopted, particularly if coated with amniotic fluid (a method traditionally used by shepherds).

After birth there is an increase in the proportion of mitral cells in the mother's olfactory bulb that respond selectively to lamb odours. Vaginocervical stimulation is also responsible for neurochemical and electrophysiological changes within the olfactory bulb that cause selective recognition of her lamb. Learning of her lamb's odours is accompanied by increased activity of a subset of mitral cells.

Maternal experience has profound effects on the speed with which ewes bond with their lambs and on neurotransmitter release within the olfactory bulb (Kendrick *et al.* 1997). Whereas almost all experienced (multiparous) ewes bond within 2 hours, only 15% of primiparous (giving birth for the first time) ewes do, and these may take 4–6 hours to bond. Maternal experience both increases the sensitivity of the brain in all future births to the effects of oxytocin and also facilitates the learning of their offspring's individual odour (Kendrick *et al.* 1997). For both primiparous and multiparous ewes, the main olfactory system is the key sense required for maternal behaviour and learning of the lamb's odour signature, not the VNO-accessory olfactory system (Lévy *et al.* 1995).

In humans, close olfactory contact between mother and baby in the first hours after birth is important for each to learn the odour signature of the other (Winberg & Porter 1998). Olfactory recognition of the baby and responses to infant odours can be related to the hormone levels of the mother (Chapter 13) (Fleming *et al.* 1997).

9.7.2 Memory and pregnancy block (Bruce effect) in mice

In female mice who have recently mated, urinary pheromones of most other males (except the one she mated with – or his identical twin) stop the newly fertilised egg from implanting and the female returns to oestrus (Brennan 1999). The more genetically dissimilar the males, the more likely she will return to oestrus.

All male mice produce the pheromone but circuits in the female's accessory olfactory bulb recognise her mate's odour and selectively stop his pheromone excitation from releasing the hormones that would block pregnancy. She learns his odour during the first few hours after mating and retains the memory for some 30 days (but forgets during a successful pregnancy).

First in the relay for the female's long-term memory of her mate's odour is the accessory olfactory bulb (AOB), which receives its sensory input from the vomeronasal organ. The mating male's individual odours stimulate a particular range of olfactory sensory neurons in the VNO to send signals to a subpopulation of mitral cells and associated granule cells. The combination of this olfactory input with the increased levels of noradrenaline after mating is thought to prime these inhibitory synapses so that if the same individual odour is smelt again, transmission to the hypothalamus of the pregnancy-blocking signal from male urinary pheromones is stopped. For the memory to be formed, mating and then exposure to the mate's odour during the sensitive period (3–4.5 h after mating) are required.

For the strange male, the benefit of the Bruce effect is clear, as he is likely to be the father of the next litter. The evolutionary advantage for the female remains unclear and many hypotheses have been put forward (Schwagmeyer 1979; Labov 1981). One, the infanticide (by the male) hypothesis, is that early termination could be the 'best of a bad job' as in re-mating with the new male, she will now have a partner less likely to kill the young.

However, Coopersmith & Lenington (1998) found no evidence of correlation between female choice, pregnancy block and the likelihood of infanticide.

Currently there is no good evidence that the Bruce effect occurs in mice under natural conditions (Brennan 1999). An alternative view suggests that the Bruce effect may be a laboratory artefact due to enforced exposure to male pheromones and their general oestrus-inducing activity. A possible evolutionary advantage of sensitivity to male pheromones for puberty advance is that it significantly increases reproductive life by bringing first oestrus to 28 days rather than the 60 days it would be otherwise, which is close to the field lifespan of only 80 days or so (Keverne & Rosser 1986).

9.7.3 The Coolidge effect

The renewed stimulation of sexual behaviour in sexually satiated animal by a new potential mate is called the Coolidge effect (see Chapter 3). For hamster males, some studies have suggested that the main olfactory system is most important for this individual recognition (or is it just novelty that is required?) (Johnston & Rasmussen 1984). However, other studies have concluded that the VNO is involved in such recognition (Steel & Keverne 1985).

In insects, the Coolidge effect has been demonstrated in cricket females, probably mediated by cuticular pheromones (Bateman 1998). In halictine bees, the frequency of male contact with female models declines over time, but rises again if a new female is presented (Smith & Breed 1995).

9.7.4 Monogamous partner recognition

Unlike most rodents, prairie voles (*Microtus ochrogaster*) are monogamous and form long-lasting pair bonds. Both parents share in rearing the young (Chapter 6). In most but not all cases, pair-bonded females exhibit a preference for the familiar male (Carter *et al.* 1995). Mating is not required for the development of preference (cohabiting may be enough) and nor are gonadal steroids (so the memory is apparently different from that involved in the Bruce effect in mice). Neuropeptides, including oxytocin and vasopressin, and the adrenal glucocorticoid, corticosterone, have been implicated in the neural regulation of partner preferences and, in the male, vasopressin has been implicated in the induction of the selective aggression toward strangers (conspecifics other than the mate) that develops in the 24 hours after mating.

9.8 | Developmental paths or metamorphosis prompted by pheromones

Primer pheromones can play a crucial switching role in major developmental changes in animals. These include the metamorphosis of marine plankton larvae to their sessile form and phase changes in locusts (Chapter 4). Two other pheromone-based examples are discussed here: the formation of a resting stage by nematode larvae when conditions get tough, and social organisation and castes in social insects.

9.8.1 Resting stage formation in the nematode, *Caenorhabditis elegans*

The first multicellular organism to have its whole genome sequenced, the nematode worm *Caenorhabditis elegans*, is providing a model system for investigating many aspects of olfaction using genetic mutants and selective destruction of identified olfactory sensory neurons with a laser (Mori 1999). Larval development of *C. elegans* is controlled by the activities of four classes of chemosensory neurons. The switch between normal development and development into a specialised, resistant larval form called a dauer larva, which can safely withstand harsh environmental conditions, is regulated by levels of conspecific dauer pheromone (whether this is a signal from the conspecifics or simply the receiving nematode's response to compounds characteristic of high densities of conspecifics is debatable and other environmental cues such as limited food and high temperature also encourage dauer formation) (Fig. 9.11) (Thomas 1993; Schackwitz *et al.* 1996). When environmental conditions improve, the dauer larva recovers, begins feeding and soon moults to resume the normal life cycle.

Response to the dauer pheromone is dependent on the major chemosensory organ in *C. elegans*, the bilaterally symmetrical amphid sensilla, located

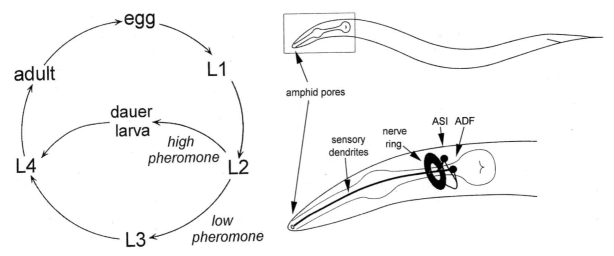

Fig. 9.11. (Left) The nematode worm *Caenorhabditis elegans* passes into an alternative resting stage, the dauer larva, if the conditions are poor, signalled by high concentrations of 'dauer' pheromone released by conspecifics. When pheromone levels drop or food becomes abundant, the dauer larva recovers and rejoins the normal cycle at the L4 stage.
(Right) (Bottom) Diagram of the position of the sensory cells (amphid pores) and two of the amphid chemosensory neurons (ASI and ADF) that control dauer formation, viewed from the left side of the animal. (Top) The whole worm with the pores indicated.
Figures from Thomas (1993).

in the tip of the anterior of the animal (Fig. 9.11). Each of these sensilla contains just 12 olfactory (or chemosensory) sensory neurons. Entry into and exit from the dauer stage are primarily controlled by different sets of the chemosensory neurons (Bargmann & Horvitz 1991). The analysis of mutants indicates that the chemosensory neurons controlling dauer formation are active in the absence of sensory inputs and that dauer pheromone inhibits the ability of these neurons to generate the signals necessary for normal development.

9.8.2 Social insects

The physiology and resulting behaviour of social insects is influenced at every stage by pheromones (Chapter 6). Many of the longer term, primer, effects are mediated by juvenile hormone (JH). For example, the age at which honeybee workers start foraging outside the nest is delayed by high levels of queen mandibular pheromone (QMP), which lower JH biosynthesis in worker bees (Chapter 6) (Pettis *et al.* 1999).

In ants, pheromones influence many developmental pathways, in many cases via effects on JH secretion. The interactions are summarised in Fig. 9.12 (Vargo 1998).

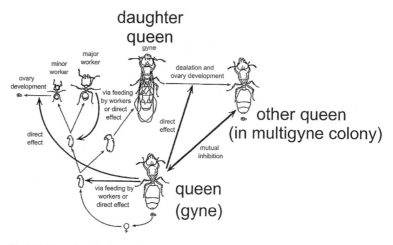

Fig. 9.12. The roles of primer pheromones in ant colonies. Only a subset of these effects have been documented for any one species. The thin lines show developmental pathways and the thick lines indicate the effects of primer pheromones from the source individuals (queens or major workers) to the target individuals. Queen-produced primer pheromones can probably influence caste determination of female larvae (which can potentially develop into a queen (gyne) or a worker), probably by affecting feeding carried out by workers, although a direct effect on larval physiology cannot be ruled out. In species with a polymorphic worker class, a pheromone produced by major workers may inhibit production of more major workers so that an equilibrium of major:minor numbers is maintained, again probably indirectly through an effect on feeding carried out by workers, although a direct effect on larval development is possible. Queens may inhibit ovary development in workers or gynes through an effect on the endocrine system. Finally, when multiple egg laying queens coexist in the nest, they may exert mutual pheromonal inhibition leading to lowered egg production. Caption and figure adapted from Vargo (1998).

Convergently evolved mechanisms, similar to those in ants, operate in termites (Chapter 6, Box 6.2). Primer pheromones act on JH synthesis to mediate caste changes in the colony (Henderson 1998). However, unlike ants, whose caste is set as larvae, termites can change caste at any stage because they are not holometabolous insects (with a pupal stage). Different thresholds for moulting into the various castes are combined with an interaction with different sensitive periods for the action of pheromone(s) on each stage (Nijhout 1994).

9.9 | Conclusion

The olfactory systems found in most multicellular organisms appear to have the same design features, ones which give olfaction a remarkable flexibility for signalling to evolve. These features include olfactory sensory neurons which each express just one or two receptor-protein types, and the linking of all sensory neurons with a particular receptor to the same few glomeruli for integration and onward projection of signals to the rest of the brain. Receptor families in different animal taxa have evolved independently from a wide range of different receptor proteins already on the surface membrane. Strikingly, in terrestrial vertebrates the VNO (vomeronasal organ) and the MOE (main olfactory epithelium) have evolved independently and the VNO has an important role in many rodent pheromone-endocrine interactions. However, this observation does not allow us to conclude that all receptors in the VNO in different vertebrates are necessarily pheromone receptors. Odorant and pheromone binding proteins may play an important role in carrying odorants to the receptors. The recent rapid progress in understanding of olfactory receptors and organisation of the olfactory system will continue as molecular biology, behaviour and neuroanatomy are combined and *in situ* probes can exploit the genomics revolution.

9.10 | Further reading

A good review of olfaction and the brain is found in Finger *et al.* (2000). Vertebrate hormones are described in Nelson (1995) and Brown (1994). Insect hormones are reviewed in Nijhout (1994).

Developments in our understanding of receptors are moving so rapidly that the best sources are recent review articles, for example, in the two journals *Cell* and *Trends in Neurosciences* and in the *Current Opinion* series.

Chapter 10

Finding the source:
pheromones and orientation behaviour

10.1 | Introduction

Finding the source of a pheromone plume in a turbulent flow is a greater challenge than finding an animal producing sound or visual signals. Nonetheless, animals responding to pheromones offer some of the most spectacular examples of long-range responses to stimuli. These include male moths attracted to females over hundreds of metres and perhaps even further. Over evolutionary time, receiving organisms have been selected to search efficiently and to find the source as quickly as possible: many odour sources (whether food or mates) do not last long – the odour source moves or another animal will get there first (Atema 1996).

Searching animals exploit the invisible odour 'landscape' created by high and low concentrations of countless molecules in overlapping plumes released by other organisms (Atema 1995; Nevitt 2000). These chemical plumes occur in a wide range of spatial scales and durations. For example, in the sea, they range from the pheromone released by a single planktonic copepod, a few millimetres long, to the odour plumes left by a tuna school 100 metres across. Odour signals have a significant but limited life before the molecules are dispersed or broken down.

How do animals distinguish chemical signals from this chemical cacophony? Signals may be identified by characteristic fine-scale changes in concentration over time or by chemical specificity (Atema 1995). Chemical specificity may come from unique compounds, but more often by unique mixtures, including common molecules in specific combinations (Chapters 1 and 3). Animals may evolve to respond to different aspects of these potential signals. Each animal species extracts its unique information from the chemical world and uniquely contributes to it (Atema 1995).

The ability of animals to use pheromones for low-cost long-distance communication may have important evolutionary implications by allowing animals at low densities to find mates. For example, in moths, it may facilitate speciation associated with larval specialisation on rare plant species (Cardé & Mafra-Neto 1997). Similarly, like other deep-sea fish, hatchet fish (*Argyropelecus hemigymnus*) live at extremely low densities and males have the biggest nasal organ relative to body size of any vertebrate, implying that olfaction is very important in mate location (Jumper & Baird 1991).

In this chapter I start by describing some of the mechanisms that animals use to find odour sources before discussing the different kinds of odour stimulus, from diffusion gradients to odour plumes, and the ways that animals have evolved to respond to these. We know most about the mechanisms used by invertebrates but the principles of olfactory orientation apply to all animals.

10.2 | Investigating orientation behaviour mechanisms

One of the most successful approaches to understanding orientation behaviour has been to explore the precise stimuli that animals use. Teleological terms, such as 'attraction' describe the endpoint or provide a handy metaphor for the 'function' or ultimate explanation of a behaviour (Kennedy 1986, 1992). However, such terms do not take us nearer the mechanisms. For example, an aggregation of animals near a pheromone source could have been attracted from a distance or instead formed by arresting (stopping) animals that came close to the source by chance. The different kinds of responses are outlined in Box 10.1.

At first sight, many animals behave as if they are aiming for a 'goal' or target but, without other evidence, it is more accurate to describe their movements as responses to the stimuli received at or up to that point (Kennedy 1986). Most orientation behaviour, chemo-orientation in particular, is based on immediate responses to stimuli. However, as important as external stimuli are behaviours initiated and modulated but not steered by the odour stimulus. These 'self-steered' behaviours are internally set (idiothetic) behaviour patterns triggered by pheromone contact (Kennedy 1986).

Dusenbery (1992, p. 416) contrasts direct and indirect guiding towards a stimulus source. The ability to detect and respond to the orientation of the odour concentration gradient is the basis for *directly guided* behavioural responses towards or away from the higher concentration. These include the various types of **taxis** responses in Box 10.1. **Kineses** are *indirectly guided* responses, in which the movements of animals are not directed by concentration

Box 10.1 | The basic mechanisms used by animals for orientation towards or away from a stimulus

Indirect guiding

Kinesis

The animal's movement is affected by the intensity of the stimulus but the direction of turns and movements *is not* related to gradients in the stimulus concentration.

Orthokinesis
The animal moves faster or slower depending on concentration.

Klinokinesis
The animal turns more or less, makes more or larger, or fewer or smaller turns, depending on concentration.

Direct guiding: directed movement towards a stimulus source

Taxis

The direction of turns and movement *is* related to the pattern of concentration. The different types of taxis reflect the different ways that animals detect the gradient or pattern of concentration, from single or paired sensors.

Klinotaxis
Directed turns based on *successive samples* of concentration in different places (one sensor moved from side to side, for 'transverse klinotaxis'; one sensor, sampling successively along the animal's path, 'longitudinal klinotaxis').

Tropotaxis
Directed turns based on *simultaneous comparison* of stimulus intensity on two points of the body, or for example the simultaneous inputs from two antennae. This is possible only where the chemical gradient is steep enough to stimulate the two sensors with different intensities.

Teleotaxis
Directed turns based on input from many receptors in a 'raster', such as vertebrate or insect eyes (and so far only demonstrated for responses to light). An extreme form of tropotaxis.

Other descriptors

Other prefixes can be added: **chemo-** (in response to chemical stimuli), **osmo-** (response to molecules in gas or water phase), **anemo-**(oriented to the wind), **rheo-** (oriented to water current flow), **meno-** (orientation at an angle to the direction of the stimulus field). For example, osmotropotaxis is the ability to detect and move up gradients of molecular concentration by simultaneous stimulation of a pair of sensors.

Largely based on the scheme of Fraenkel & Gunn (1940) since modified in many ways; reviewed by Dusenbery (1992, p. 414), Kennedy (1986), Bell (1991), Schöne (1984), Campan (1997) and other authors.

gradients but the movements may be far from random since the pheromones may trigger the self-steered (idiothetic) turning patterns (Kennedy 1986). Although these indirect guiding mechanisms are relatively simple they can work well. For example, nematodes can probably use gradients of CO_2 from plant roots to guide them from 1 to 2 m away (Dusenbery 1992) and they are used by planktonic animals, including fish larvae, to find and stay in food patches.

To determine the orientation of a concentration gradient, the animal needs to compare the stimulus intensity at different positions (Dusenbery 1992, p. 416). There are two basic methods: first, by *simultaneous sampling* of input from multiple receptors separated on different parts of organism's surface, with the spatial gradient measured by comparing the intensity at different positions (the organism needs to be large in relation to the gradient and structures such as stretched-out antennae or the head of hammerhead sharks may increase the distance between the sensors); or, second, *sequential sampling* by a single receptor moved around.

Many, perhaps most, organisms are capable of using more than one of these strategies, depending on the circumstances (Dusenbery 1992, p. 431). Designing experiments to distinguish these possible responses can be hard.

10.3 | Ranging behaviour: search strategies for finding odour plumes, trails or gradients

Animals can only respond to pheromones after entering the **active space**, defined as the zone where the pheromone concentration is at or above their threshold for detecting it (Bossert & Wilson 1963; Wilson 1970). Ranging behaviour is the initial stage of searching that brings the animal into first contact with the active space. Having contacted the chemical stimuli, the animal's behaviour changes to one guided directly or indirectly by the odours. Ranging may also be used to refind the pheromone plume if it is lost. There is surprisingly little experimental data on these search patterns and most work is based on models (reviewed by Dusenbery 1992, pp. 385ff.). The models suggest that for walking animals, searching cross-current is about 50% more efficient than searching in a random direction, as the pheromone plume presents a large target parallel to the flow. Under more natural conditions, fluctuating wind directions are likely and this will affect optimum strategies. This is shown by *Drosophila* fruit flies, which fly at right angles to the wind when the wind direction is steady, but in shifting winds fly upwind, parallel to the average wind direction (Zanen *et al.* 1994).

10.4 | Finding the source: orientation to pheromones

All life lives in a fluid medium. Although air and water differ in characteristics such as density and viscosity, the same laws apply to both media (Denny 1993; Weissburg 1997, 2000). Within either air or water, animals may respond to three very different kinds of odour stimulus. The first is *short range diffusion* of molecules from a point source, such as an injured ant releasing alarm pheromone; the second is a *trail* of pheromone laid on the substrate behind a moving animal; and the third is a *plume* of molecules carried away from the odour source by wind or water currents, in which diffusion plays almost no role. The rest of this chapter is organised around the characteristics of the three types of odour stimulus and how animals have independently evolved orientation mechanisms to cope with the different challenges these stimuli offer.

10.4.1 Diffusion very close to the source: orientation of animals to diffusing signals

A limited number of pheromone signals, particularly for animals the size of ants and smaller, act within a short range and reach the receiver by diffusion alone. This is a special case: most pheromones are carried by currents of air or water, see Section 10.4.3. From an instantaneous release of pheromone near the ground, the active space (which is the signal) expands as a hemispherical cloud of pheromone. The radius of the active space rises to a maximum, and then starts to contract and fades away to below the behavioural threshold as outwards diffusion continues.

Bossert & Wilson (1963) modelled the diffusion of pheromone molecules (Box 10.2) and showed that the active space is related first to the *quantity* of pheromone (Q), second, to the sensitivity of the receiver, its *threshold* (K), and, third, to the *diffusion coefficient* (D) of the pheromone (which is correlated with its molecular weight). In response to the selection pressures on the signalling system (Chapter 1), over evolutionary time each of these values can be changed to give an active space which is larger or smaller, quicker or slower to expand to its maximum, faster or slower to fade. The characteristics of a pheromone system can be expressed as the $Q{:}K$ ratio.

For an instantaneous pheromone signal, the maximum detection distance depends only on the $Q{:}K$ ratio. An increase in the active space can be achieved by either increasing the amount of pheromone released (Q) or lowering K (more sensitive reception), or both. Increasing the sensitivity is more likely as increases of orders of magnitude in sensitivity of chemoreceptors are possible, whereas producing more pheromone to have the equivalent effect can be difficult in air (Hölldobler & Wilson 1990, p. 244).

For continuously emitted signals, maximum range and duration are directly proportional to the $Q{:}K$ ratio and inversely proportional to the

Box 10.2 | Diffusion models for pheromones at close range

Bossert & Wilson (1963) proposed models to calculate the active space of diffusing pheromone signals, in the special case of no currents and small scale. Q, the quantity of pheromone, is measured in numbers of molecules released in a burst, or in numbers of molecules released per unit time (Hölldobler & Wilson 1990, p. 244). K, the threshold sensitivity, is measured in molecules per unit of volume. D is the diffusion constant in cm^2/s. These versions of the equations are for pheromone put on a spot on the ground: the molecules can only spread as a hemisphere up from the surface so Q is doubled (for release in three dimensions replace '$2Q$' by 'Q' throughout).

The radius (R) of active space at time t is given by:

$$R(t) = \sqrt{4Dt \log\left(\frac{2Q}{K(4\pi Dt)^{3/2}}\right)}$$

$$\text{for } 0 \leq t \leq \frac{1}{4\pi D}\left(\frac{2Q}{K}\right)^{2/3}$$

$$= 0 \text{ otherwise}$$

The second equation above provides an upper limit on time and radius when the intensity (concentration) is below threshold everywhere.

The radius of the threshold hemisphere increases through time to a maximum R_{max}:

$$R_{max} = \sqrt{\left(\frac{2Q}{K}\right)^{2/3} \times \frac{3}{2\pi e}} = 0.527\left(\frac{Q}{K}\right)^{1/3}$$

at time $t_{R_{max}}$:

$$t_{R_{max}} = \frac{1}{4\pi De}\left(\frac{2Q}{K}\right)^{2/3} = \frac{0.0464}{D}\left(\frac{Q}{K}\right)^{2/3}$$

It then fades out completely:

$$t_{fade\text{-}out} = \frac{1}{4\pi D}\left(\frac{2Q}{K}\right)^{2/3} = e.t_{R_{max}} = \frac{0.126}{D}\left(\frac{Q}{K}\right)^{2/3}$$

The equations can be rearranged to give estimates of K or Q as required from the data available.

diffusion coefficient (D) of the pheromone. For example, a shorter fade-out time, making the signal more sharply pinpointed in space and time, can be achieved by lowering the quantity of pheromone (Q) or raising the threshold, K (making the receiver less sensitive), or both (Hölldobler & Wilson 1990, p. 244).

10.4.1.1 Air vs. water

As the maximum range of a diffusing chemical signal is independent of the diffusion rate (D), the range of signals is approximately the same in air and water. However, in water the signal takes far longer to travel and then fade;

as the timescale is inversely proportional to D, the kinetics affecting rise-time and fade-out will be slowed 10 000 times in water as typical diffusivity values (in m^2/s) for small molecules are 2×10^{-5} in air, but 2×10^{-9} in water (Dusenbery 1992, p. 62). For the same substance to give the same times to maximum radius and fade-out in water, the Q:K ratio would need to be about a million times greater in water than air (Wilson 1970). That is, the amount of pheromone released (Q) would have to be a million times greater or the sensitivity threshold lowered by the same amount. Wilson suggests that by using highly soluble polar molecules (for example, anthopleurine, Chapter 8), such increases in Q might be possible in water.

Animal size will affect communication strategies. Smaller animals, 1–2 mm in diameter, live in quite a different fluid world (Box 10.3) in which diffusion can dominate. However, most larger animal species rely instead on natural currents, not diffusion, to carry their messages to the receiving animal (Section 10.4.3). Animals that live within the boundary layer (see Fig. 10.6) may be able to exploit natural currents if they can release their pheromone into the faster flowing fluids above, by lifting their pheromone gland up off a leaf (in air) or squirting pheromone out with some force (in water) (Conner & Best 1988; Vogel 1994). It is energetically too expensive to propel water or air very far but some animals briefly create their own current for short-range signalling. For example, male lobsters jet their urine signal out of their dens towards females (Section 3.10.1). Similarly, when male moths reach a female, they direct their pheromones towards her by vigorously fanning their wings (Fig. 1.2).

10.4.1.2 Ant alarm pheromones – a case study of a diffusing signal

The alarm signal of the harvester ant (*Pogonomyrmex badius*) would have a maximum radius of about 6 cm, given its Q:K and D (Bossert & Wilson 1963). This would be reached in 13 seconds and the signal would fade in 35 seconds (ideal for an alarm pheromone, with a quick rise time, short range, and rapid fade).

The alarm pheromone of *P. badius* has only one component (4-methyl-3-heptanone), but many social insects have multi-component alarm pheromones (Chapter 8). The various components may simultaneously alert, attract and evoke aggression (Hölldobler & Wilson 1990, p. 261). Is there one active space or several (one for each component)? The African weaver ant (*Oecophylla longinoda*) has four major alarm pheromone components which diffuse out from the source to create overlapping independent active spaces expanding at different rates and eliciting different behaviours (Fig. 10.1) (Bradshaw *et al.* 1975, 1979). Thus the work so far on ant alarm pheromones suggests that in the nest and close to the substrate outside it, ants are working within the boundary layer of slow air next to the surface and diffusion dominates so

Box 10.3 | Pheromone signals in the plankton – 3D-trails in water

Copepods are small (0.1–10 mm) aquatic crustaceans that live in the ocean, often separated by thousands of body lengths (Weissburg et al. 1998). In order to increase the chance of finding mates and prey they aggregate, often using chemokinesis to remain within richer patches (Chapter 4).

Planktonic animals of this size live at a scale dominated by high viscous forces and laminar flow, likened to swimming in treacle (Denny 1993; Vogel 1994). At these low Reynolds numbers (Re), a term used to describe characteristics of flow speed, viscosity, size and shape, diffusion dominates and the pheromone trails left by moving female copepods persist as trails for seconds or minutes, like vapour trails behind aeroplanes. The trail that a female copepod leaves increases her effective size to 130 times her body length, thus also increasing the area encountered by the searching male (see figure below). Male *Temora longicornis* exploit these unique 3D-pheromone trails, detecting trails up to 10 seconds old and pursuing females up to 100 body lengths away down the trail (Doall et al. 1998; Weissburg et al. 1998). The males may use receptors on their paired antennae for tropotaxis, like an ant or a snake, but following a 3D-trail. Klinotaxis may also be used (together with detection of the flow patterns (Yen et al. 1998)). Males also seem able to detect in which direction to follow the trail.

Modelling of diffusion ranges and costs of locomotion suggests that mate location using diffusable pheromones in water is worthwhile only above a critical size of animal, and within the size range 0.2–5 mm (Dusenbery & Snell 1995). Thus planktonic copepods, which are within this size range, do use pheromones but most rotifers, which are smaller, do not.

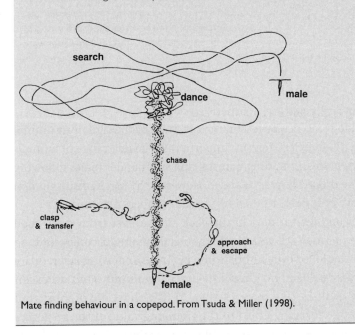

Mate finding behaviour in a copepod. From Tsuda & Miller (1998).

that each pheromone component has its own active space. However, long-range pheromones such as moth sex pheromones are carried by the wind and the different components of moth pheromones share the same active space (see Section 10.4.3.3).

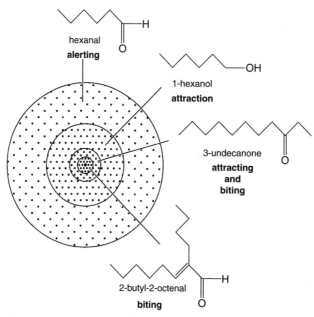

hexanal
alerting

1-hexanol
attraction

3-undecanone
**attracting
and
biting**

2-butyl-2-octenal
biting

Fig. 10.1. Where diffusion dominates, in still air close to the ground or inside an ant nest, the active space depends on the diffusion speeds of molecules (as well as the sensitivity of the receiver). The calculated active spaces at ant level for the four major components of the alarm pheromone of a major worker of the weaver ant (*Oecophylla longinoda*) 20 s after deposition at a central point of a flat surface in still air (Box 10.2). Different components elicited different behaviours and had separate active spaces represented here by the concentric rings. The active spaces are shown as circles but they are actually overlapping hemispheres that spread above ground from the point source. Diagram from Chapman (1998) using data from Bradshaw *et al.* (1975).

10.4.2 Following trails laid on a substrate

Trails consist of odour molecules left on the substrate by the signalling animal as it moves along (Chapter 7). They are important for a wide range of animals from snails to vertebrates, including hamster and salamander males following female pheromone trails (Chapters 3 and 9). However, the mechanisms used in trail following have only been studied in a small number of species.

Ants use pheromones to mark trails back to the nest from good food sources. The pheromones are volatile low molecular weight compounds so the signal is detectable for only a short time before evaporating away. A fire ant (*Solenopsis saevissima*) trail on glass lasts just a few minutes but other ant species may have trails lasting days (Chapter 7). The short life of most trail pheromones matches foraging effort to the changing value of the resources at the trail end: when the ants no longer find food at the end of the trail they no longer mark the trail on their return and thus the trail fades away. If the molecules were too persistent, trails would lead to resources long gone.

Ants following a trail respond to the airborne molecules of pheromone diffusing from the marks rather than with a contact response to the trail

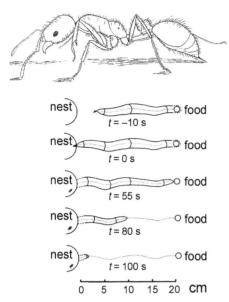

Fig. 10.2. The odour trail of a fire ant (*Solenopsis invicta*) laid on glass. Pheromone diffusing from the trail forms a semi-ellipsoidal active space. This space, and therefore the entire recruitment signal, fades after about 100 s. The times shown here are given from the moment a worker reaches the nest after laying a trail from a food source 20 cm away. The trail is shown in this model as continuous. In nature it is irregularly segmental, but the dimensions and fade-out times remain nearly the same. Ant from Wilson (1962). Lower part of figure after Wilson & Bossert (1963).

itself: fire ants can be drawn along in response to molecules diffusing away from a droplet of pheromone held on a glass rod above them (Wilson 1962). Over the trail there is a semi-ellipsoidal active space (Fig. 10.2) (Bossert & Wilson 1963). As ants travel through this 'vapour tunnel', they sweep their antennae from side to side (Hölldobler & Wilson 1990, p. 268). The ant uses tropotaxis, comparing stimulus inputs from left and right antennae (Fig. 10.3) (Hangartner 1967; Calenbuhr & Deneubourg 1992; Calenbuhr *et al.* 1992). However, when one antenna is removed the ant can still follow the trail, presumably by klinotaxis, although it does overshoot more. Both tropotaxis and klinotaxis are implicated in different species of termites, which have evolved trail laying and trail following behaviour independently of ants (Leuthold 1975, cited in Calenbuhr *et al.* 1992). Termites respond to trails of sternal gland marks by olfaction, like the ants above, but termite responses to labial gland pheromone involve contact as does trail following by tent caterpillars (Chapter 7) (Roessingh *et al.* 1988; Kaib 1999).

Trail following is important for snakes and some lizards for finding mates and food. Snakes following a trail show a characteristic rapid tongue-flicking, touching the ground about once a second (Ford 1986). Like ants, snakes may use tropotaxis to follow the trail: the deeply forked tongue provides two sensory inputs for simultaneous comparison, just like a pair of

Fig. 10.3. (a) Normal trail following in the ant *Lasius fuliginosus* using both antennae; (b) with one antenna is removed, the ant repeatedly overcorrects on the opposite side; (c) with crossed antennae, the ant has much more difficulty in following the trail, but still manages, presumably using an alternative mechanism. Figure from Hangartner (1967).

insect antennae (Ford 1986; Schwenk 1994). Unlike ants they sample molecules directly from the trail on the substrate as well as from the air. However, snakes and lizards do not have taste buds on the tongue. Instead, as the tongue goes back into its sheath, the tongue tips brush separate pads on the floor of the mouth which are then raised to take the molecules into the paired vomeronasal organ (VNO) openings in the roof of the mouth (Chapter 9). When both VNO openings are blocked, snakes cannot trail and when the VNO opening on one side is blocked, snakes tend to turn towards the unblocked side in a circus motion characteristic of tropotaxis (Waters 1993).

Male garter snakes (*Thamnophis sirtalis parietalis*) spread the tongue tips widely to increase the chance of detecting a chemical gradient in a single flick for tropotaxis (Ford 1986). As long as both tips touch the trail the snake

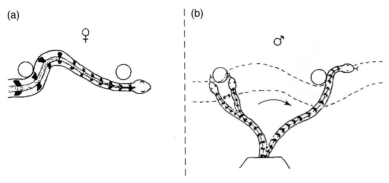

(a) ♀

(b) ♂

Fig. 10.4. Hypothesised mechanism for male garter snakes (*Thamnophis sirtalis parietalis*) to determine direction from a female snake's trail. (a) As the female garter snake pushes against objects (here, pegs) to move herself forward, she leaves traces of skin pheromone on their anterior surface. (b) Behaviour of a male garter snake when he encounters a female trail. The male tongue-flicks both sides of the pegs, detecting the pheromone on the anterior surface. He then proceeds in this direction, using chemotactic orientation on the non-directional pheromone trail she left on the substrate (dashed lines). Figure from Ford & Low (1984).

moves directly along it. If one tongue tip touches outside the trail, the snake swings his head back towards the trail before the next tongue flick. If both tips leave the trail, the snake stops and swings its head from side to side, in behaviour more like klinotaxis, tongue-flicking until contact is made with the trail again. All snakes, but only some lizard, have forked tongues. Lizard species that follow scent trails to hunt prey have forked tongues but sit-and-wait predatory lizard species do not (Schwenk 1994).

10.4.2.1 Trail orientation: which way to go

Few animal species can detect the direction of a trail. Tropotaxis does not provide any information on which direction the trail goes, as the longitudinal gradient would be too small to detect. There is no evidence that ant trails are polarised or that ants can determine which way is home from the pheromone trail alone (Hölldobler & Wilson 1990, p. 269–271). Instead, for ants the most important cues to the direction of home along the pheromone trail are probably visual, by light compass or landmark orientation for example (Wehner 1997). Some snails may produce polarised trails but the mechanism is not known (Wells & Buckley 1972). Planktonic copepods produce polarised trails (see Box 10.3).

Snakes seem to be able to determine trail direction. Male garter snakes can correctly follow the direction of a female's trail but only if she has moved on a rough surface, not a smooth one (Ford & Low 1984). The male can detect which side of the pegs the female pushed against while moving forward (Fig. 10.4). Although males can follow female trails left on a smooth paper surface, they are just as likely to go in either direction.

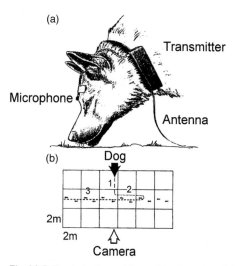

Fig. 10.5. Dog behaviour while tracking footprints. (a) The position of the microphone and transmitter while the dog was tracking. (b) How the footprints were placed in the 2 × 2 m grid, and one example of a dog's path during the searching phase (1), the deciding phase (2) and the tracking phase (3). Only 18 of the 40 squares are shown. Figure from Thesen *et al.* (1993).

The ability to detect the direction of track would also be very useful to predators. Tests of modern dog breeds suggest that they still have this ability (Thesen *et al.* 1993). Four trained German shepherd dogs always followed a person's track in the correct direction (Fig. 10.5). The tracking behaviour was divided into three phases. During the *searching* phase (equivalent to ranging) they moved quickly, sniffing 10–20 times per breath. Once the dogs had found the trail, in the *deciding* phase, they sniffed for a longer period and slowed down, sniffing at two to five footprints. In the *tracking* phase, once the direction had been established, the animals moved quickly. If odours are their cue, remarkably the dogs are detecting a difference in the concentration of scent in the air above two consecutive footprints, made 1 second apart, up to 20 minutes earlier (Thesen *et al.* 1993).

10.4.3. Orienting to an odour plume in air or water

The majority of pheromone communication probably occurs outside the smooth boundary layers of still air or water in the millimetres next to the surface, and quite beyond the physical scale at which diffusion dominates (Fig. 10.6) (Denny 1993; Vogel 1994). Far from being a smooth concentration gradient from the source, odour plumes are turbulent and unpredictable (Fig. 10.7). Gradients along plumes are too small to detect so animals cannot evaluate progress towards an odour source by the increase or decrease in odour concentration. Instead, the majority of signals have to be deciphered from a turbulent flow of molecules carried between animals in wind or water currents. Nonetheless, despite the formidable difficulties of following

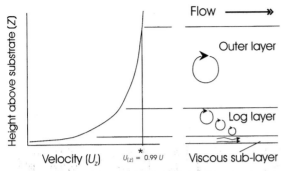

Fig. 10.6. Boundary layers and turbulence. The boundary layer is defined, arbitrarily, as extending to the height where the flow velocity (U_z) reaches 99% of the free-stream velocity (U), marked with a star on the x-axis. The flow profile of a turbulent boundary layer is qualitatively depicted in the right panel, which shows the generation of large eddies in the outer layer, momentum transfer and subsequent generation of smaller eddies in the log layer, and laminar flow in the viscous sub-layer. Turbulent boundary layers can develop over any surface immersed in flow. Figure from Weissburg (1997).

Fig. 10.7. Side view of the turbulent flow of a dye plume in sea water at 1 cm/s over gravel. Scale bar 2 cm. Figure from Weissburg & Zimmer-Faust (1993).

the track of a fluctuating meandering pheromone plume, animals *can* find the odour source.

10.4.3.1 Plume structure – what are the available stimuli?

Turbulent odour plumes form as air or water currents disperse odour molecules from their source. Odour plumes are of course normally invisible but swirling smoke clouds from a chimney provide a good visual analogy of the important features (Murlis *et al.* 1992). The smoke forms a meandering cloud that snakes downwind. If you get closer, you can see the fine-scale structure within the clouds, with filaments of high concentration interspersed with cleaner air.

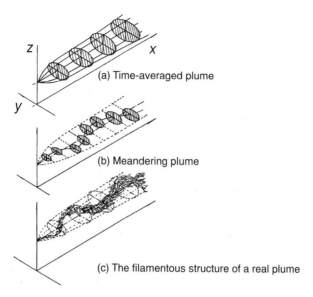

Fig. 10.8. Models of odour plumes have changed in recent years, reflecting increasing understanding of actual plumes. The first models proposed (a) a time-averaged Gaussian plume. A more accurate model (b) includes the meandering caused by changing wind direction in a plume model with concentration in each disc distributed normally about the meandering centre line. The structure of a real plume (c) in air or water has a filamentous structure of odour-filled filaments separated by clean medium. Figure from Murlis *et al.* (1992).

As a cloud of odour molecules moves from the source, turbulence tears apart the cloud into elongated odour-containing filaments, each only a few millimetres wide, separated by 'clean' water or air (Fig. 10.8) (Murlis 1997). This fine filament structure is central to the responses evolved by orienting animals. The turbulent effects are greater than diffusion (which is comparatively slow) and an important consequence is that a plume is far from a uniform cloud of pheromone drifting downwind, rather it is composed of filaments that remain relatively concentrated. Thus the pheromone concentrations within the filaments will be above the response threshold much further downstream than a diffusion model would predict – but in a spreading plume, far downstream, the odour filaments may be widely spaced.

Two characteristics of odour plumes combine to give the intermittent signal that presents so many challenges to a receiver downstream. The first is the fine filament structure and small-scale eddies. The second is the changing current direction, on a scale of seconds or minutes, which leads to the plume wandering across the landscape (made visible in field experiments with neutrally buoyant balloons or soap bubbles). The result is that a detector (or animal) at a point downstream receives a highly intermittent signal with two timescales of variation: of minutes or seconds with signal, and as much as 80% of the time without, as the plume meanders over the

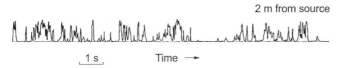

2 m from source

1 s   Time →

Fig. 10.9. As an odour plume meanders across the landscape, it will sometimes pass over an animal at a particular spot but moments later will have meandered away. The structure of plumes has been investigated by using tracers and detectors, here ionised air and a detector 2 m downwind of the source. When the plume hits the detector there are rapid fluctuations in signal as individual filaments are detected, followed by silence as the plume meanders away. Figure from Murlis & Jones (1981).

Fig. 10.10. In water, the neurotransmitter dopamine has been used as the tracer, detected by an artificial sensor with the same dimensions as the sensory hairs (diameter 30 μm) on the lobster antennule shown here *in situ* (Atema 1996). Photograph by J. Atema.

detector and away again; and on a finer scale of milliseconds while the plume is over the detector, as individual odour filaments in the plume touch the detector (Fig. 10.9). Because of water's greater viscosity, relevant scales of turbulence for plumes in water are orders of magnitude smaller than in air (Vogel 1994; Weissburg 1997). Incidentally, the turbulence in air or water makes it unlikely that *releasing* pheromone in pulses, as one moth species does (Conner *et al.* 1980), could be detected any distance downwind, although the pulses could theoretically be detected further away (Bossert 1968). The latter may be the reason why goldfish females release pulses of urine when attracting males (Appelt & Sorensen 1999).

Detecting the fine structure of odour plumes at the spatial and temporal resolution relevant to animals has been a major challenge to scientists. Animals are many orders of magnitude more sensitive than the most sensitive instruments, with animal sensory cells responding at a millisecond frequency. The way round this is to add an artificial tracer that can be detected by physical or chemical instruments (Figs. 10.9 and 10.10) (Finelli *et al.* 1999).

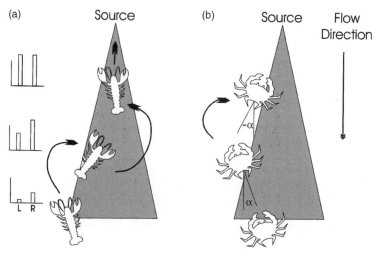

Fig. 10.11. Lobsters in turbulent environments, and estuarine crabs in smooth-flowing creeks, use different mechanisms to orient up odour plumes (shaded) to find the source. Lobsters (a) use tropotaxis, simultaneous bilateral comparisons of odour intensity to turn into the plume, and sequential comparison of odour intensity, klinotaxis, to progress towards the source within the plume. Bar diagrams show the outputs of left and right sensors at each position. The plume is supposed to be more like the turbulent plume in Fig. 10.7. An estuarine crab (b) shows positive rheotaxis to the flow and uses a binary comparison of odour presence–absence to determine that it has left the plume. The crab judges its exit angle (α) relative to the flow direction and uses this information to re-enter the plume at the same angle at which it exited the plume. Figure from Weissburg (1997).

Advances in electrical recording from insect antennae (Chapter 2) in the field confirm that the insect does receive the kind of intermittent signal predicted by tracer studies (van der Pers & Minks 1997).

10.4.3.2 Animals moving in contact with the substrate

Walking and crawling animals in many habitats (on land as well as in water) orient to odour plumes (Bell 1991). As the molecules move downstream in the flow, going upstream will bring you towards the source. Animals in contact with the substrate can detect deflection of their mechanoreceptors by the current, and then respond with rheotaxis (rheo = current) in water and anemotaxis (anemo = wind) on land. Some species can use a chemotactic response to odour, via klinotaxis or tropotaxis, as well as or instead of rheo- or anemo-taxis triggered by presence of odour. Orientation by lobsters, which live in turbulent rocky-shore habitats, and by estuarine crabs in smooth-flowing creeks, provide contrasting examples. Neither seems to use self-steering (idiothetic) responses, which are conspicuous in insect flight up plumes (Section 10.4.3.3).

American lobsters (*Homarus americanus*) orient up odour plumes by bilateral inputs from their pair of antennules in part by tropotaxis (comparing left and right inputs) (Fig. 10.11); with only one antennule they can still locate an odour

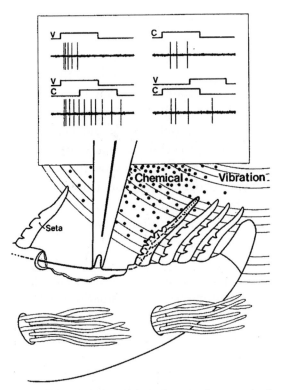

Fig. 10.12. Extracting orientation information from complex flows may be more effective if two senses are used together: for example, 'eddy-rheo-chemo-taxis', using eddies detected simultaneously by chemo- and mechano-receptors (Atema 1996, 1998). Crayfish leg sensory hairs show just such multisensory integration. Recordings from a sensory-hair nerve show the action potentials (spikes) with different combinations and orders of stimuli. The steps in the traces above the oscillograms show the start and end of chemical (C) and vibratory stimuli (V). When the vibration comes before the chemical stimulus, more spikes are evoked than when these stimuli are presented individually. When the sequence is reversed, the chemical stimulus inhibits responses to the vibratory stimulus. Figure after Hatt & Bauer (1980) in Stein & Meredith (1993).

source but the process takes much longer, and the path is convoluted (Atema 1995). When not engaged in chemotactic search, lobsters walk twice as fast.

What cues in the plume might the lobsters be using? Grasso *et al.* (1996) used a pair of artificial sensors 3 cm apart (the separation of the antennules), to investigate the characteristics of plumes (Fig. 10.10). Odour patches hit one sensor about 0.5 seconds before the other, giving the direction the plume was moving. The sensors also revealed a pattern of concentration gradients and micro-flows *within* the patches that varied predictably in different parts of the plume and nearer the source. The lobsters could be using the internal fine structure of odour plumes as cues to finding the odour source by exploiting 'eddy rheo-chemo-taxis' (Fig. 10.12). One way of testing these predictions is to give real lobsters 'virtual-reality chemical goggles', little pumps which deliver precise odour pulses to each antennule. The experimenter can guide the lobster, 'by the nose', up the virtual plume (Atema 1996).

Lobsters do not seem to use rheotaxis, perhaps because the flow patterns of their wave-swept rocky-shore habitats are too complex (Atema 1995). However, in contrast, blue crabs (*Callinectes sapidus*) in estuarine creeks do use odour-triggered rheotaxis to orient up odour plumes (Zimmer-Faust *et al.* 1995; Weissburg 1997; Finelli *et al.* 2000) (Fig. 10.11). In the tidal creeks the odour plumes revealed by fluorescein dye showed sharp boundaries, although, within the plume, finer-scale measurements show the characteristic intermittent signal from filaments of odour. Crabs contacting the odour plume turned up-current, guided by mechanoreceptors on the antennae, using positive rheotaxis switched on by the odour stimulus. However, unlike insects (which use self-steered control of lateral movements), if crabs wandered laterally across the current, they kept within the plume by chemotactic responses, turning back if the chemoreceptors on some legs hit clean water. Such responses are possible because the odour plumes in the estuarine creeks are stable and have sharp boundaries, in contrast to the more turbulent and unpredictable odour plumes to which lobsters and flying insects must orient.

10.4.3.3 Swimming and flying animals

Flying or swimming animals orienting up an odour plume cope with even greater problems than animals in contact with the substrate because the animal floating in mid-current cannot use mechanoreceptors to know if it is making progress up-current – their entire measurable world is drifting with them. Imagine swimming in a river: with your eyes shut you cannot tell if you are making any progress against the current; only seeing landmarks on the bank lets you do this. In a similar way, moths and other animals need visual feedback: for the moth, seeing the movement of the ground below tells it how much it is progressing upwind (optomotor anemotaxis). How animals orient in mid-water ocean depths, with no visual feedback, is a mystery, but one hypothesis is that animals could exploit 'eddy rheo-chemo-taxis' (Section 10.4.3.1) (Dusenbery 1992; Atema 1996).

Swimming and flying animals orienting to an odour source show similar characteristic zigzag tracks, whether they are salmon following the odour of their home stream, seabirds flying upwind to locate food, or male or female moths flying upwind to an odour source (Fig. 10.13) (Arbas *et al.* 1993; Nevitt 1999a). However, these zigzag tracks may reflect superficially similar but different behavioural solutions, evolved independently, to the problems posed by orientation to plumes (Vickers 2000).

The best-studied responses of flying animals to pheromones are undoubtedly those of male moths responding to female pheromones (Baker & Vickers 1997; Cardé & Mafra-Neto 1997; Vickers 1999). The turbulent, filamentous

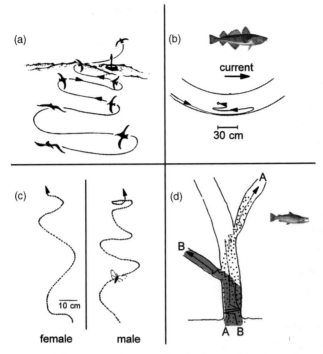

Fig. 10.13. Swimming or flying animals orienting up odour plumes show similar zigzag tracks (all viewed from above). However, they may be using different mechanisms. Figure after Arbas *et al.* (1993).
(a) Diagram of the flight pattern of procellariiform birds, such as albatross, approaching a source of food odour. Original figure from Hutchison & Wenzel (1980).
(b) A fish, the cod (*Gadus morhua*), swims down current until it detects food odour released from the pipe (o) and doubles back against the current to find the odour. Original figure from Pawson (1977).
(c) Flight tracks of male and female tobacco horn moths (*Manduca sexta*) flying upwind to odour sources (for the male, the source was female sex pheromone; for the female, a tobacco leaf). After original figure from Willis & Arbas (1991).
(d) Salmon imprinted on their separate home-stream odours (represented by filled dots in A and shading in B) swam upstream. Near the mouth of the stream, the fish swam from bank to bank, before positive rheotactic swimming (upstream) where the odour was distributed bank to bank, and zigzag swimming where both scented and unscented water plumes existed. Original figure from Johnsen (1981).

nature of odour plumes plays a crucial role in upwind orientation, as predicted by Wright (1964). Having evolved in response to turbulent plumes, moth orientation behaviour now depends on it: they are unable to progress upwind in a uniform pheromone cloud (Kennedy *et al.* 1981).

Upwind orientation by moths is the product of two behaviours, both stimulated by pheromone: first, a brief surge of upwind flight on contact with a pheromone filament (Baker 1990; Kaissling & Kramer 1990) and, second, a lapse into a pattern of crosswind flight, known as casting (self-steered counter-turning) when the filament is ended (Baker 1990). Upwind flight is thus made up of repeated sequences of surge and cast as the moth encounters the filaments of the plume (Fig. 10.14) (Baker 1990). If the male

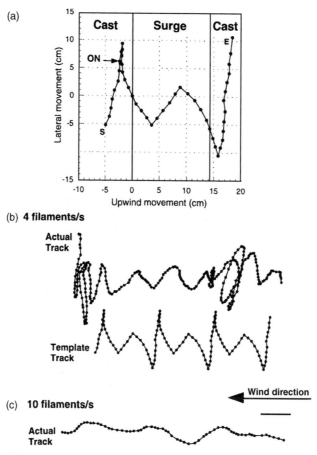

Fig. 10.14. Upwind flight of moths can be explained by short surges to contacts with pheromone filaments and casting to clean air. (a) The cast-surge-cast template derived from the average track of several males responding to a single pulse of pheromone.
(b–c) Bird's-eye-view of tracks of a male moth (*Heliothis virescens*) in response to plumes consisting of pulses (filaments) at two different frequencies. Intervals between dots along a track are 1/30 s.
(b) The upper trace is the flight track of a male responding to a plume generated at a threshold frequency of 4 pulses (filaments)/s appears to be comprised of reiterated individual responses to encounters with single filaments of pheromone. Below is a theoretical track derived from multiple iterations of the cast-surge-cast template (a), linked end to end.
(c) With higher pulse (filaments) rates the track is almost straight as contact with clean air between pheromone filaments occurs too briefly to trigger casting.
Figure from Vickers (1999) after Vickers & Baker (1994).

encounters the next pheromone-laden filaments before he has switched to casting behaviour, a straighter flight will result. Different species have different times before switching to casting. Species that switch quickly, such as *Grapholitha molesta* (Baker & Haynes 1987) will have zigzagging flights, whereas species slower to switch, such as *Heliothis virescens* (Vickers & Baker 1996), tend to have straighter flights because they are more likely to hit the next pheromone filament before switching.

The casting behaviour increases the chance of regaining contact with the shifting plume. The casting turn on loss of pheromone is self-steered, *not* chemotaxis up a chemical gradient at the edge of the pheromone plume (Kennedy 1992, p. 144–149). Male moths are not comparing inputs from their paired antennae: with only one antenna they can still fly up the centre of an odour plume just as well (Vickers & Baker 1991).

Male moths make split-second orientation responses based on both the fine temporal structure and the chemical content of filaments in pheromone plumes (Vickers *et al.* 1998). Male moths can fly up a plume made up of alternate puffs of their own species' sex pheromone and puffs of the pheromone of a sympatric species, even if the puffs are separated by just 1 mm and by at most 0.001 seconds (Baker *et al.* 1998). This is achieved by the male having the sensory cells for key components of his own species' pheromone *and* sensory cells for antagonist components (characteristic of sympatric species, Chapter 3) in the same sensory hairs on the antenna (Chapter 9): the male will only make a surge upwind to a filament containing the compounds that it should. This is how male moths find their female despite the chemical cacophony: even though plumes from females of different species will be intermingling at a coarse scale, at a fine scale the filaments remain separate (Liu & Haynes 1992). Female moths seem to fly upwind to plant odours using the same mechanisms as males use towards sex pheromones (Fig. 10.13) (Willis & Arbas 1991).

Throughout the rest of this book there are numerous examples of walking, swimming, and flying vertebrates that find mates or food by orienting up odour plumes. However, there are few detailed studies that probe the mechanisms that vertebrates may be using. Tantalising clues that vertebrates can respond to odour filaments at the fine scale shown by invertebrates comes from preliminary work on chemo-orientation to odour plumes in young American eels (*Anguilla rostrata*) (Oliver *et al.* 1996). The eels were positively rheotaxic, swimming up the flow, but when they contacted odour filaments (made visible to the human observer by red dye), they turned along the filament even if this took them across the main current.

One of the most exciting and challenging future research areas in olfaction will be to investigate vertebrate orientation mechanisms. Are vertebrates big moths or do we do something quite different?

10.5 | Conclusion

The olfactory world is a cacophony of overlapping chemical signals yet animals have evolved exquisite ways of extracting information that will lead them to mates, food and other resources. Ranging behaviour is the initial stage of searching and brings the animal into first contact with the active space, in which it can detect the signal. Mechanisms for orienting to the olfactory sign include **taxis** (guided by gradients) and **kinesis** (not directly guided).

Within either air or water, animals may respond to three very different kinds of odour stimulus: short range diffusion, a trail or a plume. The orientation of most animals occurs in turbulent plumes. Walking animals can detect currents and orient up them. Flying and swimming animals need visual or other feedback to orient upstream. Zigzag tracks of animals working up an odour plume are very similar across the animal kingdom.

10.6 | Further reading

Dusenbery (1992) provides a very good account of chemical signals, ranging and searching behaviour. Kennedy (1992) has an interesting view on the design and interpretation of experiments on orientation behaviour. Schöne (1984) and Bell (1991) review orientation and searching behaviour in general. Vogel (1994), Schneider & Moore (2000) and Denny (1993) discuss the flow of fluids and the effect this has on chemical signals carried in water or air. A special issue of the *Biological Bulletin* (Volume 198, Issue 2, April 2000) was devoted to chemical communication in ecology.

Chapter 11

Breaking the code:
illicit signallers and receivers of semiochemical signals

11.1 | Introduction

Potentially, any pheromone or chemical recognition system described in the other 12 chapters in this book could be exploited by other organisms, whether conspecifics or predators and parasites. Exploitation can take many forms, from eavesdropping on a pheromone signal, for example spiders (*Habronestes bradleyi*) attracted to fighting ants by ant alarm pheromone (Allan *et al.* 1996), to the active producion of chemicals that mimic those of a prey species, such as bolas spiders, which produce moth sex pheromones to lure male moths within striking distance. However, not all relationships involve exploitation: mutualistic relationships also involve chemical cues (see Table 11.1 for explanation of the terms used for odours in these different contexts).

Given the subtleties revealed in other chapters about chemical communication *within* species, we should expect that interspecific interactions will be no less extraordinary. Many of the examples are of insect interactions, particularly those between ants and their guests and parasites. This is largely because we know most about chemical communication in insects. Investigation of the chemical ecology of other animals in similar detail is likely to reveal a near ubiquitous role of chemical cues in other interspecific relationships. Chemical detection and interaction of partners and enemies may turn out to be the rule rather than the exception.

The attraction of invertebrate predators such as parasitoid wasps or predatory mites by semiochemicals produced by plants under insect or mite attack has excited great interest. Excellent reviews of these 'tri-trophic' systems can be found in recent sources so are not covered here (Dicke & Sabelis 1992; Turlings *et al.* 1995; Pickett 1999).

The design features of olfactory systems, described in Chapter 9, make evolution of interspecific eavesdropping of olfactory cues ('code breaking')

Table 11.1. *Classification of semiochemicals by cost and benefit to emitter and receiver*

Semiochemical[a]	Effet on emitter	Effect on receiver	Communication
Allomones	+	−	deceit, propaganda
Kairomones[b]	−	+	eavesdropping
Synomones	+	+	mutualism

[a]Definitions of allomones, kairomones and synomones are given in Chapter 1.
[b]A chemical used intraspecifically as a pheromone is termed a kairomone when eavesdropped by a predator.
After Alcock (1982).

both possible and likely. These characteristics include a basic broad sensitivity to any odour, including novel ones, and the ability to learn or evolve responses to any odour that provides selective advantage. Some specialist eavesdroppers may be as, or more, sensitive to the pheromone as the legitimate receiver. Production of chemicals to mimic those used by prey species is facilitated by the basic biochemical machinery that all animals share by common descent and which can evolve to produce new end products.

Like other interactions between organisms, eavesdropping and deception involving chemical communication can create an arms race, with strong selection pressures leading to changes, for example, in pheromone blends and the behaviour of the exploited organism, followed by counter-evolution by the exploiter (Table 11.2) (Zuk & Kolluru 1998; Haynes & Yeargan 1999).

11.2 | Eavesdropping

Any broadcast signal can be intercepted by illicit, eavesdropping receivers.

11.2.1 Egg dumping – parasitism or mutualism?
In some insect and fish species with parental guarding of eggs, the odours of recently deposited eggs or semen are exploited by other animals as a cue to add their own eggs to the pile. This leads to alloparental care in which parents look after eggs and young that are not their own.

An example of this is intraspecific egg dumping by females of the aubergine lace bug *Gargaphia solani* (Monaco *et al.*, 1998). In this species, the female stands over her eggs and defends these and first-instar nymphs from predators (see Chapter 8). The odour of newly laid eggs attracts other females to approach and add their eggs, which the mother allows.

Another example includes the numerous instances of alloparental care in fish; this is so common that it usually occurs whenever parental care has

Table 11.2. *Predator and parasitoid adaptations for exploiting sex pheromones of victim species*

Victim's signal	Exploiter	Exploiter adaptation	References
Female sex pheromone	Parasitoid wasps on moth eggs (*Trichogramma evanescens, T. pretiosum*)	Arrestment of flight in presence of prey pheromone (more advantageous than flying towards pheromone because pheromone is not exactly where the eggs are); preferential searching for prey eggs on underside of leaves, where they are deposited	(Noldus *et al.* 1991)
	Braconid wasps (*Praon volucre, P. abjectum, P. dorsale*)	Ability to recognise sexual female aphids, which may be the last chance for parasitoid to find suitable host for overwintering	(Hardie *et al.* 1991)
Aggregation pheromone	Clerid beetle (*Thanasimus dubius*)	Ability to recognise a variety of prey pheromone blends (expressed as high local variation in response to blends)	(Herms *et al.* 1991)
	Clerid beetle (*Thanasimus formicarius*)	Antennal olfactory sensory neurons (OSNs) as sensitive as prey OSNs to prey pheromone (see Chapter 9)	(Hansen 1983)
Male sex pheromone	Vespid wasp (*Vespula germanica*)	Switch from olfactory detection in morning when victim lekking peaks, to visual detection later in day when victim females are ovipositing	(Hendrichs *et al.* 1994)

After Zuk & Kolluru (1998).

evolved in fish (see review by Wisenden 1999). One well-studied association is between the green sunfish (*Lepomis cyanellus*) and the redfin shiner (*Notropis umbratili*) which deposits its eggs in the nests of male sunfish. Chemicals are important proximate cues. Hunter & Hasler (1965) were able to attract male and female shiner fish with milt (semen) and ovarian fluids of the sunfish host. If predators are around, sunfish nests with shiner nest associates produced four times as many sunfish fry as those without (Johnston 1994). Other experiments show that the benefit is mutual, as shiner nest associates benefit from parental care by the hosts.

A common feature of both the *Gargaphia* insect and the fish systems is that the young are precocial (independent from birth and able to feed themselves) so looking after additional eggs or young is not costly. Instead, in these species egg dumping could be thought as a mutualism, since the additional eggs create a protective buffer zone around the *Gargaphia* female's own eggs (Hamilton 1971). Once the eggs hatch, the dilution effect will reduce the individual risk of loss to predators for the guarder and the dumper alike.

11.2.2 Predators and parasitoids

Some of the first examples of chemical eavesdropping were discovered by accident, when traps baited with bark beetle pheromones attracted predators and parasitoid wasps as well as the intended bark beetles. The predators use the bark beetle pheromones as **kairomones** to find their prey (Table 11.1). A kairomone is a chemical emitted by an organism which attracts exploiters of another species. Any kind of pheromone can be eavesdropped. Surprisingly, there are some rare examples of pheromones being eavesdropped by visual cues. The protein-rich urine marks of voles are conspicuous in the ultraviolet part of the spectrum visible to birds: the kestrel *Falco tinnunculus* and the rough-legged buzzard *Buteo lagopus* may be able to use the vole scent marks as a cue when searching for profitable hunting and breeding areas (Viitala *et al.* 1995; Koivula & Viitala 1999).

11.2.2.1 Sex pheromones

Many predators and parasitoids use their victims' sex pheromones to locate them (Zuk & Kolluru, 1998). For example, even we can smell the pheromone coming from leks of Mediterranean fruit fly males (*Ceratitis capitata*) and yellowjacket wasps (*Vespula germanica*) use this odour to locate them (Hendrichs *et al.* 1994; Hendrichs & Hendrichs 1998).

Parasitoid wasps that parasitise moth eggs use female moth pheromones as an indirect way of locating oviposition sites (as female moths tend to call from on or near the host plant). Moths lay at night and many parasitoids hunt during the day. Adsorption of pheromone on the plant by the calling moth at night provides a 'bridge-in-time' when it is slowly released during the day (Noldus *et al.* 1991). The parasitoid wasp *Telenomus euproctidis* has a more direct way of finding the eggs. Having located the moth female by her pheromone, the parasitoid hitches a ride on the moth until the moth female oviposits. The parasitoid then jumps off and parasitises the newly laid eggs (Arakaki *et al.* 1996). *T. euproctidis* also demonstrates that eavesdropping evolves in response to local opportunities: populations of the wasp in different parts of Japan use different, allopatric species of moth as hosts and respond only to the sex pheromones of the moth species common at that locality (Arakaki *et al.* 1997).

Predator and parasitoid pressure can lead to changes in sex pheromone blend and calling behaviour. The southern green stink bug *Nezara viridula*, for example, which is native to the Old World but is now a cosmopolitan pest, produces male sex pheromones with different blends in different places (Aldrich 1995). The ratio of *trans*- to *cis*-(Z)-α-bisbolene epoxide from Japanese *N. viridula* is roughly 1:1 but populations from Mississippi, California, Europe, Brazil and Australia use ratios of 2–4:1. Some of the changes in North American populations may be due to selection pressure from the tachinid parasitoid fly *Trichopoda pennipes*.

Another possible evolutionary response to parasitoid pressure is a change in calling behaviour. Males of the spined soldier bug (*Podisus maculiventris*) release pheromones to attract females, but unfortunately these pheromones also attract specialised tachinid parasitoid flies that lay eggs on the male (Aldrich 1995). Unlike most other bugs, *P. maculiventris* males have pheromone glands that can be closed (perhaps evolved in response to parasite pressure). Some *P. maculiventris* males, attracted to the pheromone released by calling males, pursue a 'satellite male' strategy, staying 'silent' themselves and attempting to intercept females responding to the calling male (Aldrich *et al.* 1984). Silent satellite males may get fewer matings but they may also be parasitised less.

11.2.2.2 Territorial marking pheromones

East African klipspringer antelope (*Oreotragus oreotragus*) mark their territories with a resinous secretion from their antorbital gland, smeared onto twigs (Chapter 5), and *Ixodes neitzi* ticks aggregate on these territorial marks (Rechav *et al.* 1978). The ticks respond to phenolics, leached out of the scent marks by rain, by climbing up the twig. For parasites, territorial marks are an ideal place to wait as the territory owner will return repeatedly to renew the mark and conspecifics will also approach the marks to sniff them. Responses by North American ixodid ticks to pheromone secretions of their deer hosts are the likely cause of clustering of ticks along deer trails (Carroll *et al.* 1996).

11.2.2.3 Alarm pheromones

In Central America, female phorid flies (*Apocephalus paraponerae*) are attracted to fighting or injured individuals of their host, the giant tropical ant *Paraponera clavata* (Feener *et al.* 1996). The flies, which lay their eggs in the ant's head (which eventually falls off, hence the common name for these phorids of 'ant-decapitating flies'), are attracted by the ants' mandibular secretions, which contain 4-methyl-3-heptanone and 4-methyl-3-heptanol. Male flies are also attracted by these compounds as they are assured of finding females on the ant battleground.

Eavesdropping may be one reason for the evolution of fish alarm pheromones that are released when the victim is already in the predator's jaws. The alarm pheromone could attract other predators whose arrival interferes with the first predator, giving the victim a chance to escape (Chapter 8).

11.2.2.4 Aggregation pheromones

Host tree volatiles and aggregation pheromones released by the first bark beetles arriving on a suitable tree attract conspecifics, leading to an accelerating mass attack of the tree (Chapter 4). These pheromones are eavesdropped both by other bark beetle species and by a host of natural enemies (Fig. 11.1) (Birch 1984; Borden 1985). Competing bark beetle species may turn back. For example, whichever of two *Ips* species (*I. paraconfusus* and *I. pini*) arrives first in effect claims the whole tree with its pheromones (see Chapter 12). Individuals of either species have lower reproductive success in mixed than in pure aggregations.

Eavesdropping responses by predatory beetles and parasitoid wasps and flies can be dramatic: for example, 86 000 predatory beetles (*Temnochila chlorodia*) along with 600 000 of their prey *Dendroctonus brevicomis*) were caught in traps baited with the *D. brevicomis* aggregation pheromone (Wood 1982). This intense predation pressure may select for pheromone blend changes (Aldrich 1999) (and examples such as these highlight the potential for resistance to the control of moth pests using pheromones for mating disruption, Chapter 12). The pine engraver (*Ips pini*) has two main pheromone races with different male blends, the 'New York' and 'California' races. While the divergence in *I. pini* pheromone races may be largely due to separation into east and west coast populations during the Pleistocene ice age (Cognato *et al.* 1999), local predator pressure may play a part. In Wisconsin (which has the 'New York' east coast race, which uses a 1:3 blend of (R)-(−) and (S)-(+)-ipsdienol), Raffa & Klepzig (1989) found that the common local predatory beetles (*Thanasimus dubius* and *Platysoma (Cylistix) cylindrica*) were most attracted to the reverse ratio (Fig. 11.2a). In California, where *I. pini* use > 90% (R)-(−)-ipsdienol, the local predatory beetle *Enoclerus lecontei* was similarly attracted to the 'out of town-blend' (Fig. 11.2b) (Raffa & Dahlsten 1995). In each place it seems that predation has selected for *I. pini* with a blend that offers a temporary escape from the predators, in a co-evolutionary arms race.

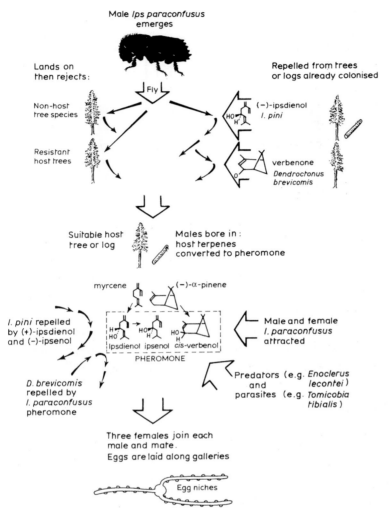

Male *Ips paraconfusus*
emerges

Lands on
then rejects:

Non-host
tree species

Resistant
host trees

Fly

Repelled from trees
or logs already colonised

(−)-ipsdienol
I. pini

verbenone
*Dendroctonus
brevicomis*

Suitable host
tree or log

Males bore in :
host terpenes
converted to pheromone

myrcene   (−)-α-pinene

I. pini repelled
by (+)-ipsdienol
and (−)-ipsenol

Ipsdienol ipsenol *cis*-verbenol
PHEROMONE

Male and female
I. paraconfusus
attracted

D. brevicomis
repelled by
I. paraconfusus
pheromone

Predators (e.g. *Enoclerus
lecontei*)
and
parasites (e.g. *Tomicobia
tibialis*)

Three females join each
male and mate.
Eggs are laid along galleries

Egg niches

Fig. 11.1. Almost every behaviour in the adult stage of the bark beetle *Ips paraconfusus* is mediated by pheromones. Some of the interactions are intraspecific whereas others are the responses of competing bark beetle species and predators and parasitoids. Figure from Birch (1984).

It is revealing that predators are not turned back by the changes in pheromone blend (late in the bark beetle aggregation on a tree) that signify to conspecific bark beetle males that the tree is 'full' and it is not worth their while to continue arriving. For example, *Dendroctonus frontalis* males are inhibited by the male-produced verbenone and *endo*-brevicomin 'anti-aggregation' pheromones (Aldrich 1999) but the predator *Thanasimus dubius* continues to arrive undeterred – a 'full' tree offers more to eat (Salom *et al.* 1992).

By definition, aggregation pheromones attract conspecifics of both sexes, which distinguishes them from sex pheromones that only attract the

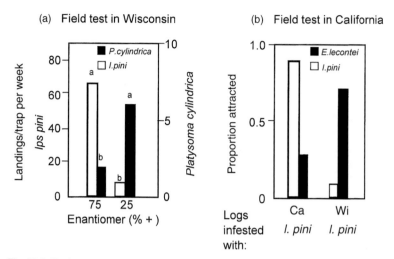

Fig. 11.2. Predation pressure may select for local populations of prey that use pheromone blends not preferred by local predators, leading to temporary escape from high predation. This was revealed in bark beetles by testing predator and prey responses to local and 'out-of-town' blends of prey pheromones.

(a) Responses of the pine engraver bark beetle (*Ips pini*) and the predatory beetle *Platysoma cylindrica* to chiral blends of ipsdienol. Figure after Raffa & Klepzig (1989) in Aldrich (1999).

(b) Proportions of *I. pini* and the predatory beetle *Enocleris lecontei* from California (Ca) that were attracted to logs infested with *I. pini* from California versus logs infested with *I. pini* from Wisconsin (Wi). Figure after Raffa & Dahlsten (1995) in Aldrich (1999).

opposite sex. However, many aggregation pheromones may really be *intra*specific eavesdropping of sexual communication directed at the opposite sex but also responded to by the same sex looking for mates, as for the spiny soldier bug (Section 11.2.2.1, Chapters 3 and 4).

11.2.3 Prey responses to predator pheromones

Many species show escape or avoidance responses to the odours of their predators. For example, in streams, after sensing the odour of the predators, herbivorous mayfly (*Paraleptophlebia adoptiva*) nymphs take more evasive action towards predatory stonefly (*Acroneuria carolinensis*) nymphs (Ode & Wissinger 1993). Many vertebrates, from fish to snakes to mammals avoid the odours of their predators (Weldon 1990; Nolte *et al.* 1994). Any cues that give early warning of hunting predators will have survival value for prey but few responses are to odours with a demonstrated role in the predator's own intraspecific communication. However, two examples where this is the case come from social insects: Japanese honeybees respond to the giant hornet marking pheromone when it marks the bee nest for attack (see Chapter 8) and the stingless bee *Trigona angustula* reacts to the pheromones of the robber bee *Lestrimelliata limao* (Section 11.5).

11.3 | Chemical communication in mutualisms

Just as interactions within ant colonies are largely mediated by pheromones, mutualisms between ants and their 'domesticated animals' rely heavily on chemical cues. Chemical cues are also important in the symbiosis between anemones and anemone fish.

11.3.1 Ants and aphids, leafhoppers, scale insects

The mutualistic association between ants and honeydew-producing homopterans (aphids, leafhoppers and scale insects) is well known: the ants gain a rich sugar source and the homopterans benefit by defence against predators and parasitoids (Nault & Phelan 1984). The close associations that have evolved between some ant species and homopterans have led to changes in the behaviour of both partners. Whereas aphids of non-ant-associated species tend to drop from the leaf (to escape the predator) when they detect the alarm pheromone released by an aphid under attack (Chapter 8), aphids of ant-associated species remain feeding and rely on their ant guards to protect them (Nault *et al.* 1976). The suppression of the dropping response is even stronger if ants are present. For their part, the ants are attracted by aphid alarm pheromone, attacking any predator menacing their 'sheep'.

11.3.2 Ants and lycaenid butterfly caterpillars

Chemical communication is the key to the close relationship between ants and caterpillars, found in almost half of the 4000 or so species of lycaenid butterflies (Fiedler *et al.* 1996). The relationships range from mutualisms (the majority) in which both species benefit (the ants guard the caterpillars from parasitoids and predators in return for sugar and/or amino acid secretions (Pierce 1989)) to parasitism (see Section 11.6.2.2).

Adoption by ants is facilitated by the specialised skin glands of lycaenid caterpillars; which secrete adoption or appeasement substances. Ants antennate these glands, which secrete amino acids in some species (Pierce 1989) and mimics of ant brood pheromone in others (Fig. 11.3) (Thomas *et al.* 1989). The signal seems to be fairly general in some species as a single caterpillar may be attended by ants from different ant species belonging to different subfamilies.

Most lycaenid caterpillars also produce a second signal, from eversible tentacle organs that release compounds that seem to be ant alert or alarm pheromones, and stimulate the ants to activity (Henning 1983). The tentacles are most frequently everted when caterpillar–ant interactions start, when caterpillars are travelling or when they are disturbed, and these are

Fig. 11.3. Mature caterpillar of the South-East Asian lycaenid butterfly *Spindasis lohita*, tended by a worker of its specific attendant ant species *Cremogaster dohrni artifex*. The caterpillar offers a nutrient-rich secretion droplet from the dorsal nectar organ and simultaneously everts the whitish membranous tentacle organs which release a pheromone. Photograph © K. Fiedler.

the times when the caterpillar is likely to deliver a drop of liquid rich in sugar and/or amino acids from its dorsal gland to the attendant ants. Caterpillar behaviour seems aimed at regulating the number of attending ants (Axén *et al.* 1996).

11.3.3 Ants and fungus

Leaf-cutting ants (Attini) have an obligatory symbiotic relationship with their fungus. The ants bring selected leaves to the fungus garden (North *et al.* 1997). The fungus provides enzymes that break down plant tissue and possibly detoxify some insecticidal plant defence compounds, enabling the ants to use a wider range of plant species. However, plants also differ in their suitability for the fungus; intriguing results suggest that plant selection by foraging ants is influenced by the response of the fungus to the plants brought in earlier (North *et al.* 1999). When foraging *Atta* workers carried in orange peel with added fungicide, leading to 'stressing' of the fungus, the ants later avoided orange peel even if it was free of fungicide – they had learned to associate the odour of the orange with food that damaged the fungus – rather like rats, which avoid food that has made them ill previously.

11.3.4 Sea anemones and anemone fish

Coral reef anemone fishes are well known for their ability to live unharmed among the stinging tentacles of sea anemones (Fig. 11.4). The 28 species of

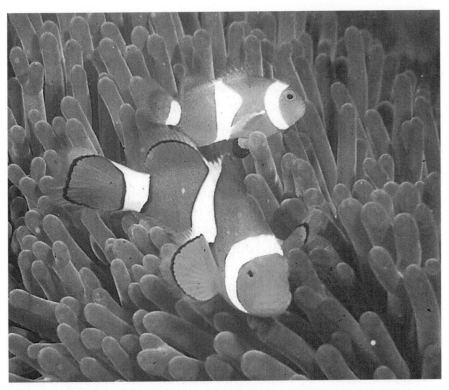

Fig. 11.4. Anemone fish *Amphiprion percula* and their anemone *Heteractis magnifica*, Lizard Island, Australia. Photograph by J. K. Elliott.

anemone fish (or clownfish: *Amphiprion* and *Premnas*) are obligate symbionts of 10 species of sea anemones (Fautin 1991; Elliott *et al.* 1999). The anemone fish species differ in their specificity: some are associated with only one anemone species, recognised by chemical cues (Arvedlund *et al.* 1999), but a few can live in any anemone species.

A protective mucous coat allows the fish to contact the tentacles of their host anemone without being recognised as 'non-self', and being stung (Elliott *et al.* 1994). However, whether the immunity comes from molecular camouflage, mimicry, or stealth (not having the sting trigger-compounds in their mucus) is not clear. Some anemone fish species appear to steal their identity from their host, like some ant guests (Section 11.6.2.1), by picking up anemone substances during the tentative 'acclimation' behaviour when they are first introduced to the anemone. However, there may be genetic factors involved as some anemone fish species can join their anemone without acclimation behaviour. Sea anemones also have symbioses with anemone shrimps, mediated by chemical cues (Guo *et al.* 1996).

11.4 | Deception by aggressive mimicry of sex pheromones

Illicit signallers can exploit the strong selection pressures on males to respond to sex pheromones. While some illicit signallers are conspecific males (see Chapter 3), the most spectacular examples are of exploitation by orchids and bolas spiders.

11.4.1 Pollination by sexual deception

Instead of producing nectar, many Mediterranean and Australian orchids use aggressive mimicry of the female sex pheromone of local solitary bees and wasps to lure males to attempt to mate with the flower, thus ensuring that they pick up the pollen packet (see reviews by Stowe 1988, Borg-Karlson 1990, Dafni & Bernhardt 1990) (Fig. 11.5). These associations have evolved multiply and independently on different continents. Each orchid species mimics receptive females of only one or a small range of pollinator species so that as the male hymenopteran continues his search for real females, he is likely to transfer the pollen to another flower of the same orchid species. When the orchids mimic the long-range pheromone released by female hymenopterans, the males fly a zigzag path from downwind. The deception is often completed by mimicry of contact pheromones and often also by visual and tactile cues. For example, Schiestl *et al.* (1999) showed that the female sex pheromone of the solitary bee *Andrena nigroaenea* is almost exactly mimicked by the European orchid *Ophrys sphegodes*, which relies on this deception for pollination (Fig. 11.6).

11.4.2 Bolas spiders

The females of some spider species, from at least two independent evolutionary lines, lure male moths by synthesising and releasing the moths' specific female sex pheromones (Stowe 1988). The majority are bolas spiders, which gain their name from their web, which is reduced to a sticky ball on a thread, held by one of their forelegs (Fig. 11.7). When a male moth is lured by the pheromone to approach the spider, she draws back this foreleg and swings the bolas at the moth. If the bolas makes contact, the moth rarely escapes.

Each bolas spider species specialises on a few species of moths (Yeargan 1994). For example, two noctuid moth species, one active early at night and the other late at night, account for more than 90% of the prey caught by the North American bolas spider *Mastophora hutchinsoni* (Haynes & Yeargan 1999). Each of these moth species uses completely different pheromones, yet individual *M. hutchinsoni* frequently catch both moth species on the

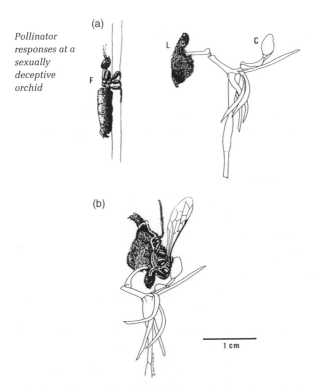

Pollinator responses at a sexually deceptive orchid

Fig. 11.5. (a) The flower of the Australian orchid *Drakaea glyptodon* (top right) mimics the look and pheromone of wingless female thynnine wasps *Zaspilothynnus trilobatus* (top left): C, column (the fused stamen and style bearing the pollinia); F, female wasp; L, the labellum, twisted to form a landing platform and mimicking the female).
(b) When a male wasp is deceived by the orchid flower and attempts to fly off with the 'female', he is catapulted into the pollinia (the orchid's pollen packet), which glues to his body. Drawings by G. B. Duckworth in Peakall (1990).

same night, suggesting that the spiders can produce diverse chemical attractants.

11.5 | Propaganda

The powerful releaser effects of alarm pheromones (Chapter 8) on social insect colonies can be exploited by other species to devastating effect. For example, slave-making ants produce and release mimics of the alarm pheromones of victim species to spread confusion during the takeover of nests (after which they carry back pupae from which workers will emerge to live and work in the slave-maker's nest). Workers of the slave maker ant *Formica subintegra* have enormously enlarged Dufour's glands loaded with approximately 700 μg (almost 10% of the ant's body weight) of a mixture of decyl, dodecyl, and tetradecyl acetates (Fig. 11.8) (Regnier & Wilson

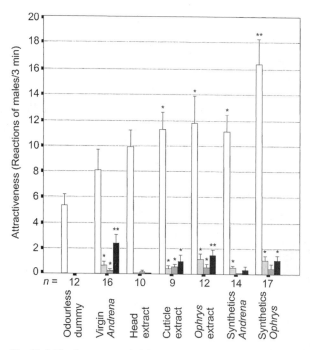

Fig. 11.6. The European orchid *Ophrys sphegodes* almost exactly chemically mimics the female pheromones of the solitary bee *Andrena nigroaenea*. This was tested in the field by offering male bees odourless dummies (chemically extracted and dried *A. nigroaenea* females) scented with different samples and synthetic mixtures reproducing the bee or orchid odours. Attractiveness is given as means ± SE of approaches to the dummy (white bar), pouncing on the dummy (light grey), alighting (dark grey) and copulation attempts (black). Asterisks indicate significant differences between the reaction types of the odourless dummy group and each of the other test groups. Figure from Schiestl *et al.* (1999).

1971). The Dufour's gland secretions are sprayed at the defenders during slave raids. These high concentrations attract other *F. subintegra* workers but panic and disperse the defenders, which cannot help but respond to them as alarm pheromones.

South American robber bees such as *Lestrimelitta limao* make similar use of 'propaganda' chemicals (Sakagami *et al.* 1993). Robber bees live by robbing the nests of various species of *Trigona* stingless bees (to which they are closely related), carrying back stolen honey and pollen. The raid begins when scout *L. limao* workers lay an aerial trail of citral (a mixture of the monoterpenes neral and geranial) to the *Trigona subterranea* nest. During the initial fighting, the robber bees release massive amounts of citral (normally used in small quantities by *T. subterranea* as its own trail and alarm pheromone), which cause the *T. subterranea* workers to retreat far into the nest or even to flee. The same chemicals attract more robber bees to the nest. However, one *Trigona* species, *T. angustula*, resists robbing because it

Fig. 11.7. Bolas spiders (*Mastophora bisaccata* is shown here) aggressively mimic the sex pheromone blends of their moth prey. When an attracted male moth approaches within striking distance, the spider draws back the sticky ball of glue, the bolas, which is suspended on a short thread, and swings it at the moth. Photograph by K. F. Haynes and K. V. Yeargan.

uses benzaldehyde rather than citral for its alarm pheromone, so its defence is not disrupted by the robber bees' propaganda – moreover, benzaldehyde is released specifically when *T. angustula* detects the characteristic pheromones (citral and 6-methyl-5-hepten-2-one) of *L. limao* scouts (Wittmann *et al.* 1990).

Some small invertebrates release large quantities of the ants' alarm pheromone when attacked by ants, to create a diversion which allows the animal to escape (see Chapter 8) (Blum 1985; Howard & Akre 1995). Examples include at least nine species of phalangids (harvestmen) in two genera.

Plants can exploit aphid alarm pheromones for their own defence. Wild potatoes (*Solanum berthaultii*) produce aphid alarm pheromones in glandular leaf hairs (Fig. 11.9) (Gibson & Pickett 1983). The aphid *Myzus persicae* was repelled at a distance of 1–3 mm from the leaves by the high

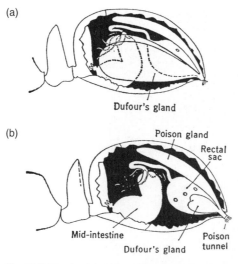

Fig. 11.8. The glandular source of propaganda substances in a slave-maker ant. (a) The abdomen of the *Formica subintegra* worker is partly filled with an enormously enlarged Dufour's gland, which carries large quantities of acetates capable of alarming and dispersing colonies of slave species. (b) The abdomen of *F. subsericea*, a more typical member of the genus, and sometimes a slave species, is shown for comparison. Figure from Regnier & Wilson (1971).

concentration of its alarm pheromone, E-β-farnesene, released from the leaf surface.

11.6 | Specialist relationships of predators, guests and parasites of social insects

Social insects have evolved into the supreme users of chemical communication (Chapters 6, 7 and 8). Indeed complex sociality depends on it: only pheromone communication and chemical cues could enable colonies containing millions of workers to act coherently.

One of the most important uses of chemical cues is for colony recognition, which allows workers to exclude unrelated individuals (Chapter 6). Social insects recognise strangers, even conspecifics from other colonies, and evict them forcibly from the nest or kill them. Nonetheless, because social insect colonies offer such rich resources of food and vulnerable brood and a sheltered environment to any organism that can breach the defences, an astonishingly large and varied zoo of invertebrates has evolved ways to break the codes. These are the myrmecophilous or termitophilous invertebrates ('ant-' or 'termite-loving' respectively – 'loving' only in the sense of being found in close association with social insects, as the relationship may be parasitic or predatory). They range from millipedes, spiders, crickets and beetles to a host of other inverte-

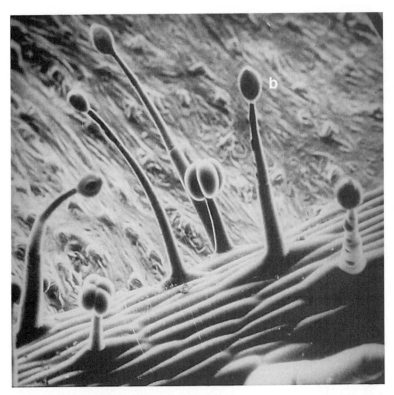

Fig. 11.9. Scanning electron micrograph of the leaf surface of the wild potato (*Solanum berthaultii*) showing the tall glandular leaf hairs (b), which contain the aphid alarm pheromone, E-β-farnesene (Gibson & Pickett 1983). Photograph courtesy of J. A. Pickett.

brates (Hölldobler & Wilson 1990; Howard & Akre 1995). The human equivalent might be to share our homes with alligators, which we insisted on feeding ahead of our own children and which ate our children unbeknown to us (Wheeler in Hölldobler & Wilson 1990). Animals can be accepted even if they look nothing like ants, so long as they smell right. Once accepted, the guests can solicit food from workers or even eat the host brood, while hidden in an olfactory cloak of respectability. Many guest species have glands that offer secretions that ants apparently find irresistible (Fig. 11.10).

Perhaps the most integrated social parasites of all are social insect species, which occur among all the groups of social Hymenoptera (ants, bees and wasps), and take over other colonies. In nearly all cases the social parasites are a closely related sympatric species of their victims. In the most extreme forms the parasitic species has no workers and cannot survive without the aid of the host species. Howard & Akre (1995) and Lenoir *et al.* (2001) give fascinating accounts of chemical cues in the lives of parasitic and slave-making species and their hosts.

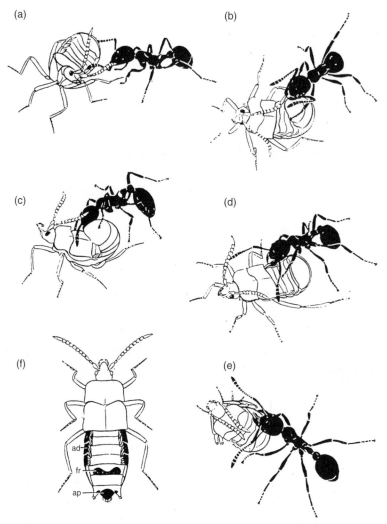

Fig. 11.10. The staphylinid beetle *Atemeles pubicollis* (f) is a social parasite that uses chemical signals to gain entry into colonies of its host, the ant *Myrmica*. In (a) the beetle taps an ant with its antennae and then in (b) offers the ant the tip of its abdomen. A secretion is produced from the beetle's 'appeasement glands' (ap), which the ant licks; this seems to suppress the normal aggressive response of the ant to intruders. In (c), the ant tries to reach the tergal rim/plate of the beetle. In (d) the beetle stretches out its bent abdomen, and in (e) the ant is prompted to pick up the beetle and carry it into the brood chamber by a host-specific attractant secreted from the 'adoption glands' (ad) in the tergal rim/plate. The secretion from the 'fright gland' (fr) is only used for defence against an unfriendly ant. Figure from Hölldobler 1970. Drawing by Turid Hölldobler.

11.6.1 Reading the signals: following trail pheromones

Many guest and specialist predatory species follow pheromone trails of their hosts or prey, sometimes with greater sensitivity than the social insect species laying the trail. These specialist myrmecophiles include other beetles, crickets, cockroaches and millipedes (Hölldobler & Wilson 1990).

Some vertebrates can also respond to ant trail pheromones. For example, the insectivorous blind snakes *Leptotyphlops dulcis* follow the trails of their ant prey, as do Australian blind snakes (Watkins *et al.* 1967; Webb & Shine 1992).

11.6.2 Intruders escaping detection: chemical camouflage and mimicry

There are two ways of blending in to a social insect colony: first, camouflage ('stealing a uniform' by picking up the surface recognition compounds from the host) and, second, chemical mimicry (by synthesising a 'counterfeit passport') (reviewed by Dettner & Liepert 1994, Howard & Akre 1995 and Lenoir *et al.* 2001; incidentally, in their account, Dettner & Liepert use the term 'camouflage' to indicate blending into the background environment rather than looking like the hosts). Guest species may use either or both strategies.

11.6.2.1 Camouflage: stealing the cuticular hydrocarbon 'uniform'

The simplest method, largely used by species that do not enter the nest, is to use the 'wolves in sheep's clothing' strategy. Larval lacewings (*Chrysopa slossonae*) carefully pluck pieces of the waxy 'wool' produced by their woolly-aphid prey and stick it onto their own bodies (Eisner *et al.* 1978). Ants guard the aphids but thus disguised, the lacewing larvae can feed on the aphids, undetected by the ant 'shepherds'.

Guest species more integrated into social insect society must not only match species-specific cues but also the colony-specific odours (Chapter 6). Mutual grooming of host colony members is an important mechanism as it allows exchange of the shimmering pool of hydrocarbons on the surface of the insects. The parasitic ant *Formicoxenus* gains its cuticular camouflage in this way, enabling it to live in the nests of its ant host *Myrmica* (Lenoir *et al.* 1997). Adoption experiments showed that the host cuticular hydrocarbons were acquired by *Formicoxenus* during the first days of adult life – in part through licking the host *Myrmica* workers, a behaviour which gives *Formicoxenus* the common name 'shampoo ant'.

Acquiring cuticular hydrocarbons from its ant hosts allows the scarab beetle *Myrmecaphodius excavaticollis* (syn. *Martinezia dutertrei*) the flexibility to colonise a variety of *Solenopsis* fire ant species. When the adult beetles first enter a nest they are immediately attacked and most are killed (Vander Meer & Wojcik 1982). Only the beetles' strong exoskeleton and passive defence enables some of them to survive long enough to acquire the species-specific and colony-specific odours – inadvertently the ants provide many of these colony-specific hydrocarbons when they liberally apply postpharyngeal gland secretions during the attack. Once annointed, the beetle moves freely in the ant nest, being fed by its ant hosts and eating the ant brood.

Camouflage exploits the way that social insects themselves acquire their colony odours (Chapter 6). As well as picking up the colony-specific hydrocarbon mix from nestmates, the new adult ant learns the odour of its nestmates. An ant will take as nestmates the ants around when it emerges from its pupa (which is why 'slavery' in ants can work – a pupa taken from another species will adopt the species of nest it emerges in, even joining in raids against its source colony).

11.6.2.2 Chemical mimicry by counterfeiting recognition cues

An alternative strategy is to synthesise cuticular hydrocarbons to match those of the host. Radiolabelling experiments show that the termitophilous beetle *Trichopsenius frosti* synthesises its own cuticular hydrocarbons to match those of its termite host *Reticulitermes flavipes* (Howard *et al.* 1980). As a result, the beetles are accepted and fed by the termites. Not only do the beetles synthesise the species-specific hydrocarbons used by the host species, but different beetle species synthesise the cuticular hydrocarbons that are most like the termite caste they spend most time with (queens for *T. frosti*, workers for the other *Trichopsenius* species).

The larvae of the syrphid flies *Microdon* spp. are fully integrated into the nests of many different ant species. This integration is achieved by synthesising a set of cuticular hydrocarbons which seem to signify 'ant larva' at least for the genus of their ant host, so the hydrocarbons can be a passport to a range of ant hosts (Howard *et al.* 1990a, b). One *Microdon* species (*M. mutabilis*) in Europe has, paradoxically, a wide range of host ant species across its distribution but extreme local adaptation of each *M. mutabilis* population not simply to one species of ant host, but to an individual host population and possibly even to local strains or family groups (Elmes *et al.* 1999). It seems the syrphid has become an extreme specialist, synthesising cuticular coatings for the egg that give immunity only to a single nest of ants or their close relatives: if the eggs of a female *M. mutabilis* are placed in the entrance of a nest of the same ant species just a few kilometres away they are quickly destroyed. Elmes *et al.* point out the similarity of this specialisation to the co-evolution of hosts and parasitic diseases.

Many guests use a combination of techniques: synthesising some recognition cuticular hydrocarbons (particularly the more general ones, which may vary less between species and colonies) and, once safe in the nest, acquiring the colony-specific blend by mutual grooming (Howard & Akre 1995). Some palearctic lycaenid butterfly species (*Maculinea* spp.) that have become nest parasites of their ants show this pattern. The *Maculinea* caterpillar feeds in flowers for the first three instars then it crawls down to the ground (Elmes *et al.* 1991). The caterpillar synthesises the cuticular hydrocarbons to match the cues of an ant larva so *Myrmica* ants pick up and bring the caterpillar into the nest. However, the caterpillar grows best if by chance it is picked up by the right ant species for that *Maculinea* spp. The caterpillar of the most specialised of the butterfly species, *Maculinea rebeli*, which gets fed like a cuckoo by the ant workers (in contrast to the less efficient predation on ant brood used by some other *Maculinea* spp.) synthesises a cuticular hydrocarbon uniform that has the closest match to its host species (Thomas & Elmes 1998; Akino *et al.* 1999). Once in the nest the caterpillars pick up additional hydrocarbons from their ant hosts for a near-perfect match. There may be a trade-off between specialisation to increase the penetration of host societies (for greater reward) and a reduction in the range of hosts that can be exploited (Thomas & Elmes 1998).

11.7 | Conclusion

Why do male moths still respond to the lures of bolas spiders and why do ants feed cuckoo caterpillars? If the deceiver is at relatively low frequency, then the cost of not treating the signal as genuine is too high if legitimate use of the signal increases reproductive success or survival. As in other cases of mimicry, as long as bolas spiders are comparatively rare compared with genuine female moths, male moths must respond to female moth sex pheromone – even if this sometimes leads to a sticky end.

Similarly, sexual selection drives displays towards greater conspicuousness but also lays the signaller vulnerable to increased predation risk (Darwin 1871). Where different populations are affected by different predators, local selection by predators could lead to divergence in sexual signals – as perhaps may be happening in the male pheromone of the green stink bugs *Nezara viridula* (Section 11.2.2.1); this has the potential to lead to speciation (Verrell 1991).

As 'code-breaking' is a capability built into all olfactory systems, and chemical signals dominate the lives of most animal species, chemical eavesdropping is likely to be the norm. Just as every animal has its parasites, every pheromone communication system is likely to be exploited.

11.8 | Further reading

Dettner & Leipert (1994) and Howard & Akre (1995) offer wide ranging reviews – especially strong on the surface hydrocarbon story in social insects and guests, a theme also explored for the ants in Hölldobler & Wilson (1990) and Lenoir *et al.* (2001). Haynes & Yeargan (1999) also give a wide-ranging review limited to arthropods but including other modalities such as sound and vision. Stowe (1988) gives a general discussion of chemical mimicry, touching on the relationship with other forms of mimicry. Zuk & Kolluru (1998) specifically cover eavesdropping and deception involving sexual signals.

Chapter 12

Using pheromones:
applications

12.1 | Introduction

The importance of pheromones in the natural behaviour of animals has long been recognised and, long before it was known what pheromones were, people used them to manipulate the behaviour of animals. For example, traditionally, shepherds have encouraged a ewe to adopt a strange lamb if her own died at birth, by covering the strange lamb with the skin of her dead lamb (see Chapter 9).

The clear potential for applied uses of pheromones was an early encouragement to research them. At the turn of the twentieth century and in its first few decades, the potential of synthetic chemical signals to control insect pests was anticipated in both North America and in Europe (Chapter 1) (see reviews by Hecker & Butenandt 1984, Roelofs 1995, Plarre 1998). There is now increasing use of pheromones to affect the behaviour of domesticated animals, from cows to bees, as well as use as 'greener' alternatives to pesticides, largely for the control of insect pests. However, no matter how elegant the science, pheromones will be exploited only if they are commercially viable (Winston 1997). In this chapter I will concentrate on such examples, although I will also mention some promising leads. Human odours will be discussed in the next chapter (Chapter 13).

12.2 | Pheromones used with beneficial and domestic animals

12.2.1 Reproduction of farm animals
Both priming and signalling effects of sex pheromones are used for the manipulation of reproductive behaviour in domestic mammals (Izard 1983;

Booth & Signoret 1992; Houpt 1998). Modern farming practices, including isolation of the sexes, crowding and artificial insemination (AI), may block pheromone mechanisms evolved before domestication (Izard 1983). A greater understanding of pheromones could lead to improvements both in reproductive efficiency and animal welfare.

12.2.1.1 Priming

The effects of pheromones to bring forward, delay, or synchronise puberty and oestrus in mammals (Chapter 6) have been widely used with domestic farm animals (Booth & Signoret 1992; Vandenbergh 1999c).

Priming pheromones and puberty

A brief daily exposure to a boar (adult male pig), or his odour advances puberty in juvenile female pigs by at least a month; the priming pheromone is probably 3α-androstenol produced by the boar's saliva (Booth & Signoret 1992). Exposure to the boar can counter the delaying effect on puberty of the confinement of young female pigs in intensive pig production systems (Izard 1983). Priming pheromones could be cheaper to administer, involve less labour on the farm and cause fewer human health concerns than the hormone treatments currently used to cause earlier puberty in both pigs and cows.

Priming pheromones and ending seasonal anoestrus

Female sheep undergo seasonal anoestrus during the winter and spring (Booth & Signoret 1992). In spring, introduction of a male to previously isolated females results in synchronised oestrus 18–25 days later (by a priming effect of secretions on his wool). In sheep farming, this synchronised oestrus results in a desirable clumping of lambing, which helps labour management and means that the lambs can be marketed as a uniform group (Izard 1983). The use of pheromone exposure to give a predictable and synchronised oestrus in cows, pigs and sheep means that artificial insemination can be used more easily.

Priming pheromones and the postpartum interval

To ensure efficient reproduction in pigs and cattle, farmers must rebreed sows and cows as soon after they give birth as possible (Izard 1983). The introduction of a boar to a group of sows and their litters brings forward their next oestrus. The effect is due to priming pheromones: the time to next oestrus in pigs after pregnancy is reduced from 27 days to 10 days after weaning if they are sprayed with the commercial 5α-androstenone pheromone (Boar Mate^{TM}) within 2 days of weaning (Hillyer 1976, cited in Booth & Signoret 1992).

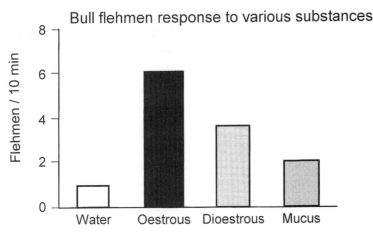

Fig. 12.1. Bulls and other ungulate males show a specific behaviour, flehmen, of lip curl and drawing in air, when exposed to female sex pheromones (Fig. 9.7). The behaviour may serve to bring odours into the vomeronasal organ. Bulls show significantly more flehmen when exposed to urine from an oestrous cow and can identify females coming into oestrus while farmers cannot do this easily. Research is investigating the pheromonal cues that the bull uses. Figure redrawn from Houpt et al. (1989).

12.2.1.2 Signalling

In many ungulates, females signal when they are coming into oestrus: using pheromones and distinctive behaviour, they solicit approaches by the male (Chapter 3). The farmer using artificial insemination for animal breeding also needs to choose the time with the best chance of fertilisation. The time window may be narrow: in cows it is a 12–22 hour period on the day of oestrus. Dogs can be trained to recognise the changes in odour of cow urine to detect oestrus and trained rats are being used in ongoing research to find the pheromone cue used by bulls (*Bos Taurus*) (Fig. 12.1) (Dehnhard & Claus 1996).

Under modern farm management, a pheromone signal (the salivary steroid pheromones 3α-androstenol and 5α-androstenone) from the boar that would normally elicit behavioural signs of oestrus in sows may be missing. Most oestrous sows will respond to pressure on the back, with a stereotyped 'lordosis' or 'standing reflex' response with the ears cocked, an immobile posture and back down, which would allow the male to mount and mate (Booth & Signoret 1992). About a third of sows in oestrus do not react to pressure alone. More than half of these will however respond to the pressure test if they are sprayed with synthetic 5α-androstenone (Boar Mate™). Farmers thereby miss fewer oestrus females who could have been inseminated. Incidentally, female pigs are five times more sensitive to the male pheromone than males (Dorries *et al.* 1995) so it is not surprising that sows have been traditionally used to find the fungal delicacy of truffles (*Tuber melanosporum*), which contain 3α-androstenol (Chapter 13) (Claus *et al.* 1981).

12.2.2 Aquaculture

The use of fish sex pheromones to control reproduction in fish aquaculture has a promising future but research into commercial applications is in its early stages (Patiño 1997). For example, using female pheromones to induce sperm production in male carp would be cost effective and less stressful on broodstock than the hormone injections sometimes used currently (Chapter 1) (Stacey *et al.* 1994).

Many hatchery-reared fish released into the wild do not survive long because they do not associate predators with danger. Brown & Smith (1998) describe how naïve juvenile rainbow trout (*Oncorhynchus mykiss*) could be taught to associate predator odour (from the northern pike, *Esox lucius*) with the chemical alarm signal from trout skin extract (Chapter 8) before being released to stock rivers.

12.2.3 Pheromones, conservation, and animal welfare

A knowledge of reproductive pheromones, largely gained from studies of domesticated and laboratory animals, is important for the success of captive breeding programmes for rare species. For example, an understanding of chemical cues in kin recognition and their role in mediating incest taboos (Chapter 6, Box 6.1) may explain why a rare lowland gorilla female was not interested in males it had lived around since birth, but was successfully mated with a male from another zoo (Pfennig & Sherman 1995) (odour cues are also important in human mate choice, see Section 13.3.3). Pheromone cues for aggregation and social organisation of animals, particularly in relation to Allee effects at low population levels (Chapter 4), could be an important tool in conservation efforts for the maintenance and reintroduction of rare species (Stephens & Sutherland 1999).

With captive animals we may act inadvertently in ways that cause greater stress because we do not understand their olfactory world. For example, partial cage cleaning, by replacing the sawdust substrate without completely cleaning the cage of odour marks, can increase aggression in caged groups of male laboratory mice because social odour cues are disrupted (Gray & Hurst 1995).

Like many other mammals, cats and dogs use urine marks extensively for communication (Chapter 5) and some of these activities can cause problems for pet owners (Houpt 1998). Spray marking in the home by adult neutered male cats can occur 12 times a day; a pheromone, synthetic cat cheek gland pheromone (Feliway™) may be used to reduce the frequency of spraying (Frank *et al.* 1999). Boar Mate™ (Section 12.2.1.1–2) can be used to reduce aggression among newly mixed pigs (McGlone & Morrow 1988) and has been used to reduce aggression by pet minipigs (K. A. Houpt, pers. comm.).

12.2.4 Manipulating the behaviour of beneficial insects

Pheromones orchestrate or modulate every aspect of honeybee life (Chapter 6) and offer many opportunities for intervention (Free 1987; Winston & Slessor 1998; Pettis *et al.* 1999). For example, swarming, which results in decreased honey yields, can be delayed by adding controlled-release sources of synthetic queen mandibular pheromone (QMP) to the hive (Winston & Slessor 1998).

QMP can be sprayed on fruit trees or other crops to increase bee visits: almost twice as many bees visited the sprayed areas as unsprayed areas. This leads to higher pollination rates (often a limiting factor) and increased yield of some fruits, for pears by about 6% and for cranberries 15% by weight, giving increases in grower income of US$1000–4000 per acre (Winston & Slessor 1992, 1998).

Other honeybee pheromones are also being used commercially. Worker bee Nasonov pheromone has been proposed as another pollination attractant, although many field trials have failed to show significant yield increases (Pettis *et al.* 1999). Worker bee Nasonov pheromone is routinely used to bait swarms into trap hives to help bee-keepers (Free 1987, p. 129) (see also monitoring schemes, Section 12.3.1).

12.3 | Pheromones in pest management

The greatest and most successful applications of pheromones are for insect pest management, with significant cost and environmental benefits to the farmer, the consumer and society. Pheromones are the safest of all currently available insect control products (Minks & Kirsch 1998). There are many successful schemes using pheromones for the direct control of insect pests at least as effective as the conventional pesticides they have replaced (Cardé & Minks 1995). Pheromones for many insect pests have been identified. A web site, 'Pherolist', for example, cites more than 670 genera from nearly 50 families of Lepidoptera in which female sex pheromones have been identified (Arn *et al.* 1995). Although most uses have been in agriculture and forestry, manipulation of vector insects with behaviour-modifying chemicals looks increasingly possible in medical and veterinary entomology.

The main ways of exploiting an understanding of pheromones to control pests are monitoring, mating disruption, 'lure and kill' or mass trapping, and other manipulations of pest behaviour. Some of these techniques have been applied to control other animal pests, including vertebrate herbivores such as deer.

A major strength of pheromones is their effectiveness as part of integrated pest management (IPM) schemes, because of their compatibility with biological control agents and other beneficial invertebrates such as bees and spiders. Pheromones fit neatly into the *virtuous* spiral, for example in green-house IPM, where the use of one biological control agent such as a predatory spider mite encourages (or requires) moves away from conventional pesticides for other pests (van Lenteren & Woets 1988).

12.3.1 Monitoring

An important use of pheromones is for baiting traps for monitoring populations of insect pests of crops, orchards (Wall 1989), stored products (Phillips 1997; Plarre & Vanderwell 1999) and forestry (Borden 1993, 1997). Pheromone-based monitoring provides one of the most effective survey methods for detecting the presence and density of a pest species. Thanks to the specificity of insect pheromones, almost all animals attracted to the trap will belong to that species. For example, the rapid spread of the larger grain borer beetle (*Prostephanus truncatus*) across Africa after accidental introduction in the 1980s, has been monitored with pheromone traps baited with the male aggregation pheromone (Chapter 4). Similarly, the spread of Africanised honeybees in North America has been monitored by placing swarm traps baited with the Nasonov pheromone of worker honeybees (Fig. 12.2) (Chapters 4 and 6) (Schmidt & Thoenes 1987).

Fig. 12.2. Placing a swarm trap, baited with Nasonov pheromone, in a tree to attract honeybee swarms as part of the monitoring and control of Africanised honeybees in the USA (Schmidt *et al.* 1989). Photograph by J. O. Schmidt.

Pheromone traps can be used to monitor the number of pests so that pesticides can be targeted when and where they are actually needed, rather than sprayed just in case or at regular intervals, whether or not pests are present. This can give significant reductions in pesticide use and improved pest control (Wall 1989).

Pheromones can also be used to monitor vertebrate populations. The responses of rodents to scented traps can be used to infer the status and density of the population. For example, adult muskrats (*Ondatra zibethicus*) in a densely populated breeding population avoided musk-scented traps, but scented traps placed in an unexploited area did attract members of a pioneer population (Müller-Schwarze 1990).

The scent marks (Chapter 5) left by the animals themselves can also be used for density estimates. Visible scent marks such as the faecal piles of muskrats, scent mounds of beavers (*Castor canadensis*), carnivore droppings on conspicuous features or along trails, and urine marks in the snow, have all been used to estimate how many individuals are living in a given area (Müller-Schwarze 1990).

12.3.2 Mating disruption

The aim of mating disruption is to stop fertilisation of eggs by preventing adult males and females finding each other. When larvae, such as caterpillars, cause the damage, anything that reduces numbers in the next generation of larvae is a potentially effective control agent. For most moths, which rely on pheromones for the sexes to find each other (Chapters 3 and 10), this can be achieved by flooding the air in the host crop with synthetic pheromone. Area-wide treatments that involve every farmer over a region are the most effective, as shown by successful schemes against the pink bollworm moth in cotton and the codling moth in orchards (Table 12.1) (Cardé & Minks 1995).

Mating disruption is one of the most significant successful uses of insect pheromones and has often come to the rescue when pests have become resistant to conventional pesticides. For example, mating disruption using sex pheromones was able to almost eradicate populations of the light brown apple moth (*Epiphyas postvittana*) that were resistant to organophosphate pesticides (Fig. 12.3) (Suckling *et al.* 1990).

A variety of slow-release formulations have been developed to release small quantities of volatile pheromone over the months of the insect pest season. These include polyethylene rope (sealed tubes containing pheromone that diffuses through the tube walls), hollow fibres, rubber septa or PVC beads impregnated with pheromone, and microcapsules (Howse *et al.* 1998). A recent approach is to release pheromone from aerosol 'puffers' at regular intervals during mating periods (Mafra-Neto & Baker 1996; Shorey & Gerber 1996).

Table 12.1. *Examples of successful mating disruption*

All are in commercial use except the nematode control.

Crop or resource	Pest	Pheromone formulation	Countries	Notes	References
	Moths				
Apples and pears	Codling moth (*Cydia pomonella*)	Polythene rope (sealed tube containing pheromone)	Washington State, USA	Area-wide treatment	Gut & Brunner 1998
Apples	Light brown apple moth (*Ephiphyas postvittana*)		New Zealand, Australia		Suckling *et al.* 1990
Apricots and peaches	Oriental fruit moth (*Grapholitha molesta*)		South Africa, Australia, USA, Europe		Rothschild & Vickers 1991
Cotton	Pink bollworm (*Pectinophora gossypiella*)	Polythene rope	Arizona, California, USA	Area-wide treatment, almost complete suppression of bollworm	Staten *et al.* 1997
	P. gossypiella	Microencapsulation, polythene rope	Egypt	Pheromones now used instead of pesticides	Howse *et al.* 1998, p. 324
Tomatoes	Tomato pin worm (TPW) (*Keiferia lycopersicella*)	Hollow fibres	Mexico		Trumble 1997
Grapes (vineyards)	European grape moth (*Eupoecilia ambiguella*) Grape vine moth (*Lobesia botrana*)	Open-mouthed tubes (ampullae)	Europe		Arn & Louis 1997
	Nematode				
Soybean	Soybean cyst nematode (*Heterodera glycines*)		USA		Meyer *et al.* 1997

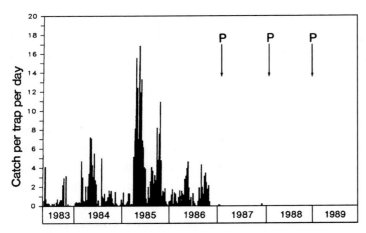

Fig. 12.3. Pheromone trap catches of the light brown apple moth (*Epiphyas postvittana*) at Moutere Bluffs, New Zealand, before and after the use of mating disruption (P, pheromone) to control insecticide resistant insects. Figure from Suckling *et al.* (1990).

Another, complementary, way in which mating disruption reduces crop damage is by reducing or eliminating pesticides, thereby keeping more natural enemies alive, so the few caterpillars present have a greater chance of being killed by predators or parasitoids. In addition, pollinating insects stay alive, which increases fruit set.

These effects are demonstrated in an integrated pest management (IPM) scheme for tomatoes grown for export in Mexico (Trumble 1997). The tomato pin worm (TPW) moth (*Keiferia lycopersicella*) had become a major pest as it developed resistance to a wide variety of pesticides. In an attempt to control the moth, 20–45 applications of broad-spectrum insecticides were applied per season and insecticide costs rose. Worse, Mexico's one billion dollar tomato industry was in jeopardy as two major problems developed. First, tomatoes were being rejected because of high pesticide residues and, second, farmers entered the pesticide treadmill of secondary pests, such as the leafminer fly (*Liriomyza sativae*), created from previously minor pest species when their predators were killed by the pesticides targeted against the original pest.

The IPM programme combines mating disruption of TPW with the control of other moth pests. The other moth species are unaffected by the TPW mating disruption, so they are controlled by abamectin (a moth pesticide with low mammalian toxicity), *Bacillus thuringiensis* (an insect pathogen specific to moths), and release of *Trichogramma* parasitoid wasps, which parasitise moth eggs (Trumble 1997). The TPW female sex pheromone, (*E*)-4-tridecenyl acetate, in a hollow fibre formulation, is applied by hand to about 1000 release sites, giving 10 g pheromone per hectare, about 3–4 times per planting.

In the autumn planting, low TPW larval populations mean that there is no difference in damage to fruit in IPM and conventional treatments but costs are lower on the IPM programme so the net profits are higher (Fig. 12.4c,d). The fruit damage in winter and spring plantings was much higher on the conventionally treated plots (70–90%) than on the IPM plots (33–35%). The reduction of damage was because fewer moth eggs were laid (Fig. 12.4a) and there was a higher egg parasitism rate in the IPM plots (Fig. 12.4b). Over the year, a grower following the IPM programme would make an additional US$3.5 million dollars on 1000 hectares over that from the conventional pesticide regime (Fig. 12.4d). The new methods also meet the desire of consumers for reduced pesticide use. In addition, there should be fewer problems with pesticide resistance in the future, and there will be health benefits for farm workers, their families, and the environment.

While mating disruption can be very effective, how it works is still not understood in detail. The potential mechanisms include: (1) sensory adaptation (of the olfactory sensory neurons, Chapter 9) or habituation (in the central nervous system); (2) false-trail following (competition between natural and synthetic sources); (3) camouflage of natural plumes by ubiquitous high levels of synthetic pheromone; (4) imbalance in sensory input by massive release of a partial pheromone blend; and (5) the effects of pheromone antagonists and mimics (Cardé & Minks 1995; Sanders 1997). If we understood these better we might be able to explain why some formulations are successful and others not, and why results may vary between places and years (Cardé & Minks 1995). Most, but not all, trials of mating disruption of moth pests suggest that the full blend offers the most effective disruption, at the lowest dose (Minks & Cardé 1988). The very fine discrimination of odour filaments in time and space that moths are able to achieve helps to explain why antagonists (anti-pheromones) have not been successful. The correct blend needs to hit the moth antenna as a unit to elicit the full behavioural response (Chapters 9 and 10).

Moths are not the only pests that could be controlled with mating disruption. Parasitic nematodes are major pests in agriculture and a major cause of human and animal disease. Female sex pheromones might offer control. For example, the sex pheromone of the soybean cyst nematode (*Heterodera glycines*), which is vanillic acid, has been used in successful field trials of mating disruption (Meyer *et al.* 1997).

12.3.3 Lure and kill (attracticide) and mass trapping

The aim of 'lure and kill' or mass trap pest control is to reduce the pest population by attracting pests with pheromones and then either trapping or killing responding individuals (Table 12.2). With 'lure and kill' (also called attracticide), pest animals attracted to the pheromone source pick up a lethal dose of pesticide or a pathogen (Howse *et al.* 1998, p. 300). Mass trapping simply confines the animals in a trap.

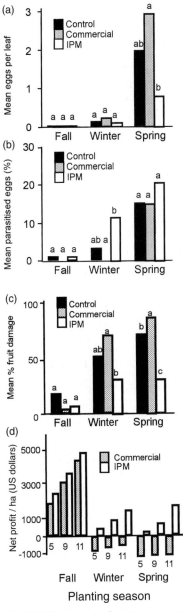

Fig. 12.4. The integrated pest management (IPM) programme, including mating disruption of the tomato pin worm (TPW, *Keiferia lycopersicella*), for tomatoes in Sinaloa, Mexico, over three seasons. The reduction of damage (c) in IPM plots (with mating disruption) was due to fewer moth eggs laid (a) and a higher percentage of parasitism of moth eggs by parasitoid wasps (b). In the autumn planting, low TPW larval populations meant there was no difference in damage to fruit in IPM and conventional (commercial) treatments (c) but costs were lower on the IPM programme so the net profits (d) were higher. Numbers directly below the bars in (d) indicate prices (US dollars) for a standard carton of tomatoes. Letters above bars indicate significant differences within plantings at the $p < 0.05$ level (ANOVA, Fisher's Protected LSD). Percentage data were transformed by the arcsine square root transformation prior to analysis. Figure redrawn from Trumble (1997).

Table 12.2. *One of the simplest uses of pheromones is for luring pest animals to the source and then either trapping them or letting them contact a killing agent*

Most of the examples in the table are at the earliest stage of development as the pheromones have not been identified. More background on the biology of each type of pheromone and many of the animals will be found in other chapters in the book (see index).

Pest	Lure	Trap or killing agent	Resource affected by pest	Reference
Vertebrates				
Marine lamprey (*Petromyzon marinus*)	Attractant pheromones from larvae	?	Fish (Gt Lakes, N. America)	Li *et al.* 1995
Brown tree snake (*Boiga irregularis*)	?Sex pheromone	?Pathogen	Native birds (Guam Island)	Mason & Greene in March-lewska-Koj *et al.* (2001)
Crustacea				
Ectoparasitic copepod (*Lepeophtheirus salmonis*)	Sex	?	Marine salmon farming	Pike & Wadsworth 1999
Arachnida				
Ticks				
African bont ticks (*Amblyomma hebraeum*)	Sex + attraction-aggregation attachment pheromone	Pesticide impregnated tail lures	Cattle (Africa)	Norval *et al.* 1996
Dog tick (*Dermacentor variabilis*)	Sex + attraction-aggregation-attachment + mounting pheromone	Pesticide impregnated decoys	Cattle, dogs (N. America)	Sonenshine *et al.* 1992
Mites (Acari) *Varroa jacobsoni* (an ectoparasite)	Kairomone in brood cells of host	?	Honeybee	Trouiller *et al.* 1992

Table 12.2. (cont.)

Insecta

Ants

Leaf-cutter ants *Atta*	Trail pheromones	Pesticide baits?	Citrus trees	Robinson et al. 1982
Termites	Trail and gnawing pheromones	Pesticide baits?	Crops, wooden buildings	Kaib 1999

Insect disease vectors

Mosquito (*Culex*)				
Sand fly (*Lutzomyia*)	Oviposition pheromone	Pesticide or entomophagous fungus	Health of people, livestock	McCall & Cameron 1995; Pickett & Woodcock 1996
Blackfly (*Simulium*)				

Other insects

Boll weevil (*Anthonomus grandis*)	Male aggregation pheromone (Grandlure)	Pesticide	Cotton	Hardee & Mitchell 1997
Flour moth (*Ephestia kuehniella*)	Female sex pheromone	Pesticide	Stored food products (warehouse)	Trematerra 1997
House fly (*Musca domestica*)	Female sex pheromone ((Z)-9-tricosene)	Pesticide	Nuisance, health of people, livestock	Chapman et al. 1998

Both 'lure and kill' and mass trapping rely on the specificity of pheromones to attract only members of a single pest species. This means that the whole crop or forest does not need to be sprayed, thus helping to save beneficial insects that would normally be killed with area-wide pesticide sprayings.

In forestry, 'trap trees' are a traditional variation on lure and kill, with aggregation pheromones (Chapter 4) attracting, for example, the bark beetle *Ips typographus* to a log or tree which is then cut up and destroyed, or that tree only is dosed with pesticide (Drumont *et al.* 1992).

The effectiveness of attracticide or mass trapping depends on which sex(es) are attracted by the pheromone (whether it is just the males, females or both), the proportions of the population that can be attracted/caught or killed, the mating system of the pest, their dispersal patterns, and their potential rates of increase. With a high potential rate of increase, trapping may be ineffective. For example, mass trapping exercises in Europe have caught enormous numbers of bark beetles but control is hard to demonstrate (Schlyter & Birgersson 1999). If the lures only attract males, almost all the males in a population must be caught, as those remaining can usually mate with many females.

12.3.4 Using deterrent odours

Another approach to pest control uses the responses of animals to their own pheromones or to those of their predators to 'persuade' animals to go somewhere else to lay eggs, nest or feed.

12.3.4.1 Host-marking pheromones and insect herbivores

Synthetic host-marking pheromones (HMP) (sometimes called oviposition deterring pheromones, ODP) could be used to protect crops by persuading female insects that the plants or fruit are already occupied with conspecific eggs. Flies, beetles and butterflies (Chapter 4) use HMPs. While some HMPs have been identified and field trials showed good control of cherry fruit fly (*Rhagoletis cerasi*) with its HMP (Katsoyannos & Boller 1980), the *R. cerasi* HMP molecule is complex and currently too expensive to produce commercially (Landolt & Averill 1999). As with other effective treatments it may be necessary to offer untreated 'sacrificial' trap rows, which are later destroyed, so that females can find somewhere to lay their eggs. Without this, despite the pheromone, they may eventually lay eggs on the HMP-protected crop.

12.3.4.2 Predator odours to control vertebrate herbivores

Many vertebrate herbivores, such as deer and rodents, are repelled by the odours of predatory carnivores, in particular urine, gland secretions and droppings (see reviews by Weldon 1990, Müller-Schwarze 1990). Many of the active compounds are sulfur containing and from the carnivore anal glands. Snowshoe hares fed less on pine seedlings treated with a sulfur compound (3-propyl-1,2-dithiolane) from stoat anal glands (*Mustela erminea*) (Sullivan & Crump 1984). Deer are repelled by lion dung extracts (Abbott *et al.* 1990) and this may be a general response to predator excreta. For example, the repellency of coyote (*Canis latrans*) urine to four rodent species seems to be related to the sulfurous metabolites of meat digestion that are excreted in the urine (Nolte *et al.* 1994).

12.3.4.3 Conspecific scent marking to prevent colonisation
Scent marking forms a very important part of territorial behaviour in mammals. However, in most species the scent marks do not act as 'keep out' signals (Chapter 5). Nonetheless, for some species these may have some application as control agents. For example, beavers (*Castor canadensis*), which have recovered their numbers in the USA to reach pest status, may be deterred by synthetic scent marks so long as there are other sites available to colonise (Chapter 5) (Welsh & Müller-Schwarze 1989). A greater understanding of badger marking behaviour could help current investigations in the UK into the possible role of badgers in the transmission of tuberculosis and, in particular, the role of scent marks in distributing populations after some groups of badgers have been removed.

12.3.5 Alarm pheromones
Alarm pheromones of aphids have been used commercially to increase the effectiveness of conventional pesticides or biological control agents such as the fungal pathogen *Verticillium lecanii* (Howse *et al.* 1998, p. 348). The synthetic alarm pheromones and the increased activity of the aphids in response to their alarm pheromone (Chapter 8) increases mortality because they contact more insecticide or fungal spores (Pickett *et al.* 1992).

12.3.6 Push-pull or stimulo-deterrent diversionary strategies
Pheromones including host marking pheromone, feeding deterrents and trap crops (later destroyed) can be used to manipulate the behaviour of pests by a push-pull or stimulo-deterrent diversionary strategy (SDDS) so that they cause less damage or are easier to control (Pickett *et al.* 1997). For example, valuable trees can be protected from bark beetles by making them appear to be resistant non-host trees already fully occupied by conspecifics or competing species (the push) and by at the same time baiting trap trees with attractive pheromones (the pull) (Fig. 12.5) (Chapters 4 and 11) (Borden 1997).

Predators and parasitoids could also be manipulated by semiochemicals. Many insect predators and parasitoids eavesdrop the pheromones of their

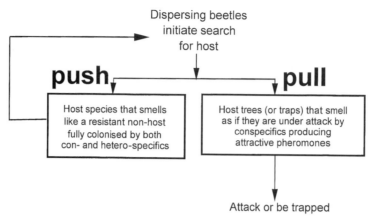

Fig. 12.5. Bark beetles respond to positive and negative semiochemical cues when choosing trees to attack (Chapter 4). These responses can be exploited by 'pushing' beetles away from a tree or group of trees at risk of attack, by spraying these trees with the semiochemicals that indicate to the beetles that the trees are unsuitable non-hosts or resistant to attack, or are already full of beetles. Nearby sacrificial trees or traps are baited with attractive semiochemicals, indicating them as ideal hosts only sparsely occupied, to offer a lethal alternative that 'pulls' the beetles away. Figure redrawn from Borden (1997).

prey, using them as kairomone cues to locate the prey (Chapter 11). Kairomones have not yet been widely exploited commercially but one promising idea is to spray synthetic aphid sex pheromones to attract parasitoid wasps into the crop to parasitise the aphids (Powell 1998).

12.3.7 Understanding barnacle settlement and how to stop it
Barnacles and other encrusting ('fouling') organisms slow down ships and cost maritime activities dear. However, the metal-based antifouling coatings, in particular those containing organotins, cause significant environmental problems and international bans are pending (Clare 1998). The search for non-toxic alternatives has renewed interest in the cues that barnacle larvae use when settling from the plankton into aggregations (Chapter 4) (Clare & Matsumura 2000). Certain lectins, natural sugar-binding proteins, can inhibit the effect of the adult settlement pheromone (Matsumura *et al.* 1998a).

12.3.8 Plant breeding and biotechnology for self-protecting plants
Theoretically, plants could be bred to be self-protecting. Wild potato species produce aphid alarm pheromone in glandular hairs on the leaf (Chapter 11). Could these and other insect pheromones be engineered into plants? For example, plants could be engineered to produce their own mating disruption levels of moth sex pheromones or high concentrations of host-marking pheromones on their fruits (Pickett 1985).

12.3.9 Primer pheromones
Pheromones may provide a way of controlling important hymenopteran pests, for example by using queen pheromones to suppress queen reproduction in

fire ants (*Solenoposis invicta*) (Chapter 6) (Vargo 1998). Similar approaches may be possible with termites when their pheromones are better understood (Chapter 6) (Kaib 1999). In future, locust pheromones might offer ways of disrupting gregarisation in the early stages of population build-up, before locust swarms form (Chapter 4) (Hassanali & Torto 1999).

The primer pheromone influences on reproduction in many rodents (Chapter 9) also suggest that there should be rich possibilities for pest management, but I am not aware of commercially available treatments.

12.4 | Pest resistance to pheromones?

There have been few studies of the long-term effects of mating disruption on pheromone signalling in pest moths. Haynes *et al.* (1984) found no change in emission rates or blend in the pink bollworm moth (*Pectinophora gossypiella*) after 3–5 years of mating disruption. However, pheromones are unlikely to be immune to the development of resistant strains of pests – in this case individuals that are able to mate despite the high concentrations of the artificial pheromone in the crop. Resistance could come from changes in behaviours, such as the use of different mating sites, use of other mating cues, or selection for the dispersal of mated females (McNeil 1992); however, it is most likely to come via selection for mutant females producing a different pheromone blend to that used for mating disruption and for males able to respond to this. Resistance to mating disruption pheromones could also come from males responding to other chemical cues (already produced by the females and perceived by males but not currently part of the pheromone complex) in place of the original components (Pickett 1992). We already have good evidence that natural populations of insects have evolved regional differences in the pheromone blend used by a species (Chapters 3, 4, 7 and 8), which shows that change is possible. An illustration of the potential for signal change is the mutation in pheromone blend of the cabbage looper moth (*Trichoplusia ni*) observed by chance in a laboratory culture (Chapter 3). What we can predict is that insects are likely to surprise us.

12.5 | Commercialisation – problems and benefits of pheromones

Given the power of pheromones for insect pest control there still remains the question of why they have not been used more. In common with much of IPM, the most serious challenges to getting greater use of pheromones in pest control are economic and political (Silverstein 1990; Dent 1991; Wyatt 1997).

The first problem is economic. Large agrochemical companies are unlikely to develop commercial pheromone technologies for more than a very few major crops, largely because of the problem of recovering high development costs (Jutsum & Gordon 1989). The difficulties include small markets, formulation, marketing, limited patentability, plus a greater realisation of how subtle and complex pheromones can be, and competition with the companies' own conventional products (Arn 1990; Silverstein 1990). The specificity of pheromones, which is their strength for minimal impact, is a commercial disadvantage as each species has its own pheromone, and this situation is further complicated by geographical races using different pheromone blends (Chapter 3).

Most of the small companies involved in commercialisation of pheromones do not have the resources for basic research, so much of the research and extension work will need to be government supported. However, as Jones (1994) points out, it is ironic that just when politicians are legislating for reductions in conventional pesticide use (see below), they seem most keen to reduce the worldwide capacity for scientific research into alternative pest control methods at government and academic institutions.

Gaining acceptance for pheromone-based pest control, like the adoption of other new IPM methods, requires considerable effort (Wyatt 1997). Adoption will be more likely if farmers find that there *must* be change, for example in response to resistance problems with conventional pesticides. Getting acceptance of ultimately better but less certain pest control methods is hard. Pheromones need as much selling as pesticides, perhaps even more.

Costs of active ingredients have been a limiting factor in some mating disruption schemes. Developments in synthetic chemistry, in particular chiral synthesis, may bring down costs. The use of biotechnology for pheromone synthesis might lead to cheaper production of chirally correct pheromones (Appendix A2) (Pickett 1992; Hick *et al.* 1997).

Second, political changes at the level of consumers and governments will affect the development of pheromones in pest control. Changing consumer attitudes to pesticide use are already improving the climate for alternatives to conventional pest control and the use of pheromones is likely to benefit (Coats 1994; Trumble 1997; Winston 1997). Government influences through legislation include promising solutions such as the Swedish tax on pesticides, which provides money for pheromones and other alternatives to pesticides. Making it easier to register pheromone products would also make a big difference (Minks & Kirsch 1998). Legislation restricting pesticide use is increasingly important. For example, tougher legislation has dramatically changed the patterns of pesticide use in California (Trumble 1997). Government subsidy for pheromone treatments, such as those to wine-grape growers in Germany and Switzerland, can also help (Arn & Louis 1997).

12.6 | Conclusion

Each insect pheromone symposium since the 1960s has emphasised the potential for pest control but only some of that is now being realised. However, we should remember that it is still less than 40 years since the first chemical identification of an insect pheromone and we have made massive strides since then (Wyatt 1997). So far the greatest successes have been with moth and beetle pheromones. The central role of pheromones in the biology of most insects, crustaceans, fish and mammals suggests there is still a vast potential for intervention with pheromones: we have only just begun.

12.7 | Further reading

The role of olfaction in domestic mammals is well reviewed by Houpt (1998). The chapters by Owen (O. T.) Jones in Howse *et al.* (1998) give an excellent recent review of applied pheromone use in insect pest management. Winston (1997) reviews alternatives to conventional pesticides. Pheromones are important in most insects, not just moths, and the chapters in Hardie & Minks (1999) cover these sometimes neglected groups.

Chapter 13

On the scent of human attraction:
human pheromones?

13.1 | Introduction

As we are mammals, it is highly likely that many behavioural and physiological aspects of human biology are influenced by pheromones. However, only recently has the importance of chemical cues for humans become clear, so I start by investigating the evidence that odours are important for humans. I then look at how these natural odours are produced and perceived. Human responses to perfumes and artificial scents would form a book on their own (see further reading).

Humans are just one of some 200 living species of primate (Dixson 1998). While we and our nearest relatives, the great apes (the gorillas, orang-utans, bonobos and chimpanzees), may not use odour communication in quite the same ways as the Old World and New World monkeys, chemical communication is still important to us (Fig. 13.1) (Table 13.1). Unlike other primates, humans and great apes have axillary (armpit) scent glands. In terms of numbers and sizes of sebaceous and apocrine glands (see Box 13.1), humans have to be considered the most highly scented ape of all (Stoddart 1990). These scent glands give off a plethora of natural products that envelope the body with a complex, and probably individually distinctive, volatile label (Schaal & Porter 1991).

Olfaction is the suppressed sense in contemporary Western society. We give it the least attention and seem to value it the least of our senses. However, its importance is felt when lost. Oliver Sacks quotes a man who lost his sense of smell after a head injury: 'when I lost [my sense of smell] it was like being struck blind. Life lost a good deal of its savor – one doesn't realise how much 'savor' is smell. You *smell* people, you *smell* books, you *smell* the city, you *smell* the spring – maybe not consciously, but as a rich unconscious

Fig. 13.1. Olfaction is important in social and sexual behaviour for most primates. This figure shows mutual naso-genital investigations in the white-cheeked mangabey (*Cerocebus albigena*). Male chimpanzees and gorillas, our nearest relatives, both investigate the odour of the vagina of prospective mates. Figure redrawn from Gautier & Gautier (1977) in Stoddart (1990).

background to everything else. My whole world was suddenly radically poorer.' (O. Sacks, *The man who mistook his wife for a hat*, quoted in Classen *et al.* 1994, p. 1.) In Süskind's imaginative novel *Perfume* (1986), the sinister main character has an extraordinarily sensitive sense of smell and memory for odours, but no scent of his own.

The small relative size of the olfactory bulb in the brain means that humans (and the great apes) are described as 'micro-osmic', but the sense of smell is more complicated than simple brain proportions (Schaal & Porter 1991). Small olfactory bulbs do not show that smell is unimportant, because the number of neurons they contain may be less important than their integration: complex olfactory messages are also decoded in other brain regions in addition to the olfactory bulb (Keverne 1983).

In humans, as in other mammals, it seems that most odour preferences are learned and thus individual and cultural experience are important factors, although, like colour blindness, the range of odours that *can* be detected may be genetically determined (Section 13.5) (Chapter 9). Soussignan *et al.* (1997) used the facial expressions of newly born babies (neonates) to judge their responses to odours (for example, smile or disgust). The responses of neonates to some odours are very different to those of adults and this continues into early childhood. For example, young children to the age of 4–5 years tolerate odours such as butyric acid (rancid butter odour), which older children and adults find unpleasant or disgusting (Schaal 1988). Socially appropriate responses to some odours may be learned.

Table 13.1. *Occurrence of specialised glands and of scent-marking displays in primates*

Primate group	Number of genera studied	Functional vomeronasal organ	% Genera with specialised skin scent glands	Main positions of skin scent glands	% Genera which mark using glands	% Genera mark using urine
Prosimians (e.g. lemurs)	17	yes	94	Circumgenital/ ventral thorax	94	76
New World monkeys (e.g. marmosets)	16	yes	94	Circumgenital/ ventral thorax (all)	88	69
Old World monkeys (e.g. baboons)	14	no (?)	14	Few have skin glands	14	0
Apes and man	5	no (?)	100	Sternal (chest) glands: (*Hylobates, Symphalangus, Pongo*) Axillary (armpit) glands (*Pan, Gorilla, Homo*)	0	0

After Dixson (1998). Vomeronasal information from Preti & Wysocki (1999).

13.2 | Cultural and social aspects of odours and humans

> Smell is not just a biological and psychological phenomenon: smell is a cultural, hence a social and historical phenomenon (Classen *et al.* 1994, p. 3).

Evidence that olfactory cues are important to humans comes from detailed case studies by clinical psychologists and psychiatrists, together with comparative data from anthropologists, human ethologists, and sociologists (Schaal & Porter 1991). Different societies may have different universes of odour. These, in turn, give rise to different osmologies (words in different languages for smells that also reflect cultural significances) (see Classen *et al.* 1994, for a fascinating discussion).

Over the last 300 years, cultural norms in the Western world have moved towards intense suppression and masking of natural body odours and control over overt appreciation (Schaal 1988; Classen *et al.* 1994; Corbin 1996). However, these attitudes are not entirely recent as the Ancient Greeks also commented on bad smelling armpits. Smell can be political: those too poor to keep clean easily have been looked down on as the 'great unwashed'.

In some non-Western cultures, which could be described as 'smell-seeking', attitudes to the smells of others are very different. For example, among the Amazonian Yanomamö or the New Guinean Kaum-Irebe, rubbing another's body and then sniffing one's fingers are common acts (Schaal & Porter 1991).

Even today in Western society, body odours hold a special place in sexual interaction. Some contact (personal) advertisements request partners to remain unwashed. In the nineteenth century, Napoleon is said to have written to Empress Josephine asking her 'not to wash' ('Ne te lave pas. J'arrive') during the 2 weeks before they would meet again (Ackerman 1990, p. 9).

Body odours (either natural or added to culturally, for example by perfumes) are important to people in both Western and non-Western societies. The underlying motives and techniques for suppressing our own naturally produced body odours and adding artificial odours vary a great deal between cultures – or even within them (Schaal & Porter 1991). The functions of added odours include, first, as a label of same group/other group. Second, seductive perfumes are in use almost everywhere: in Melanesia, men wear a strong musky aromatic leaf during dances and, of course, perfume advertisements fill Western magazines (references in Schaal & Porter 1991); further examples are given by Stoddart (1990) and Classen *et al.* (1994).

Like so many aspects of our behaviour, our current condition is so strongly influenced by social evolution that it is hard to disentangle cultural influences from the biological. Even a modern cross-cultural approach may not yield the answers, as no present-day society necessarily shows the ancestral patterns of odour production and response. The challenge is to discover the biological odours and responses that are dependent on experience, and potentially common to all humans. The idea that odours might affect our emotions or subconscious, and is not entirely under our control, is scary to modern sensibilities (but paradoxically, at the same time it is often the claim of perfume advertisements). What is the evidence that human odours are biologically important to us?

13.3 | Evidence that olfaction is important in human behaviour and biology

Chemical cues are important in two main areas of human behaviour and biology: first, for recognition of kin and familiar people and for distinguishing biological categories such as male and female; and, second, as we are still discovering, olfaction plays an important role in reproductive behaviour and biology.

13.3.1 Recognition
For a species once thought to have a reduced sense of smell we are surprisingly good at recognising familiar humans by smell. A clue to the individuality of our odours is given by the ease with which dogs are able to tell us apart by smell (unless we are identical twins on identical diets). Only relatively recently have human abilities to do this been more systematically investigated – with surprising results (see review by Schaal & Porter 1991).

The typical experiment asks the people whose abilities are being tested to sniff the smell that has collected on a standard piece of clothing (Doty 1985). This is usually a clean T-shirt provided by the experimenters, which the 'odour donor' had worn for a time, perhaps overnight. The odour donors are usually given instructions including which soaps to use and strong smelling foods to avoid, to try to minimise the number of environmental odours that could affect the tests. The use of T-shirts or odour extracts of various kinds makes it easier to standardise experiments and make replicates possible. Odour donors do not then have to be present during the testing, removing the risk that responses will be influenced by the audio, visual, or behavioural characteristics of the odour donors.

The odours we produce reflect our internal physiological and metabolic state in a complex interaction between our genetic make-up and the environment, including diet. The odours carry messages that give clues to our age, gender, reproductive phase and health status (Schaal & Porter 1991). Many of these will be important in the recognition of potential mates and in mate choice (Section 13.3.3).

Male and female humans, once they have reached puberty, can often be distinguished by their odours. However, the results of experiments are rather mixed and, sometimes, correct identifications are made *less* often than by chance. People often make errors, with stronger smells often assumed to have come from males even if this was not the case (Doty 1985). In these and many other odour tests, women tend to be correct more often than men.

Various tests of the ability to recognise one's own T-shirt by odour have been made. For example, between a quarter and a third of people can correctly pick out their own T-shirt by smell out of a choice of ten worn T-shirts in two out of three tests (Hold & Schleidt 1977; Schleidt 1980). About the same proportions of people can correctly identify their partner by smell.

13.3.1.1 Parents and offspring, including babies

Some of the most important work on human recognition by odour has been done on the interactions between mothers and newborn babies (neonates) (see reviews by Schaal & Porter 1991, Porter & Winberg 1999).

Newborn babies learn the specific odours of their mother in the first hours after birth and by 3 days old will turn towards a pad carrying the odours of their mother's breast rather than odours of another lactating mother or a clean pad. The baby also learns the individual odour of its mother's axillae (armpits), and will turn towards these odours in tests. Babies move their head and arms less and suck more when they are exposed to their mothers' odours.

The olfactory basis for the individual recognition is likely to include odours related to the major histocompatibility complex (MHC) (Box 3.1). There may be long term effects of this early olfactory experience on adult mate choice (Section 13.3.3.1).

Naturally occurring maternal odours common to all lactating mothers, including secretions from the skin glands in the areola around the nipple, may have a role in guiding the neonate to the nipple and thus contribute to early nipple attachment and sucking. These olfactory cues are disrupted by washing. If offered a choice, newly born babies spontaneously turn towards and suckle on the unwashed breast (Varendi *et al.* 1994). While the baby is breast feeding or sleeping, the mother is also learning its odours. Mothers can recognise their babies by odour alone within the first 6 days even if they have had only a few hours' contact (Schaal *et al.* 1980; Porter *et al.* 1983). Later experiments by Kaitz *et al.* (1987) showed that all mothers with an unimpeded sense of smell could recognise their baby by smell after they had had an hour's contact. In the hours after giving birth many mothers rate the odours of their babies pleasing and find baby odours in general pleasing. In first-time mothers this, and success at recognising the odours of their own baby, are positively correlated with their level of the adrenal hormone, cortisol (Fleming *et al.* 1997). Cortisol may simply be an easily measured marker for other, as yet unidentified, hormone changes underlying the changes in perception (Chapter 9). Olfactory recognition is not confined to the time after birth. Mothers can recognise their older children by smell too.

The comfort given by 'security blankets' and other favourite 'attachment objects' carried round by young children is largely due to their familiar smell (attachment objects may be rejected if washed). Young children with sleeping difficulties may be calmed by the axillary odours of their mother (Schaal 1988).

Other members of families can also recognise each other by smell (Porter *et al.* 1986). In a two-choice discrimination test, most fathers correctly recognised the T-shirt worn by their own infant, less than 72 hours after birth. Aunts and grandmothers also reliably identified their neonatal relatives by odour. Children aged 3–8 similarly were able to identify correctly T-shirts worn by their similar-aged siblings.

The growing understanding of olfaction in mother–neonate interactions has important practical clinical implications (Winberg & Porter 1998). For example, removing breast odours by washing may reduce the olfactory cues important at the initiation of breastfeeding. Cleanliness is not necessarily a virtue. Similarly, close olfactory contact between mother and neonate in the hours after birth may be important for both to learn the odour signature(s) of the other, at a time when the mother and baby may be physiologically primed to learn each other's odours (Chapter 9).

Early recognition of newborn infants may be facilitated by the similarity of their odours to those of other family members to whom they are related (Schaal & Porter 1991). Many animals can use such 'phenotype matching' (Box 6.1) to recognise genetically related but unfamiliar animals. This ability was tested in humans by asking strangers to match the T-shirt odours of children and parents, which they were able to do successfully. It was assumed that if strangers could do this, family members would be able to. Cues such as diet in common between child and parent were ruled out as factors.

13.3.2 Odours and memory

Memories are powerfully associated with odours. Perhaps the most famous example in literature is Proust's account, in his novel *Swann's way* (the first volume of *Remembrance of things past*, published in 1913), of the way the taste and smell of a madeleine biscuit dipped in linden tea triggered intense joy and memories of the hero's childhood. Odour memory (of odours and the memories associated or evoked by odours) may be a separate memory system from visual-verbal memory (Herz & Engen 1996). Laboratory tests suggest, however, that odour-associated memories are not recalled more accurately. Rather, odours seem to be better cues to memory because they evoke more emotional recall, which is felt to be more real than other associative memory stimuli such as visual stimuli (Herz 1998). Recently, synthetic odours used in a museum to make the exhibit of tenth century Viking York more authentic gave an opportunity to make 'field tests' of memory associations (Aggleton & Waskett 1999). Even though the average time since their last visit was more than 6 years, people were able to answer more questions about the museum if the odours of the exhibits were presented during the questionnaire. Short-term memory of visual stimuli is superior to that of odour recognition tests. However, some 4 months later, the accuracy of odour recognition is greater than picture recognition (Engen & Ross 1973). Schaal & Porter (1991) wonder if the memory for individual odours might be a long-lasting mechanism for social recognition. Adult humans, for example, are able to identify their adult siblings correctly by odour even if they have not been in contact for up to 30 months.

13.3.3 Human mate choice

It is perhaps not surprising, given the subtlety of odour discriminations that we can make, that smell appears to be important in mate choice by humans. We use chemical senses, both taste and smell, during courtship and sex (in some languages, to kiss means to smell) (Penn & Potts 1998a).

The most searched-for human smell is a releaser pheromone to make the wearer irresistible to potential partners. There are a number of reports, mostly dating back to the 1970s, of the effects on human behaviour of male pig pheromone (Chapter 12). These experiments include, for example, the rating of photographs while exposed to 3α-androstenol (men rated the photographs of women more positively) and waiting room seats sprayed with 5α-androstenone (women tended to choose these, apparently). These same stories seem to get recycled in the popular media almost every year. The experiments were prompted by the discovery of androstenone in human armpits (Section 13.4.1). The experiments are interesting, and describe effects that may one day be shown to be robust. However, there are weaknesses in the methods of the experiments, for example experimental design or the lack of neutral odour controls (see Doty 1985). A more natural experiment on the attractiveness of 3α-androstenol showed no effects (Black & Biron 1982).

Various kinds of mate choice could be based on odour (Chapter 3). The development of scent glands in humans at puberty, in particular the axillary glands in our armpits, and the differences in rates of secretion in males and females, all point to these glands being sexually selected. In many mammals, females can use odour to choose dominant males, whose high androgen levels have produced the largest odour glands (Chapter 3). Theoretically, human females might be attracted to dominant males, those with the smelliest armpits, the main androgen-dependent glands in humans. However, we have no evidence of this in present-day humans (Section 13.4.1.1). Humans might use odour cues to avoid potential mates with diseases (bodily odours are used by doctors and dentists for diagnosis, Section 13.6.2). Odour cues related to symmetry might give similar information (Section 13.3.3.2). Finally, mates might be chosen for optimum outbreeding, avoiding close kin. These possibilities are discussed in Section 13.3.3.1.

Alas, despite the attraction of the idea of the irresistible smell, and the many products marketed on the strength of it, there is no good evidence yet that any smell will guarantee – or even increase – success (Preti & Wysocki 1999) (Section 13.6.4). However, evidence of odour-influenced mate choice in humans is just as interesting, particularly work on the possible role of the major histocompatibility complex (MHC), described in the next section.

13.3.3.1 The major histocompatibility complex (MHC) and human mate choice

Odours associated with the enormously variable major histocompatibility complex (MHC) of the immune system are used by mice to avoid mating with close kin (Chapter 3). The mice can distinguish otherwise genetically identical mice differing at only one MHC locus. Remarkably, humans can distinguish the odours of these same mice (Gilbert *et al.* 1986). Humans also have an MHC immune system and rats can learn to distinguish the urine odour of people who differ at the MHC (Ferstl *et al.* 1992), which in humans is called HLA (Human Leucocyte Antigen) because the molecules are on leukocytes, a type of white blood cell. As humans have both variability in MHC odours and the ability to detect them, do we use them in mate choice?

T-SHIRT ODOUR EXPERIMENTS

A first investigation of MHC and human mate choice examined the responses of Swiss female students to the odour of a cotton T-shirt that had been worn by a male student (Wedekind *et al.* 1995). Wedekind and colleagues tissue-typed the women and men to determine their HLA (MHC) group and compared the responses when the female had a similar or dissimilar MHC to the male who had worn the T-shirt (Fig. 3.6). Women *not* taking the oral contraceptive pill preferred the odour of T-shirts previously worn by MHC-dissimilar men. In contrast, women who were taking the pill preferred T-shirts worn by MHC-similar men. This result was not due to perceptual differences of the intensity of odours. Preferred odours by women both taking and not taking the pill were rated just as intense as less preferred odours (Fig. 3.6); this finding is important because strong odours often repel in this kind of experiment. When asked who the odour of the T-shirt reminded them of, women were more likely to say 'of a current or previous mate' if the T-shirt had the odour of an MHC-dissimilar male.

A second experiment, designed slightly differently, but with the same protocol for T-shirts, tested the responses of both genders to male and female odours (Wedekind & Füri 1997). Women not on the pill and men preferred the odour of MHC-dissimilar individuals, and the preference was negatively correlated with the degree of MHC similarity between the smeller and the T-shirt wearer. The preference seemed to be for difference in MHC, not for any particular complementary MHC combination. Again, in contrast, women taking the pill preferred MHC-similar odours.

The oral contraceptive pill acts by simulating pregnancy hormonally. The reversed preference of women taking the pill suggested to Wedekind and colleagues that during pregnancy they may prefer close kin, just as female mice tend to rear their young cooperatively with MHC-similar females (Chapter 3).

The first experiments have been criticised (see Hedrick & Loeschcke 1996, and Wedekind's reply) but Wedekind is cautious in his interpretation of the experiments. He argues only that they show that females and males can discriminate MHC-related odours. The 'reminding of past or present partners' data suggests (but no more) that these preferences are acted on.

In a study of perfumes and MHC, Milinski & Wedekind (2001) found that people with given MHC tended to choose the same perfume ingredients, and that these were different from those chosen by people with a different MHC. This was only true when the subjects were asked to choose a perfume for themselves, not when asked what perfume would they like a partner to wear. Milinski & Wedekind wonder if perfumes are selected 'for self' to amplify in some way the body odours that reveal a person's immunogenetics.

Women's choice of MHC-related male odours is influenced by the HLA (MHC) alleles they inherit from their fathers (but not their mothers) (Jacob *et al.* 2002). In this study, women were able to detect differences of one HLA allele among male odour donors with different MHC genotypes. They preferred odours from donors that had one HLA-allele in common with themselves, rather than none in common or identical HLA-alleles. The HLA-alleles inherited from their fathers were more important in determining odour preference than exposure to HLA-associated odours of their family. A fascinating subject has been opened up here and it would repay further investigation (see review by Penn 2002). As Wedekind & Füri (1997) conclude, 'no one smells good to everybody: it depends on who is sniffing whom, and it is related to their respective MHCs'.

EVIDENCE FROM MARRIAGE – POPULATION DATA

We would expect that if mates were chosen to be different at the MHC, there would be fewer same-MHC type matings than expected at random. However, the statistical demonstration of this can be difficult because the high polymorphism in the MHC makes nearly all mating combinations rare (Hedrick 1999). Non-random mating for MHC has been found in some mouse experiments (Chapter 3), but the evidence in human populations is contradictory (Edwards & Hedrick 1998; Penn 2002).

In one study, of 411 couples in a Hutterite population (a northern US and western Canadian group isolated by religious beliefs), couples were less likely to share MHC haplotypes (genotype) than expected, even after statistically controlling for cultural incest taboos (Ober *et al.* 1997). However, a second study, of 194 couples from 11 South Amerindian tribes, found no evidence for non-random MHC-dependent mating preference (Hedrick & Black 1997).

Ober *et al.* (1997) found that in married couples, each partner was slightly less likely to have the other's maternal MHC than expected by chance. When young, mice learn the MHC odours of their parents and siblings and when adult they prefer mates that are different (Chapter 3). If MHC-based choice in adult humans *does* occur, could it be as a result of olfactory imprinting by babies at their mother's nipple? An olfactory mechanism might also explain the 'kibbutz' or Westermarck effect in which people do not pair with people they have grown up with in the same 'family' or close social unit (see Schneider & Hendrix 2000).

One selective advantage for choosing a dissimilar partner is that there is a higher successful implantation rate if the foetus and mother differ in MHC (Chapter 3). This was revealed for a human population by an analysis of pregnancy outcomes in the Hutterite population mentioned above (see reviews by Ober 1999 and Penn 2002).

13.3.3.2 Symmetry and odour

Across the animal kingdom, females may use body symmetry as a sensitive indicator of male quality (see Chapter 3 for discussion). In humans, men who are more symmetrical apparently have more sexual partners, are thought more facially attractive and have more extra-pair-copulations (affairs) (references in Gangestad & Thornhill 1998). In a double-blind test, Gangestad and Thornhill asked women to rate the pleasantness of T-shirts worn by different men. Normally ovulating women preferred the scent of symmetrical men during their period of peak fertility. Such women showed no preference for symmetry or asymmetry during periods of low fertility in their cycle. Interestingly for the idea of an attractive human pheromone, women rated the

T-shirts worn by symmetrical men as more attractive than a clean, unworn T-shirt. It is too early to know what these effects might be due to.

13.3.4 Advertisement of oestrus or concealed ovulation?

In most mammals, hormone cycles lead females to be sexually receptive during discrete intervals known as oestrus, signalled by changes in behaviour, smells and visual signs (Hrdy & Whitten 1987). In some primate species, olfactory signals are important, in others there are visual signals such as brightly coloured sexual skin produced during oestrus. However, in a number of primate species, including humans, there is no signal of oestrus visible to human observers and females are sexually receptive throughout the cycle. These species, which include humans themselves, are termed 'concealed ovulators' (Sillén-Tullberg & Møller 1993). However, Dixson (1998, p. 352) argues many primate species described as 'concealed ovulators', *do* have rhythmic changes in sexual activity through the ovarian cycle which are signalled by olfactory cues, subtle visual cues or result from increases in female initiation of mating at those times. Thus to males of these species, for example vervets, marmosets and tamarins, oestrus is not concealed (even if it is from the human observer).

Is ovulation in humans really concealed from the male (and the woman herself?). There are no obvious visual signs but work in the 1970s suggested that cyclic changes in vaginal odour might be a signal in rhesus monkeys and, since we are also primates, perhaps in humans too.

In many monkey species, the common chimpanzee and gorilla, a male will intensively sniff the vagina of a prospective mate, insert a finger and sniff it before mating (Dixson 1998, p. 106). In the early 1970s releaser sex pheromones in rhesus monkeys were reported. The pheromones, named 'copulins', were said to be C_2–C_5 aliphatic acids, found in the vagina of the female around ovulation, which attracted males and stimulated male sexual behaviour (Michael & Keverne 1970). The same fatty acids are found in the vaginal secretions of some women (Section 13.6.4). However, other researchers (Goldfoot *et al.* 1976; Goldfoot 1981) questioned these results: olfactory cues were not required for rhesus monkey male response, the highest concentrations of the aliphatic acids in rhesus monkey females occurred *after* the fertile period, and adding the natural or synthetic putative 'copulin' mixtures to ovariectomised females did not increase mating by males (see reviews by Dixson 1998, Stoddart 1990, Preti & Wysocki 1999). Olfactory cues are *not* ruled out but, on current evidence, a key releaser role for vaginal odours is not supported. The earliest experiments implicating fatty acids (by Michael and colleagues) used conditioning responses with a small number of rhesus monkey males, and may have been inadvertently testing other learned responses to olfactory cues (see Michael & Keverne 1970). The story is a complicated one – not surprising since the sexual behaviour of primates is highly complex. Nonetheless, to date there is no evidence that 'copulins' exist as human pheromones. The only double-blind placebo-controlled study with 62 human couples over 3 months showed no increase in sexual activity when the synthetic aliphatic acid mixture 'copulins' rather than controls (water, alcohol or a perfume) had been applied by the woman to her chest at bedtime (Morris & Udry 1978). Disturbingly, the early Michael papers on 'copulins' are still sometimes quoted as evidence of human pheromones without mention of the associated debate.

Are there any odour cues to human ovulation? In a test of human responses to vaginal odours, Doty et al. (1975) found that although the mean results showed a pattern, there was so much individual variation between cycles that odours were unlikely to be a reliable way to identify the most fertile phase. However, as they admit, their 'sniff' test of odours by a panel of volunteers in a laboratory setting might be rather out of context. More objective physiological responses of males might be a better way to investigate the question. In many mammals, such as voles, blood testosterone rises in sexually mature males exposed to female odours (Chapter 9), and this was the measure used in an unpublished study by Jütte of responses of human males (Jütte, 1995 [not seen]). Unfortunately the stimuli were the 'copulins' mentioned above and the numbers were very small, but the approach is an interesting one.

Odour changes in axillary secretions and mouth odours across the menstrual cycle have been reported (Preti & Wysocki 1999) but until recently there was no evidence that humans could use olfactory cues to identify the ovulatory phase. Singh & Bronstad (2001) have now suggested that men may be able to distinguish odours from women in their ovulatory (late follicular) and non-ovulatory (luteal phases).

13.3.5 Priming pheromones

Primer pheromones act by evoking a cascade of effects via stimulation of the recipient's endocrine system; such effects on reproduction in mice, sheep and marmosets are well established (Chapters 6 and 9). Recently, evidence for primer pheromones in humans has been strengthened.

13.3.5.1 Menstrual synchrony
FEMALE–FEMALE INFLUENCES

Women who live together and interact over several months can, under some circumstances, develop synchronised menstrual cycles (McClintock 1971, 2000; Weller & Weller 1993, 1997).

Recent experiments have demonstrated a pheromone mechanism for menstrual synchrony: secretions produced by women at different stages in

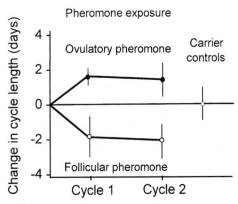

Fig. 13.2. Women produce pheromones which may affect the menstrual cycling of other women, leading to menstrual synchrony. This graph shows the effect of axillary compounds, donated by women during the follicular or ovulatory phases of their menstrual cycle, on the menstrual cycle length of recipients. Figure from Stern & McClintock (1998).

their menstrual cycle either speed up or slow down the cycles of other women. Building on earlier studies by Preti *et al.* (1986), Stern & McClintock (1998) collected armpit samples from nine healthy donor women at different phases of the menstrual cycle. The apparently odourless samples were applied daily to the upper lip of 20 recipient women, under their noses. Donor samples secreted during the late follicular phase brought forward the preovulatory surge of luteinising hormone of recipient women and shortened their menstrual cycles (Fig. 13.2). Samples collected later in the cycle (at ovulation) had the opposite effect. Two pheromones may be involved, as proposed for control of ovulation in rats (McClintock 2000): one, produced before ovulation, shortens the ovarian cycle; the other, produced at ovulation, lengthens the cycle. The experiments demonstrate that odours from other humans can have significant effects on neuroendocrine processes, in this case the menstrual cycle. The results open up the distant possibility of potential applications for control of fertility and contraception. In rats and mice, many social interactions are mediated by ovarian-dependent pheromones (Chapter 6). These include the age of puberty and inter-birth intervals. Perhaps some of these effects will be demonstrated in people in the future.

MALE–FEMALE INFLUENCES

Another influence on the female menstrual cycle comes from male axillary secretions (Cutler *et al.* 1986). Using a similar protocol for collecting axillary extracts, in this case from males, and applying them to the upper lip, the menstrual cycles of women volunteers with unusually short or long cycles became more regular and approached the normal length of 28 days or so. A control group of women given the solvent alcohol were not affected. The

Box 13.1 | Human sebaceous and sweat glands – major sources of odour

Three primary gland systems supply chemicals to the skin surface: the sebaceous glands and two types of sweat glands, the eccrine and apocrine (see figure below and Table 13.2) (Labows & Preti 1992).

The sebaceous glands are found over most of the body, especially the upper part. There may be up to 400–900 glands/cm^2 on the upper chest, back, scalp, face and forehead. Their secretions are rich in lipids.

The two kinds of sweat glands produce quite different secretions. Eccrine sweat glands are distributed over the entire body. Each human adult has about three million eccrine sweat glands, capable of secreting up to 12 litres of very dilute aqueous fluid per day, for thermoregulation. Some eccrine sweat glands respond to psychological stimuli. Eccrine sweat does not contribute much to odour. In what we colloquially describe as 'sweating' in hot weather or with hard physical exertion, almost all of the secretion comes from the eccrine sweat glands (hence the saying, 'honest sweat doesn't smell'). However, the moisture does help to create the damp conditions good for the growth of bacteria.

The apocrine glands secrete only small volumes of material (about 1–10 μl/day per gland) but are the most important for odour production. They are found primarily round the nipples, genitals and in the axillary region. They secrete into the hair follicles above the sebaceous gland, at times of emotion, stress or sexual arousal.

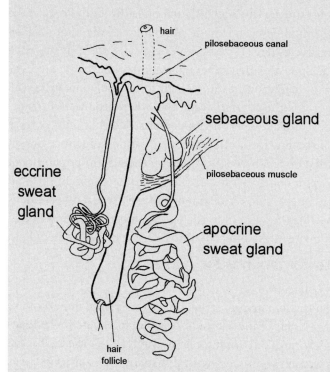

Diagram of a human hair follicle and associated glandular structures. Figure from Stoddart (1990), redrawn from Montagna & Parakkal (1974).

Sebaceous and apocrine glands are secondary sexual characteristics that develop fully in puberty, giving adult odours. The glands are androgen sensitive and are equivalent to the sexually selected apocrine and sebaceous glands in other mammals (Chapter 3).

effect seems to require physical closeness: non-intimate social contact between men and women is not enough.

WHY SYNCHRONISE?

The adaptive functions of menstrual synchrony are not clear. The advantages of synchrony of matings and subsequent births in some seasonally breeding primates might relate to predation pressure and to seasonal availability of food (Dixson 1998, p. 458). A troop with many young might move slowly, whereas a lone mother with her offspring might be left behind. Humans and apes are not seasonal breeders, however, and synchronisation of cycles in free-ranging gorillas or chimps has rarely been described. Other reasons for menstrual synchrony in humans can be speculated upon. First, menstrual synchrony would affect the operational sex ratio (the ratio of females in breeding condition to sexually active males), bringing it closer to 1:1. If all females ovulate at the same time, there is less opportunity for any one male to gain the majority of matings. Each male would, therefore, be forced to link to one or a small number of females, as seen in many nesting bird species with a short mating season. This may increase paternal investment, to the advantage of the woman. Second, the advantage of synchronised births might be the opportunity for communal rearing (Kiltie 1982; Lewis & Pusey 1997).

13.3.5.2 Age of female puberty?

Effects of primer pheromones on age at puberty in some prosimian primates seem likely but none has been proved (Dixson 1998, p. 460). For example, in a survey of female age at puberty in humans, Surbey (1990) concluded that physical exposure to a *related* man delays menarche (puberty), whereas physical exposure to an unrelated man (such as a stepfather) has the opposite effect.

13.4 | Candidate compounds for human pheromone odours

As described in the sections above, odours are important in many aspects of human mate choice and social interactions. Many of our body odours are produced by bacteria and other micro-organisms on the skin acting on largely odourless secretions. The characteristic odours of different parts of our body (Box 13.1, Tables 13.2 and 13.3) come from the interaction

Table 13.2. *Human secretory glands*

Gland	Distribution	Main secretions	Control	Notes
Eccrine sweat gland	Whole body surface	Aqueous secretion, reflects concentrations of dilute plasma	Thermoregulation, emotion/stress	
Apocrine sweat gland	Armpits (axillae), around nipples, around anus and on genital area	Fatty acids on carrier proteins, steroids	Continuous but especially if emotion/stress/sexual arousal	Secretes at and after puberty, androgen dependent
Sebaceous gland	Hair follicles, upper lip, male and female genitalia, scalp, forehead	Sebum (lipids such as cholesterol, cholesterol esters, long-chain fatty acids, squalene, triglycerides)	Continuous secretion	Secretes at and after puberty, androgen dependent

Constructed using data from Stoddart (1990) and Labows & Preti (1992).

Table 13.3. *The odour producing areas in humans*[a]

Area	Odour formation	Odour class
Scalp/hair	Yeast/bacteria on sebaceous lipids	Acids, lactones
Mouth/breath[b,c,d]	Bacteria on saliva amino acids and sugars	Acids, sulfides, indoles
Axillae/underarms (armpits)[b,c]	Bacteria on apocrine/ sebaceous secretions	Acids, steroids
Chest	Bacteria on apocrine/ sebaceous secretions	Acids, steroids
Genital/vaginal[b,c]	Yeast/bacteria on gland secretions	Acids, amines
Feet	Bacteria on eccrine secretions and lipids	Acids

[a-d] There are several metabolic disorders that may produce abnormal body odours and/or bad breath in adults (see Table 13.5). In addition, the odours in several of these body areas may be influenced by gender[b], physiological state[c], diet[d] and/or age.

After Labows & Preti (1992), with additional information from Preti & Wysocki (1999).

between secretions from different glands and the bacteria that live there (which are influenced, for example, by levels of moisture and oxygen) (see reviews by Doty 1985, Stoddart 1990, Schaal & Porter 1991, Labows & Preti 1992, Preti & Wysocki 1999). The secretion of semiochemicals at virtually all the odour-producing sites is dependent on endocrine activity, which in turn can be acutely sensitive to psychosocial events (Schaal & Porter 1991, p. 151).

Olfactory research on humans has mostly focused on the axillae, for many reasons: first, the growth of armpit hair and activity of glands in the axillae are some of the most conspicuous changes at puberty; second, axillary glands are unique to humans and the great apes (Table 13.1); third, the axillae differ between males and females in their odours (like male humans, male gorillas have a distinctive pungent odour, different from that of females, Dixson (1998)); fourth, primer, recognition and other olfactory effects can all be shown with samples taken from the upper body (in the archetypal T-shirt test), and most can be demonstrated with samples of axillary odours; and, fifth, the smells can be detected, when strong, from a distance.

The complex of sebaceous and apocrine glands, axillary hair, and associated micro-organisms in the underarm region have been termed the 'axillary scent organ' and are the source of the scent perhaps most characteristic of humans.

13.4.1 Axillae

The apocrine secretions in the axillae are odourless when first secreted, but if incubated with skin bacteria the characteristic odours emerge. The springy hair of the armpit provides a vast surface area for the bacteria to live on and an evaporating surface for odours when the axillae are exposed by lifting the arms – not unlike the hair pencils of male moths and tarsal-gland hair tufts of deer (Chapters 2 and 3). Shaving the armpit dramatically increases the length of time before odours build up (Fig. 13.3). Other ways of controlling odour by changing the axillary ecosystem include antiperspirants, which reduce the flow from eccrine glands so that there is less moisture and nutrients for bacterial growth, and compounds that change the pH to make it less favourable for bacterial growth, kill the bacteria or inhibit their enzymes (Labows & Preti 1992).

What are the characteristic odours? Much of the urine-like muskiness reported in human axillae is caused by 3α-androstenol (also found in human urine), 5α-androstenone and other steroids (Fig. 13.4) (Gower & Ruparelia 1993). Men produce up to 50 times more 5α-androstenone in their axillae than women. More is produced in the right axillae of right-handed people and similarly in the left axilla of left-handed people; the reason is not known. The steroid 5α-androstenone is also found in truffles, a gourmet fungus (Claus *et al.* 1981). This might explain its popularity – the nineteenth century French gourmand Brillat-Savarin (1825, p. 93) concluded that 'The truffle is not a true aphrodisiac; but . . . it can make women more affectionate and men more attentive'. Section 12.2.1.2 mentions the use of pigs to find them.

Fig. 13.3. The hair in the axillae provides a home for bacteria that convert the odourless secretions of the apocrine sweat glands into the characteristic armpit smells. The effect can be demonstrated by shaving one armpit of male volunteers, washing both their armpits with soap and then rinsing. A simple sniff test of each man's armpits showed that the unshaved, hairy axilla recovered its typical odour after 6 hours, whereas the shaved armpit only regained a noticeable odour after 24 hours. Figure redrawn from Shelley *et al.* (1953).

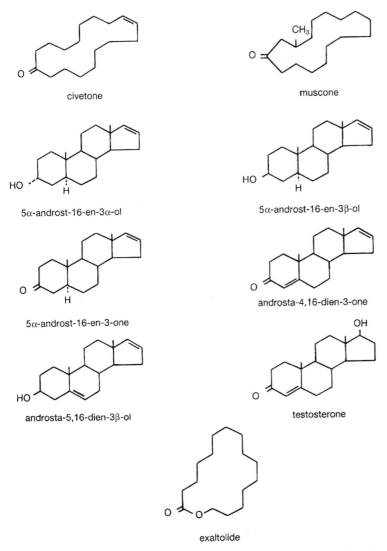

Fig. 13.4. Structures of some 16-androstenes. Testosterone and some other cyclic odorants are shown for comparison; 5α-androstenone (5α-androst-16-en-3-one in the figure) and 3α-androstenol are found in human male armpits and contribute a musky/urinous odour. Figure from Gower (1990).

However, recent organoleptic (sensory-based) and analytical chemical analysis suggests that other compounds create the axillae odours we find characteristic: volatile $C_6 – C_{12}$ branched, straight-chained, and unsaturated aliphatic acids (Fig. 13.5) (Appendices A1 and A2) (Zeng et al. 1991, 1992, 1996). In terms of general abundance, these acids, in particular (E)-3-methyl-2-hexenoic acid ((E)-3M2H or the synonym TMHA, Three Methyl Hexenoic Acid) are present at more than 100 times the amounts of the volatile steroids.

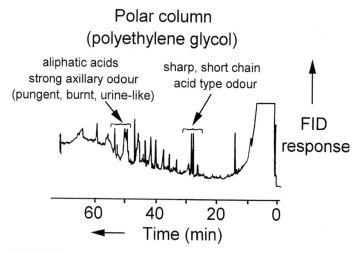

Fig. 13.5. While the main axillary odours of humans were long thought to come from steroids such as 5α-androstenone and 3α-androstenol, the results of the organoleptic (human sensory) sniff evaluation of the compounds in male axillary secretion extract suggest that the characteristic odours are associated with smaller aliphatic acids, which came out of the gas chromatography column much sooner than the steroids (which came out after more than 65 minutes on this column). FID, flame ionisation detector (see Chapter 2). Figure from Zeng *et al.* (1991).

The volatile aliphatic acids are secreted in the same odourless apocrine secretions as the steroids. Bacterial activity releases the acids from the carrier proteins they are bound to (Spielman *et al.* 1995). One of the binding proteins is apolipoprotein D (apoD), but it is glycosylated differently from the way it is elsewhere in the body, which is interesting as other members of the same protein family, the lipocalins, are implicated in specific pheromone-protein complexes in mice, hamsters, pigs and elephants (Preti & Wysocki 1999).

13.4.1.1 Which odours do what?

There are at least three functions that appear to be mediated by the complex odours produced by the axillae: first, recognition; second, potentially, mate choice; and, third, priming effects of possibly different kinds. These many effects could be signalled by different compounds in the cocktail. Currently we do not know but, for example, mate choice by females could theoretically be based on the odours of armpits reflecting androgen levels in two ways: by the size of the gland or its activity. Both of these mechanisms would lead to more odorous axillae, which might, if humans are like other mammals, have been attractive in the distant past (even if men with the most 'powerful' armpits do not win the most mates today). The species in the bacterial flora, and thus odours, might also vary with androgen levels. While the steroid odours may not be the characteristic odours of axillae, the much greater concentrations in male axillae, the variation between males and possible differences in sensitivity between males and females (Section 13.5), suggests they could have or have had this role in mate choice.

Males and females have largely the same aliphatic acids in their axillae, but the relative proportions differ (males have more (E)-3M2H than any of the straight-chain C_6–C_{11} acids, whereas the opposite occurs in females) (Preti & Wysocki 1999). While the original secretions may differ, many of the odour differences between males and females could be due to the differences in bacterial flora in their axillae. For example, men typically have higher proportions of *Corynebacterium* spp. than women and this is reflected in stronger odours. If the same odourless apocrine secretion is inoculated with different skin bacteria, very different odours are produced, which match the odours of the axillae providing the bacteria (Labows & Preti 1992).

13.5 | Perception of odours

Human sensitivities to different odours have been investigated for evidence that they are pheromones. We would expect that biologically important odours will be perceived by humans, even if unconsciously (as shown by EEG traces). The story is not clear, but there are interesting differences between the sexes (and between individuals).

For most odorants, for example floral smells, there is a bell-shaped normal distribution of the lowest thresholds of individuals, but for many odorants associated with the human body there is a bimodal distribution with individuals either being sensitive or insensitive to the odour (Labows & Preti 1992). This phenomenon is called specific anosmia, which is the inability to smell a specific odour molecule or related class of molecules at concentrations readily perceived by other individuals, despite an otherwise good sense of smell (Amoore 1977). These 'odour blindnesses' are similar to colour blindness, but whether the anosmics are different in their olfactory receptor proteins from those who can smell a particular odour is not yet known (see Chapter 9). While there are many other anosmias, those in Table 13.4 are particularly common. Labows & Preti (1992) comment that people live in different olfactory worlds.

Nearly half of the adult human population does not perceive an odour when sniffing 5α-androstenone, even at high concentrations (which is one reason why researchers looked for other odours in the axillae). In contrast, about 15% detect a subtle odour but are not offended by it and may even find it pleasant. The remaining 35% are exquisitely sensitive to 5α-androstenone, detecting less than 200 parts per trillion (10^{12}), and say the odour is like stale

Table 13.4. *Humans may be unable to smell particular compounds (specific anosmias), which include compounds found in body odours*

Odour source	Odorant	Primary odour (description)	Anosmia (%)
Axilla	Androstenone	Urinous	46–50
Axilla	Androstenol	Musky	12
Axilla	(E)-3-Methyl-2-hexenoic acid ((E)-3M2H)	Axillary	Some people, % not measured
Axilla	4-Ethylheptanoic acid	Hircine (goaty)	16
Foot	Isovaleric acid	Sweaty	3
Uraemic breath	Trimethylamine	Fishy	6
Semen	1-Pyrroline	Spermous	16

After Labows & Preti (1992).

urine or sweat. A genetic basis for the ability to smell 5α-androstenone has been shown in studies of identical and fraternal twins (Wysocki & Beauchamp 1984). The two types of twins did not differ in their response to a noxious test odour, pyridine, but identical twins had similar thresholds to each other for 5α-androstenone whereas fraternal twins showed little correlation. The influence of experience is discussed in Chapter 9.

Before puberty, most children can smell 5α-androstenone. However, during and after puberty many post-pubescent males become anosmic (Dorries *et al.* 1989). In contrast, females become more sensitive to androstenone as they get older and a larger proportion of women remain able to smell androstenone.

The differences in sensitivity of adult males and females to 5α-androstenone and related compounds found in some studies have been used as evidence that these compounds might be human pheromones. More women than men are sensitive to 5α-androstenone (see review by Gower & Ruparelia 1993) but up to a third of women are anosmic. Some studies have shown marked sex-related differences in sensitivity to the related alcohol, 3α-androstenol, and women were thought to show increased olfactory acuity just before ovulation and in the mid-luteal phase of the menstrual cycle. However, other workers have been unable to show this phenomenon. Doty (1985) reminds us that rhythms of all sorts of physiological parameters, especially of sensitivity to stimuli, vary with the menstrual cycle, including responses to many things almost certainly not sexual (Fig. 13.6).

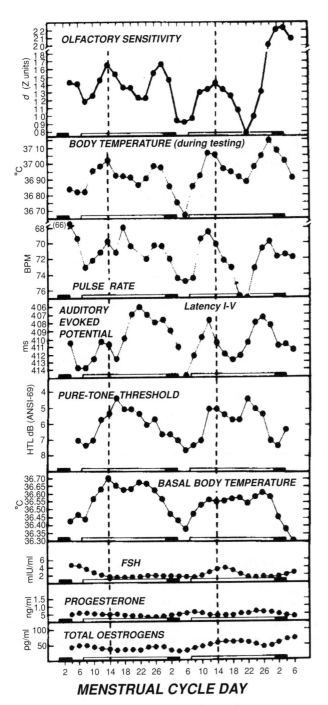

Fig. 13.6. Variation in sensitivity to putative human pheromones over a menstrual cycle may not be reliable evidence of biological activity because many other sensitivities, including hearing, also change as shown here in two consecutive menstrual cycles of a woman taking oral contraceptive medication. Dark rectangles on the abscissae signify periods of menstrual bleeding and open rectangles days during which the oral contraceptive medication was taken. Figure from Doty *et al.* (1982).

13.5.1 Do we have a functional vomeronasal organ?

In addition to the main olfactory epithelium, some mammals have an area of specialised sensory receptors for some pheromones, called the vomeronasal organ (VNO), which is described in more detail in Chapter 9 (Fig. 13.7). It is important to point out that while many sexually important pheromones in mice, other rodents and elephants, for example, are mediated by the VNO, in other mammals including other rodents, such as the hamster, some or all of the important sexual pheromones are sensed through the *main* olfactory system (MOE) (Chapter 9). For example, the male sex pheromone in pigs, 5α-androstenone, can be perceived by the female pig's MOE, and the VNO is not necessary for detection/discrimination of androstenone, although of course in unmanipulated animals it may be involved (Chapters 9 and 12) (Dorries *et al.* 1997).

Whether humans have a functional VNO is controversial (Preti & Wysocki 1999; McClintock 2000; Doty 2001; Meredith 2001). Most researchers seem agreed that there are pits in the nasal cavity, but whether these lead to a functioning VNO with active nervous connections to the brain is disputed.

A series of publications from Monti-Bloch and colleagues has maintained that the VNO is functional in human adults and that it responds to steroidal compounds different from the ones mentioned in the chapter so far (Monti-Bloch *et al.* 1998). The compounds, named 'vomeropherins' by the researchers, were isolated from human skin and were reportedly active on the VNO only. For example, one of the vomeropherins, estratetraenyl acetate (ETA) induced an electrovomerogram (EVG) in the putative VNO but did not stimulate the

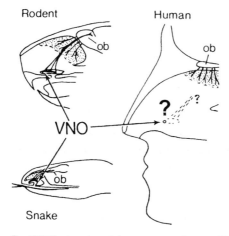

Fig. 13.7. The location of the vomeronasal organ (VNO) at the base of the nasal septum in rodents and snakes. The vomeronasal nerves project to the accessory olfactory bulb, dorsocaudal to the main olfactory bulb (ob). In humans, a vomeronasal organ is doubtful, the existence of a neural connection with the brain is uncertain, and no accessory bulb is identifiable. The stippled areas show the extent of the olfactory epithelium. Figure from Meredith (1999).

olfactory epithelium. Preti & Wysocki (1999) raise many reservations about the experiments.

The neuroanatomical and histological evidence for a VNO in humans and Old World primates is not positive (Eisthen 1992; Preti & Wysocki 1999; see also McClintock 2000, Doty 2001, Meredith 2001). Although New World primates such as the marmoset certainly have a functional VNO, it is doubtful that Old World primates do. If Old World primates do have a VNO then it would be easier to hypothesise the presence of one in humans. However, a functional VNO would need nerves from the structure to a functioning accessory olfactory bulb. This has not been demonstrated in adult humans or in any Old World primate. Instead, although a vomeronasal nerve does appear in early embryos, by 36 weeks old there is no trace of it in the human foetus, only a vestigial structure on the nasal septum (Boehm *et al.* 1994). However, some other researchers suggest there is evidence for survival of the nerve at 28 weeks (Smith *et al.* 1998) and Monti-Bloch *et al.* (1998) report other work suggesting its existence in adults. A functioning VNO would need to have olfactory sensory neurons and the presence of neurons has yet to be demonstrated in the putative VNO pit in humans. Rodriguez *et al.* (2000) described a putative VNO receptor gene expressed in the human olfactory mucosa but Kouros-Mehr *et al.* (2001) suggest that this may not be a vomeroreceptor and that all the human genes homologous with mouse VNO receptor genes are non-functional pseudogenes.

Overall, like most in the field, Preti & Wysocki (1999), Doty (2001) and Meredith (2001) are sceptical about the likelihood of a functional VNO in adult humans on current evidence. It would be very exciting if humans do indeed turn out to have an active VNO, but an important role for chemical signals in human biology does not depend on it.

Recently, two of the steroids, Δ4,16-androstadien-3-one and 1,3,5(10)16-estratetraene-3-ol, suggested by some researchers to be human 'pheromones' (above), have been shown to affect mood in humans when presented as odours (Jacob & McClintock 2000). Both steroids increased positive mood state in women but decreased it in men. These results are quite different from the sexually specific effects on the putative VNO epithelium reported by Monti-Bloch and colleagues (androstadienone, the male-typic steroid, affecting only females, and the female-typic steroid, estratatraenol affecting only males (see Jacob & McClintock 2000, for references).

13.6 | Putting human odours to use: applications

Getting rid of human odours is big business. In 1990, an estimated US$1.6 billion $[10^9]$ per year was spent on products to eliminate or mask axillary odours in the USA. However, there are uses for human odours and our reactions to them. The potential clinical importance of being aware of the role of odours for the newborn baby has already been mentioned (Section 13.3.1.1).

13.6.1 Forensic uses

Forensic uses for police detective work exploit some of the individual variation in smell between people that enables kin recognition in families (Section 13.3.1). It has long been known that dogs can distinguish people by odour, only being confused by identical twins. It was once even suggested that dogs might be used to sniff out MHC-compatible donors for organ transplants (see Section 13.3.3.1 for discussion of the human MHC) (Thomas 1974 in Penn & Potts 1998b). Dogs have remarkable abilities to track human scent trails (Chapter 10).

The acute olfactory abilities of dogs have also been used to match suspects to trace odours, left by hand prints for example, at the scene of the crime or on a weapon. An understanding of the reliability of dog decisions is important as the traces of odour are too small to be independently tested using current technologies such as gas chromatography. A particular question concerns whether dogs can generalise from a hand print, for example, to the odours of other parts of the same person. Hamsters cannot generalise across odours from different parts of another hamster unless they are familiar with the animal and have learned to associate the various odours with the individual (Chapter 6) (Johnston & Jernigan 1994). Dogs may have similar difficulties. Worryingly for use of dog identifications as evidence by the police, Brisbin & Austad (1991) found that dogs could not correctly distinguish their handler's elbow scent from a stranger's hand scent on objects. The training of dogs might inadvertently train them to respond to hand scents rather than to those of other parts of the body. However, a later study by Settle *et al.* (1994) found average success rates for a forensic task (comparing hand odour with an odour trace from another part of same person's body) greater than 80%. They recommended, though, that dogs used in criminal investigations should be selected for aptitude and the tests replicated so that conclusions are not based on tests by a single dog. Schoon (1997) has developed new ways of presenting the odour choices to the dogs that improve the reliability of dog decisions.

13.6.2 Medical uses of smell

Disease, either as a result of infection or metabolic disorder, may be reflected in unusual odours (Table 13.5) that can be detected by other people (and may influence human mate choice, Chapter 3). Smelling the patient can help in the diagnosis of many diseases and has a long history. One thousand years ago, the Arabian physician Avicenna observed that an individual's urine odour changed during sickness (Penn & Potts 1998a). Today,

Table 13.5. *Body odour is a diagnostic indicator for many human diseases*

Disease	Description of odour (and identity of key odours if known)
Infectious diseases	
Typhoid	Freshly baked brown bread
Diphtheria	Sweetish
Yellow fever	Butcher's shop
Scrofula (tuberculosis)	Stale beer
Gangrene	Obnoxious
Syphilis	Characteristic
Giardia	Rotten eggs (hydrogen sulfide)
Non-infectious diseases	
Diabetic ketosis	Breath and sweat have the fruity aroma of decomposing apples (acetone and other ketones)
Gout	Sweat has characteristic odour
Lung cancer	o-Toluidine, aniline (in breath)
Liver cirrhosis	Dimethyl sulphide, limonene (in breath)
Schizophrenia	Sweat pungent (high levels of (E)-3-methyl-2-hexanoic acid ((E)-3M2H))
Genetic metabolic disorders	
Trimethylaminuria	Like dead fish (trimethylamine, in breath and urine)
Phenylketonuria	A musty odour, resembling stale, sweaty locker-room towels (phenylacetic acid)

After Penn & Potts (1998), with additional information from Doty (1985) and Labows & Preti (1999).

an astute doctor can still make a preliminary diagnosis of many rare metabolic diseases of newborn babies by their characteristic smells alone and start life-saving therapy while the diagnosis is being confirmed by slower laboratory tests (Doty 2001). However, the use of smells by doctors for diagnosis is fast becoming a lost art. In the future, new technology may be designed to automate non-invasive diagnosis of disease from breath and other odour samples.

For some other diseases, early diagnosis can come from loss of the sense of smell. Loss of this sense can be due to many causes, including simple congestion of the nose from a cold. However, patients with early stages of Parkinson's disease and Alzheimer's disease often show partial or complete loss of smell, which distinguishes these from other neurological diseases with similar early symptoms (Doty 2001). The smell loss can be revealed by using standardised olfactory tests with micro-encapsulated 'scratch & sniff' odorants.

13.6.3 Odours as treatment/mood changers

Given the importance of odours in so many aspects of human biology, it is highly likely that both human and other odours used as treatments will have an influence on us but I do not have space to cover the growing research on this interesting topic. There is a large literature on the effects of artificial (non-human) odours, ranging from peppermint to aromatherapy, some it based on unsubstantiated claims, some of it well researched (see Van Toller & Dodd 1992, Martin 1996, Lorig 1999). For example, Redd *et al.* (1994) used a fragrance to reduce stress in patients being scanned in MRI machines. The patients reported less anxiety, although there was no change in physiological measurements such as heart rate.

13.6.4 Commercial products containing 'human pheromones'

Human pheromones to make the wearer irresistibly attractive to potential partners have long been searched for. People are willing to pay a great deal for products that make these claims, despite the lack of firm evidence (Preti & Wysocki 1999; McClintock 2000). In the 1970s 'copulins' were patented as human releaser pheromones, based on research on rhesus monkeys. These were apparently added to perfumes (Box 13.2, Section 13.3.4). Nowadays, products claiming to contain human pheromones are widely advertised on the Internet as well as in the popular press. These range from 5α-androstenone and 3α-androstenol to other steroid-like compounds.

However, Preti & Wysocki (1999) conclude that 'No peer-reviewed data supporting the presence of endogenous, human derived "pheromones" that cause rapid behavioural changes, such as attraction and/or copulation have been documented.'

Box 13.2 | Some commercial products (past and present) with alleged pheromonal properties

COPULINS

These are a mixture of C2–C5 aliphatic acids with alleged releaser pheromone activity in rhesus monkeys but they have been tested in double blind study in humans with no effect (Morris & Udry 1978).
- A French patent claims similar effects in humans (Michael 1972)
- They were added to many fragrances (c. 1970–79)

ANDROSTENONE/ANDROSTENOL

These are present in sweat and boar saliva and act as a releaser pheromone in pigs but there is no support for a releaser pheromone in humans (Black & Biron 1982).

- Androstenol was added to Andron® fragrances (Jovan; *c.* 1979–86)
- Products with androstenol or musks are available via advertisements in tabloids by or mail order
- Companies advertise on the World Wide Web pheromone/sex attractant products for men containing androstenone or androstenol as the alleged 'active ingredient', e.g. The Scent, The Secrete, Yes Pheromone, Sex Attractant for Men, each suggesting that women will be irresistibly drawn to the wearer. There are no similar products for women.

REALM MEN®/REALM WOMEN®

These are fragrances that contain 'vomeropherins'– claimed by some to be pheromones that simulate or send messages through the putative human vomeronasal organ.

'Vomeropherins' consist of volatile and non-volatile androstenes, progestins and oestrogens said to be isolated from the skin, although there are no published peer-reviewed studies describing the isolation and identification process.

It is alleged that the 'Wearer of these fragrances will feel more relaxed and self-confident, hence more attractive.'

PHEROMONE 1013

This product contains dehydroepiandrosterone (DHEA), a non-volatile steroid in ethanol.

It is alleged to make women more confident and attractive and increase their love life and is added to an individual's favourite fragrance.

PHEROMONE 10X

This product may contain volatile and non-volatile steroids, synthetic musks, lipids and silicon compounds.

It is alleged to increase attractiveness and romantic encounters and is added to men's cologne or aftershave.

After Preti & Wysocki (1999).

13.7 | Conclusion

To smell is human. We each produce a heady mixture of odours and, clearly, smells do affect our behaviour and physiology. However, we are at a very early stage in research into human olfaction and communication. If only we knew as much about humans as we do about moths and mice!

Odours and potential human pheromones are undoubtedly important, but hard to investigate. One of the hardest problems to overcome is cultural conditioning of responses to odours. Nonetheless, more conclusive studies of putative human pheromones are coming within reach as diverse techniques in chemical analysis, behaviour, molecular biology and brain imaging can be focused on the same investigation. We have strong candidate compounds to investigate: the volatile steroids (including 5α-androstenone and 3α-androstenol) and the short chain C_6–C_{12} acids such as (E)-3M2H from human armpits and many other compounds from other parts of the body. McClintock (1998) makes the important point that for pheromones to be treated as a signal by mammals such as the hamster, a complex interaction of prior experience and the context of the message is necessary (Chapter 9); we should expect human responses to be no less complex.

One of the major challenges to human pheromone research is that of designing rigorous experiments that eliminate other cues and variables. As well as the complexity of odour that being a mammal brings, humans are also complex emotionally. This makes us doubly difficult as experimental subjects. The greatest need is for well controlled experiments, run double blind so that neither subject nor experimenter knows which treatment is being tested. Full publication in the peer-reviewed scientific literature will also be needed if we are to make progress. There is also a strong need to bring the serious human odour investigators and hypothesis-testing evolutionary biologists together – so that better tests are made.

The early discoveries of primer effects on mouse fertility encouraged research into the possibilities of human equivalents. The prize of identifying the compounds responsible for the demonstrated effects in humans still waits to be claimed.

Even though we try hard to forget it, we are mammals first and foremost and, as such, odours are likely to be important means of communication. Indeed, as one of the smelliest primates, there is surely an interesting world of olfactory communication to investigate. No doubt there will be some surprises to come as we get the results.

To summarise, demonstrated effects in research on human pheromones include parent–offspring recognition and menstrual synchrony. However, firm conclusions on human pheromones in sexual attraction remain elusive despite unceasing interest from the popular press. There is a paradox in the enormous interest, yet surprisingly little solid work on this subject; when it comes to pheromones, we know much more about moths and mice than people.

13.8 | Further reading

This chapter touches on so many interesting topics that I do not have space to explore in this book. However, it is worth bearing in mind that the field has been changing rapidly in the last 5 years, in particular for data on the human MHC and the debate on whether the human VNO is functional.

On the VNO, the review by Meredith (2001) of the best and worst cases is recommended as a starting point.

Dixson (1998) describes primate sexuality, including that of humans, from a biologist's point of view. Stoddart (1990) gives a very interesting overview of human olfaction and reproduction in both biological and historical settings. Schaal & Porter (1991) give a full and enjoyable review. Classen *et al.* (1994) discuss the sociology of smell and are strong on history and social context. Patrick Süskind's novel *Perfume* (Süskind 1986) is chilling but compelling and is translated into many languages. Two popular accounts of our senses are by Ackerman (1990), who discusses all of the senses and Vroon *et al.* (1997) who describe smell only.

A number of web sites offer entry points to good sources of information about chemical ecology on the World Wide Web. These include: Monell Chemical Senses Centre, Philadelphia, USA (http://www.monell.org/); the US National Institute on Deafness and Other Communication Disorders (NIDCA), which maintains a site with general information about smell and taste disorders at http://www.nidcd.nih.gov/health/st.htm; and the Leffingwell & Associates website at http://www.leffingwell.com/, which is aimed at perfume, flavour, food and beverage companies but has useful general information on perfume and flavour chemistry and olfaction.

Appendix A1

An introduction to pheromones for non-chemists

This brief introduction to the chemical structure of pheromones is largely after Stevens in Howse *et al.* (1998). Like other organic molecules, pheromones are based on a chain of carbon and attached hydrogen atoms. The carbon backbone forms a zigzag because of the tetrahedral arrangement of the four carbon bonds (these angles are important for other characteristics of the molecular shape, see Appendix A2 Isomers).

The hydrogen atoms attached to the carbon backbone lie in two planes, above and below the paper, (represented here by bonds as solid wedges (the plane above) and as dashed wedges (the plane below)):

The structure is often simplified to show just the carbon backbone:

When other atoms such as oxygen or nitrogen, or other functional groups, are added to the chain or substituted for hydrogens or carbons, the chemical nature of the molecule changes (see Table A1.1 for common functional groups).

The naming of compounds tells the reader the length of the carbon chain and what and where the important functional groups occur. A common chemical modification of the alcohol group is the combination with an organic acid (with loss of a water) to give an ester. The naming of the

Table A1.1. *Prefixes and suffixes for common functional groups*

Functional group	Formula	Prefix	Suffix
Alcohol	-OH	Hydroxy-	-ol
Aldehyde	-CH=O	Formyl-	-al
Amine	-NH$_2$	Amino-	-amine
Carboxylic acid	-COOH	Carboxy-	-oic acid
Ester	-COOR	R-oxycarbonyl-	-R-oate
Ketone	>C=O	Oxo-	-one

From Howse et al. (1988).

molecule reflects the addition of the acid, often acetic acid, to the original chain, for example to produce dodecan-1-ol acetate:

dodecanol   acetic acid

The name also indicates the number and position of carbon double bonds (C=C). The formal name for the pheromone of the pink bollworm (*Pectinophora gossypiella*) is (Z,E)-7,11-hexadecandien-1-ol acetate (and the Z,Z isomer; see Appendix A2) (common name Gossyplure). The 'dien' tells us that there are two double bonds, the '7,11' that these occur at carbons 7 and 11:

Appendix A2

Isomers and pheromones

Pheromone synthesis is catalysed by enzymes and pheromones are detected by receptors, both of which are proteins, which recognise their substrates and signal molecules (ligands), respectively, by shape in three dimensions. Molecules with atoms in the same formula can also sometimes be put together in a number of different structures. Even more subtly, molecules with the same structure (connections between atoms) can have different shapes in three dimensions. These different variations are called isomers.

This potential variation in molecular shape, in synthesis and reception, has been acted on by natural selection so that now these variations in shape are vitally important to many species for species recognition. Each of the possible isomer types can be important biologically as in each case the shape of the molecule is different and it may stimulate a different range of olfactory receptors (Chapter 9).

There are two main different kinds of isomers: constitutional isomers have the atoms connected in different ways, whereas stereoisomers have the same connectivity but differ in the arrangement of atoms in space.

A2.1 | Constitutional isomers

A2.1.1 Functional group isomers

From the same molecular formula the atoms can be connected in ways that produce different functional groups, giving quite different chemical properties to the molecule. For example, n-butyl alcohol (below left) and ethyl ether (below right) are both $C_4H_{10}O$

A2.1.2 Positional isomers

Positional isomers differ, for example, in the position of a functional group (e.g. 2-heptanone and 3-heptanone) or the position of a double bond

Table A2.1. *Naming of isomers: the meanings of the letters and symbols*

The development of chemical nomenclature has left us with a number of different naming schemes, including some for the same differences, which can be slightly confusing. In addition, some chemical formulae allow for many possible isomers to be made from them so that the range of isomers could include, say, geometrical, positional and optical isomers among the possible molecules sharing the same chemical formula.

Symbols	Synonym	Type of isomer
E, Z	*trans, cis* (older texts)	Geometrical spatial arrangement around double bond
L, D	− , +	Optical isomers (*laevo, dextro* twist of polarised light)
S, R		Plane of bond around chiral carbon

(e.g. Z-9-tetradecen-1-ol acetate and Z-11-tetradecen-1-ol acetate, the pheromone components of the summer fruit tortrix moth, *Adoxophyes orana*):

2-heptanone   **3-heptanone**

A2.2 | Stereoisomers

Stereoisomers are compounds with identical formulae and order of connecting the atoms together, but with different shapes due to different spatial orientations of the atoms. Stereoisomers may behave almost identically in most chemical reactions but in biological systems, driven by enzymes and receptor proteins sensitive to the shape of molecules, these differences have profound effects. There are a several types of stereoisomer. These are described in Sections A2.2.1 and A2.2.2, and the nomenclature used is shown in Table A2.1.

A2.2.1 Chirality and enantiomers (optical isomers)

Chiral molecules are identical except that they are mirror images of each other and, just like your hands, cannot be superimposed on each other (chiral comes from the Greek word *cheir*, meaning 'hand'). The two mirror images are called **enantiomers**. The ability to distinguish between a pair of enantiomers requires a chiral agent; a glove is a chiral agent that distinguishes right and left hands. Enzymes and receptors are just such systems in biology. Nature is inherently chiral.

The chirality is a consequence of the tetrahedral geometry of carbon atoms: if different groups are attached to each of the four bonds of one of the carbon atoms, the molecule can be made in two different ways (or as two enantiomers, see digram below). Such a carbon atom is called a chiral carbon or chiral centre:

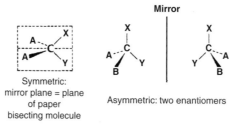

Symmetric:
mirror plane = plane of paper bisecting molecule

Asymmetric: two enantiomers

Symmetry and asymmetry in tetrahedral carbon.

A special property of chiral compounds, which led to the discovery of the phenomenon, is optical activity. Solutions of pure enantiomers rotate the plane of polarisation of light passing through them: opposite enantiomers rotate it in opposite directions. L-molecules rotate the plane of polarisation to the left, and D-molecules rotate it to the right. The abbreviations come from Latin: L (*laevo*, for 'left') and D (*dextro*, 'right'). L and D are often called − and +, respectively.

An example of our sensitivity to enantiomers is L-carvone: this gives us the smell of spearmint whereas its mirror image, D-carvone, is perceived by us as the very different smell of caraway:

mirror

L-carvone	D-carvone
smell: spearmint	caraway

A **racemic mixture** or **racemate** is an equal mix of the two enantiomers. As they cancel each other out optically, such solutions do not rotate the plane of polarisation of light, so they are optically inactive. Most chemically synthesised compounds, unless enzymes or other special steps are used in synthesis or purification, are racemic mixtures.

The L, D naming system is based on the observation of the direction in which polarised light is shifted but does not tell us the actual position of all atoms in space around a molecule. This is done by the **absolute**

configuration, which uses a rather complicated set of rules which allow the chemist to describe the orientation at each of the chiral centres as either *R* (from *rectus* = 'right') and *S* (from *sinister* = left). As the *R* and *S* are defined according to a set of nomenclatural rules, these do not necessarily predict which way the whole molecule will shift polarised light (L/D). The rules for naming are not important here, but biologically, the differences *are* important (see Howse *et al.* 1998, p. 152).

An example of stereoselectivity in a mammalian pheromone is dehydro-*exo*-brevicomin, which comes in *R,R* and *S,S* forms; only the *R,R* form is active biologically and only this form is produced by male mice (Novotny *et al.* 1995):

(*R*,*R*)-3,4-dehydro-*exo*-brevicomin

(*S*,*S*)-3,4-dehydro-*exo*-brevicomin

A2.2.2 Diastereoisomers

When a molecule has more than one chiral centre, a second form of stereoisomerism is possible, diastereoisomerism. Each of the chiral centres can be in one of the two forms, *R* and *S*. If there are two chiral centres there will be 2^2 stereoisomers, that is four. For example, 2,3-octandiol (a pheromone produced by the grape borer beetle (*Xylotrichus pyrrhoderus*) occurs in four forms: (2R,3R), (2R,3S), (2S,3S) and (2S,3R):

HO H

HO H   (2*R*,3*R*)

HO H

H OH   (2*R*,3*S*)

HO H

HO H   (2*S*,3*S*)

H OH

HO H   (2*S*,3*R*)

The four optical isomers of 2, 3-octandiol.

Some of these molecules represent mirror images of each other, for example, (2S,3S) and (2R,3R) and are called enantiomers. Other pairs of isomers, such as (2R,3R) and (2S,3R) are not mirror images and these are called **diastereoisomers**, that is, they are stereoisomers that are not enantiomers. These are common in branched compounds. In the case of the grape borer beetle, only one of the four isomers is attractive (2S,3S).

If there are three chiral centres in the molecule there will be eight isomers (2^3), as the number of optical isomers is 2^N where N is the number of chiral centres.

A2.2.2.1 Geometrical isomers

Geometrical isomers are a particular form of diastereoisomer. Many moth pheromones have particular hydrogens removed, catalysed by specific desaturase enzymes, to leave double bonds. The double bond makes the carbon chain rigid at that point. Because of the angles of the chain, there are two versions of the molecule depending on whether the remaining chains are connected to (i.e. spatially arranged around) the double bond on opposite sides (*E*) or the same side (*Z*) (from the German *Entgegen*, opposite, and *Zusammen*, together). These two forms are different chemical compounds with characteristic chemical and physical properties. You will also come across the older terms *trans* (now E) and *cis* (now Z).

(Z,E)-7,11 hexadecandien-1-ol acetate

Appendix A3

Further reading on pheromone chemical structure

The chapters by Stevens in Howse *et al.* (1998) are warmly recommended as an introduction to chemical structures for non-chemists. Although the book is principally about insect pheromones, it would also be useful to people working on vertebrates. The Leffingwell & Associates web site http://www.leffingwell.com/chirality/chirality2.htm, has molecular structures of odour molecules that you can visualise on your computer.

References

Abbott, D. H., Baines, D. A., Faulkes, C. G., Jennens, D. C., Ning, P. C. Y. K. & Tomlinson, A. J. (1990). A natural deer repellent: chemistry and behaviour. In *Chemical signals in vertebrates*, Vol. 5, ed. D. W. Macdonald, D. Müller-Schwarze & S. E. Natynczuk, pp. 599–609. Oxford: Oxford Science Publications, Oxford University Press.

Abbott, D. H., Saltzman, W., Schultz-Darken, N. J. & Tannenbaum, P. L. (1998). Adaptations to subordinate status in female marmoset monkeys. *Comparative Biochemistry and Physiology C-Pharmacology Toxicology & Endocrinology*, **119**, 261–274.

Ackerman, D. (1990). *A natural history of the senses.* New York: Random House (Phoenix Paperback).

Adams, E. S. & Traniello, J. F. A. (1981). Chemical interference competition by *Monomorium minimum* (Hymenoptera, Formicidae). *Oecologia*, **51**, 265–270.

Aggleton, J. P. & Waskett, L. (1999). The ability of odours to serve as state-dependent cues for real-world memories: can Viking smells aid the recall of Viking experiences? *British Journal of Psychology*, **90**, 1–7.

Agosta, W. C. (1992). *Chemical communication: the language of pheromones.* San Francisco: Scientific Library, W. H. Freeman.

Akino, T., Knapp, J. J., Thomas, J. A. & Elmes, G. W. (1999). Chemical mimicry and host specificity in the butterfly *Maculinea rebeli*, a social parasite of *Myrmica* ant colonies. *Proceedings of the Royal Society of London Series B-Biological Sciences*, **266**, 1419–1426.

Al Abassi, S., Birkett, M. A., Pettersson, J., Pickett, J. A. & Woodcock, C. M. (1998). Ladybird beetle odour identified and found to be responsible for attraction between adults. *Cellular and Molecular Life Sciences*, **54**, 876–879.

Alberts, A. C. (1990). Chemical-properties of femoral gland secretions in the desert iguana, *Dipsosaurus dorsalis. Journal of Chemical Ecology*, **16**, 13–25.

Alberts, A. C. (1992). Constraints on the design of chemical communication-systems in terrestrial vertebrates. *American Naturalist*, **139**, 62–89.

Alberts, A. C. (1993). Chemical and behavioral-studies of femoral gland secretions in iguanid lizards. *Brain Behavior and Evolution*, **41**, 255–260.

Albone, E. S. (1984). *Mammalian semiochemistry: the investigation of chemical signals between mammals.* Chichester: John Wiley.

Albone, E. S. & Perry, G. C. (1976). Anal sac secretion of the red fox, *Vulpes vulpes*: volatile fatty acids and diamines: implications for a fermentation hypothesis of chemical recognition. *Journal of Chemical Ecology*, **2**, 101–111.

Alcock, J. (1982). Natural-selection and communication among bark beetles. *Florida Entomologist*, **65**, 17–32.

Alcock, J. (1989). *Animal behaviour. An evolutionary approach*, 4th edn. Sunderland, Mass.: Sinauer Associates.

Alcock, J. (1998). *Animal behaviour. An evolutionary approach*, 6th edn. Sunderland, Mass.: Sinauer Associates.

Aldrich, J. R. (1995). Chemical communication in the true bugs and parasitoid exploitation. In *Chemical ecology of insects 2*, ed. R. T. Cardé & W. J. Bell, pp. 318–363. London: Chapman and Hall.

Aldrich, J. R. (1999). Predators. In *Pheromones of non-lepidopteran insects associated with agricultural plants*, ed. J. Hardie & A. K. Minks, pp. 357–381. Wallingford, Oxon: CAB International.

Aldrich, J. R., Kochansky, J. R. & Abrams, C. B. (1984). Attractant for a beneficial insect and its parasitoids: pheromone of the predatory spined soldier bug, *Podisus maculiventris* (Hemiptera: Pentatomidae). *Environmental Entomology*, **13**, 1031–1036.

Aldrich, J. R., Neal, J. W., Oliver, J. E. & Lusby, W. R. (1991). Chemistry vis-à-vis maternalism in lace bugs (Heteroptera, Tingidae) – alarm pheromones and exudate defense in *Corythucha* and *Gargaphia* species. *Journal of Chemical Ecology*, **17**, 2307–2322.

Allan, R. A., Elgar, M. A. & Capon, R. J. (1996). Exploitation of an ant chemical alarm signal by the zodariid spider *Habronestes bradleyi* Walckenaer. *Proceedings of the Royal Society of London Series B-Biological Sciences*, **263**, 69–73.

Allee, W. C. (1931). *Animal aggregations, a study in general sociology*. Chicago: Chicago University Press.

Alonso, W. J. (1998). The role of kin selection theory on the explanation of biological altruism: a critical review. *Journal of Comparative Biology*, **3**, 1–14.

Amoore, J. E. (1977). Specific anosmia and the concept of primary odors. *Chemical Senses*, **2**, 267–281.

Andersson, M. (1994). *Sexual selection*. Princeton, N. J.: Princeton University Press.

Andersson, M. & Iwasa, Y. (1996). Sexual selection. *Trends in Ecology & Evolution*, **11**, 53–58.

Andrade, M. C. B. & Roitberg, B. D. (1995). Rapid response to intraclonal selection in the pea aphid (*Acyrthosiphon pisum*). *Evolutionary Ecology*, **9**, 397–410.

Aoki, S. (1977). *Colophina clematis* (Homoptera, Pemphigidae), an aphid species with soldiers. *Kontyû*, **45**, 276–282.

Apanius, V., Penn, D. [J.], Slev, P. R., Ruff, L. R. & Potts, W. K. (1997). The nature of selection on the major histocompatibility complex. *Critical Reviews in Immunology*, **17**, 179–224.

Appelt, C. W. & Sorensen, P. W. (1999). Freshwater fish release urinary pheromones in a pulsatile manner. In *Advances in chemical signals in vertebrates*, ed. R. E. Johnston, D. Müller-Schwarze & P. W. Sorenson, pp. 247–56. New York: Kluwer Academic Publishers/Plenum Press.

Arakaki, N. (1989). Alarm pheromone eliciting attack and escape responses in the sugarcane woolly aphid, *Ceratovacuna lanigera* (Homoptera, Pemphigidae). *Journal of Ethology*, **7**, 83–90.

Arakaki, N. (1990). Colony defense by first instar nymphs and dual function of alarm pheromone in the sugar cane woolly aphid, *Ceratovacuna lanigera*. In *Social insects and the environment*, ed. G. K. Veeresh, B. Mallik & C. A. Viraktamath, pp. 299–300. Bombay: Oxford University Press.

Arakaki, N., Wakamura, S. & Yasuda, T. (1996). Phoretic egg parasitoid, *Telenomus euproctidis* (Hymenoptera, Scelionidae), uses sex-pheromone of tussock moth *Euproctis taiwana* (Lepidoptera, Lymantriidae) as a kairomone. *Journal of Chemical Ecology*, **22**, 1079–1085.

Arakaki, N., Wakamura, S., Yasuda, T. & Yamagishi, K. (1997). Two regional strains of a phoretic egg parasitoid, *Telenomus euproctidis* (Hymenoptera: Scelionidae), that use different sex pheromones of two allopatric tussock moth species as kairomones. *Journal of Chemical Ecology*, **23**, 153–161.

Arathi, H. S., Shakarad, M. & Gadagkar, R. (1997). Factors affecting the acceptance of alien conspecifics on nests of the primitively eusocial wasp, *Ropalidia marginata* (Hymenoptera: Vespidae). *Journal of Insect Behavior*, **10**, 343–353.

Arbas, E. A., Willis, M. A. & Kanzaki, R. (1993). Organization of goal-oriented locomotion: pheromone-modulated flight behavior of moths. In *Biological neural networks in invertebrate neuroethology and robotics*, ed. R. D. Beer, R. E. Ritzmann & T. McKenna, pp. 159–198. Boston, Mass.: Academic Press.

Arcese, P. (1999). Effect of auxiliary males on territory ownership in the oribi and the attributes of multimale groups. *Animal Behaviour*, **57**, 61–71.

Arn, H. (1990). Pheromones: prophesies, economics, and the ground swell. In *Behavior-modifying chemicals for insect management*, ed. R. L. Ridgway, R. M. Silverstein & M. N. Inscoe, pp. 717–722. New York: Marcel Dekker.

Arn, H. & Louis, F. (1997). Mating disruption in European vineyards. In *Insect pheromone research: new directions*, ed. R. T. Cardé & A. K. Minks, pp. 377–382. New York: Chapman and Hall.

Arn, H., Tóth, M. & Priesner, E. (1995). *The pherolist*. Internet edn: http://www. nysaes. cornell. edu/pheronet.

Arnold, G., Le Conte, Y., Trouiller, J., Hervet, H., Chappe, B. & Masson, C. (1994). Inhibition of worker honeybee ovaries development by a mixture of fatty-acid esters from larvae. *Comptes Rendus de l'Academie des Sciences Serie III-Sciences de la Vie-Life Sciences*, **317**, 511–515.

Arvedlund, M., McCormick, M. I., Fautin, D. G. & Bildsøe, M. (1999). Host recognition and possible imprinting in the anemonefish *Amphiprion melanopus* (Pisces: Pomacentridae). *Marine Ecology-Progress Series*, **188**, 207–218.

Asa, C. S. (1997). Hormonal and experiential factors in the expression of social and parental behavior in canids. In *Cooperative breeding in mammals*, ed. N. G. Solomon & J. A. French, pp. 129–149. Cambridge: Cambridge University Press.

Atema, J. (1986). Review of sexual selection and chemical communication in the lobster, *Homarus americanus*. *Canadian Journal of Fisheries and Aquatic Sciences*, **43**, 2283–2290.

Atema, J. (1995). Chemical signals in the marine environment: dispersal, detection, and temporal signal analysis. In *Chemical ecology: the chemistry of biotic interaction*, ed. T. Eisner & J. Meinwald, pp. 147–159. Washington, D.C.: National Academy of Sciences.

Atema, J. (1996). Eddy chemotaxis and odor landscapes – exploration of nature with animal sensors. *Biological Bulletin*, **191**, 129–138.

Atema, J. (1998). Tracking turbulence: processing the bimodal signals that define an odor plume. *Biological Bulletin*, **195**, 179–180.

Averill, A. L. & Prokopy, R. J. (1987). Intraspecific competition in the tephritid fruit-fly *Rhagoletis pomonella*. *Ecology*, **68**, 878–886.

Axén, A. H., Leimar, O. & Hoffman, V. (1996). Signalling in a mutualistic interaction. *Animal Behaviour*, **52**, 321–333.

Ayasse, M., Birnbaum, J., Tengö, J., van Doorn, A., Taghizadeh, T. & Francke, W. (1999). Caste- and colony- specific chemical signals on eggs of the bumble bee, *Bombus terrestris* L. (Hymenoptera: Apidae). *Chemoecology*, **9**, 119–126.

Baker, T. C. (1990). Upwind flight and casting flight: complementary phasic and tonic systems used for location of pheromone sources by male moths. In *Proceedings of the 10th International Symposium on Olfaction and Taste*, ed. K. B. Døving, pp. 18–25. Oslo: GCS A/S.

Baker, T. C. & Haynes, K. F. (1987). Manoeuvres used by flying male oriental fruit moths to relocate a sex pheromone plume in an experimentally shifted wind-field. *Physiological Entomology*, **12**, 263–279.

Baker, T. C. & Vickers, N. J. (1997). Pheromone-mediated flight in moths. In *Insect pheromone research: new directions*, ed. R. T. Cardé & A. K. Minks, pp. 248–264. New York: Chapman and Hall.

Baker, T. C., Fadamiro, H. Y. & Cosse, A. A. (1998). Moth uses fine tuning for odour resolution. *Nature*, **393**, 530.

Balthazart, J. & Schoffeniels, E. (1979). Pheromones are involved in the control of sexual behavior in birds. *Naturwissenschaften*, **66**, 55–56.

Bargmann, C. I. & Horvitz, H. R. (1991). Control of larval development by chemosensory neurons in *Caenorhabditis elegans*. *Science*, **251**, 1243–1246.

Barrett, J., Abbott, D. H. & George, L. M. (1990). Extension of reproductive suppression by pheromonal cues in subordinate female marmoset monkeys, *Callithrix jacchus*. *Journal of Reproduction and Fertility*, **90**, 411–418.

Bateman, P. W. (1998). Mate preference for novel partners in the cricket *Gryllus bimaculatus*. *Ecological Entomology*, **23**, 473–475.

Beauchamp, G. K., Yamazaki, K. & Boyse, E. A. (1985). The chemosensory recognition of genetic individuality. *Scientific American*, **253** (July), 86–92.

Beckers, R., Deneubourg, J. L. & Goss, S. (1993). Modulation of trail laying in the ant *Lasius niger* (Hymenoptera, Formicidae) and its role in the collective selection of a food source. *Journal of Insect Behavior*, **6**, 751–759.

Bell, W. J. (1991). *Searching behaviour. The behavioural ecology of finding resources*. London: Chapman and Hall.

Bengtsson, B. O. & Löfstedt, C. (1990). No evidence for selection in a pheromonally polymorphic moth population. *American Naturalist*, **136**, 722–726.

Bennett, A. T. D., Cuthill, I. C., Partridge, J. C. & Lunau, K. (1997). Ultraviolet plumage colors predict mate preferences in starlings. *Proceedings of the National Academy of Sciences of the United States of America*, **94**, 8618–8621.

Bennett, N. C., Faulkes, C. G. & Jarvis, J. U. M. (1999). Socially-induced infertility, incest avoidance and the monopoly of reproduction in the cooperatively breeding African mole-rat, Family Bathyergidae. In *Advances in the study of behavior*, Vol. 28, ed. P. J. B. Slater, J. S. Rosenblatt, C. T. Snowdon & T. J. Roper, pp. 75–114. New York: Academic Press.

Bertness, M. D., Leonard, G. H., Levine, J. M. & Bruno, J. F. (1999). Climate-driven interactions among rocky intertidal organisms caught between a rock and a hot place. *Oecologia*, **120**, 446–450.

Beynon, R. J., Robertson, D. H. L., Hubbard, S. J., Gaskell, S. J. & Hurst, J. L. (1999). The role of protein binding in chemical communication: major urinary proteins in the house mouse. In *Advances in chemical signals in vertebrates*, ed. R. E. Johnston, D. Müller-Schwarze & P. W. Sorenson, pp. 137–148. New York: Kluwer Academic Publishers/Plenum Press.

Billen, J. & Morgan, E. D. (1998). Pheromone communication in social insects: sources and secretions. In *Pheromone communication in social insects: ants, wasps, bees, and termites*, ed. R. K. Vander Meer, M. D. Breed, M. L. Winston & K. Espelie, pp. 3–33. Boulder, Colo.: Westview Press.

Birch, M. C. (1984). Aggregation in bark beetles. In *Chemical ecology of insects*, 1st edn, ed. W. J. Bell & R. T. Cardé, pp. 331–354. London: Chapman and Hall.

Birch, M. C., Poppy, G. M. & Baker, T. C. (1990). Scents and eversible scent structures of male moths. *Annual Review of Entomology*, **35**, 25–58.

Black, S. & Biron, C. (1982). Androstenol as a human pheromone: no effect on perceived physical attractiveness. *Behavioral and Neural Biology*, **34**, 326–330.

Blum, M. S. (1974). Pheromonal bases of social manifestations in insects. In *Pheromones*, ed. M. C. Birch, pp. 190–199. Amsterdam: North-Holland.

Blum, M. S. (1982). Pheromonal bases of insect sociality: communications, conundrums and caveats. *Les Colloques de l'Institut National de la Recherche Agronomique (INRA)*, **7**, 149–162.

Blum, M. S. (1985). Alarm pheromones. In *Comprehensive insect physiology, biochemistry and pharmacology*, Vol. 9, ed. G. A. Kerkut & L. I. Gilbert, pp. 193–224. Oxford: Pergamon Press.

Boake, C. R. B. (1991). Coevolution of senders and receivers of sexual signals: genetic coupling and genetic correlations. *Trends in Ecology & Evolution*, **6**, 225–227.

Boehm, N., Roos, J. & Gasser, B. (1994). Luteinizing-hormone-releasing hormone (LHRH)-expressing cells in the nasal-septum of human fetuses. *Developmental Brain Research*, **82**, 175–180.

Bonabeau, E., Theraulaz, G., Deneubourg, J. L., Aron, S. & Camazine, S. (1997). Self-organization in social insects. *Trends in Ecology & Evolution*, **12**, 188–193.

Bonabeau, E., Theraulaz, G., Deneubourg, J. L., Franks, N. R., Rafelsberger, O., Joly, J. L. & Blanco, S. (1998). A model for the emergence of pillars, walls and royal chambers in termite nests. *Philosophical Transactions of the Royal Society of London Series B-Biological Sciences*, **353**, 1561–1576.

Booth, D. W. & Signoret, J.-P. (1992). Olfaction and reproduction in ungulates. In *Oxford reviews of reproductive biology*, Vol. 14, ed. S. R. Milligan, pp. 263–301. Oxford: Oxford University Press.

Boppré, M. (1990). Lepidoptera and pyrrolizidine alkaloids – exemplification of complexity in chemical ecology. *Journal of Chemical Ecology*, **16**, 165–185.

Boppré, M. & Schneider, D. (1985). Pyrrolizidine alkaloids quantitatively regulate both scent organ morphogenesis and pheromone biosynthesis in male *Creatonotos* moths (Lepidoptera: Arctiidae). *Journal of Comparative Physiology A-Sensory Neural and Behavioral Physiology*, **157**, 569–578.

Borden, J. H. (1985). Aggregation pheromones. In *Comprehensive insect physiology, biochemistry and pharmacology*, Vol. 9, ed. G. A. Kerkut & L. I. Gilbert, pp. 257–286. Oxford: Pergamon Press.

Borden, J. H. (1993). Strategies and tactics for the use of semiochemicals against forest insect pests in North America. In *Pest management: biologically based technologies*, ed. R. D. Lumsden & J. L. Vaughn, pp. 265–279. Washington, D.C.: American Chemical Society.

Borden, J. H. (1997). Disruption of semiochemical-mediated aggregation in bark beetles. In *Insect pheromone research: new directions*, ed. R. T. Cardé & A. K. Minks, pp. 421–438. New York: Chapman and Hall.

Borg-Karlson, A.-K. (1990). Chemical and ethological studies of pollination in the genus *Ophrys* (Orchidaceae). *Phytochemistry*, **29**, 1359–1387.

Bossert, W. H. (1968). Temporal patterning in olfactory communication. *Journal of Theoretical Biology*, **18**, 157–170.

Bossert, W. H. & Wilson, E. O. (1963). The analysis of olfactory communication among animals. *Journal of Theoretical Biology*, **5**, 443–469.

Bourke, A. F. G. (1993). Lack of experimental-evidence for pheromonal inhibition of reproduction among queens in the ant *Leptothorax acervorum*. *Animal Behaviour*, **45**, 501–509.

Bourke, A. F. G. (1997). Sociality and kin selection in insects. In *Behavioural ecology*, 4th edn, ed. J. R. Krebs & N. B. Davies, pp. 203–27. Oxford: Blackwell Science.

Bourke, A. F. G. & Franks, N. R. (1995). *Social evolution in ants*. Princeton: Princeton University Press.

Bradbury, J. W. & Vehrencamp, S. L. (1998). *Principles of animal communication*. Sunderland, Mass.: Sinauer Associates.

Bradshaw, J. W. S., Baker, R. & Howse, P. E. (1975). Multicomponent alarm pheromones of the weaver ant. *Nature*, **258**, 230–231.

Bradshaw, J. W. S., Baker, R. & Howse, P. E. (1979). Multicomponent alarm pheromones in the mandibular glands of the African weaver ant, *Oecophylla longinoda*. *Physiological Entomology*, **4**, 15–25.

Brant, C. L., Schwab, T. M., Vandenbergh, J. G., Schaefer, R. L. & Solomon, N. G. (1998). Behavioural suppression of female pine voles after replacement of the breeding male. *Animal Behaviour*, **55**, 615–627.

Brashares, J. S. & Arcese, P. (1999a). Scent marking in a territorial African antelope: I. The maintenance of borders between male oribi. *Animal Behaviour*, **57**, 1–10.

Brashares, J. S. & Arcese, P. (1999b). Scent marking in a territorial African antelope: II. The economics of marking with faeces. *Animal Behaviour*, **57**, 11–17.

Breed, M. D. (1998). Recognition pheromones of the honey bee. *Bioscience*, **48**, 463–470.

Breed, M. D., Stiller, T. M. & Moor, M. J. (1988). The ontogeny of kin discrimination cues in the honey bee, *Apis mellifera*. *Behavior Genetics*, **18**, 439–448.

Breed, M. D., Garry, M. F., Pearce, A. N., Hibbard, B. E., Bjostad, L. B. & Page, R. E. (1995). The role of wax comb in honey-bee nestmate recognition. *Animal Behaviour*, **50**, 489–496.

Brennan, P. (1999). Bruce effect. In *Encyclopedia of reproduction*, Vol. 1, ed. E. Knobil & J. D. Neill, pp. 433–438. New York: Academic Press.

Brillat-Savarin, J.-A. (1825). *The philosopher in the kitchen*. Harmondsworth, UK: Penguin.

Brisbin, I. L. & Austad, S. N. (1991). Testing the individual odour theory of canine olfaction. *Animal Behaviour*, **42**, 63–69.

Brossut, R. (1996). *Phéromones. La communication chimique chez les animaux*. Paris: CNRS Editions.

Brown, G. E. & Brown, J. A. (1995). Kin discrimination in salmonids. *Reviews in Fish Biology and Fisheries*, **6**, 201–219.

Brown, G. E. & Smith, R. J. F. (1998). Acquired predator recognition in juvenile rainbow trout (*Oncorhynchus mykiss*): conditioning hatchery-reared fish to recognize chemical cues of a predator. *Canadian Journal of Fisheries and Aquatic Sciences*, **55**, 611–617.

Brown, G. E., Chivers, D. P. & Smith, R. J. F. (1995). Fathead minnows avoid conspecific and heterospecific alarm pheromones in the faeces of northern pike. *Journal of Fish Biology*, **47**, 387–393.

Brown, G. E., Chivers, D. P. & Smith, R. J. F. (1996). Effects of diet on localized defecation by northern pike, *Esox lucius*. *Journal of Chemical Ecology*, **22**, 467–475.

Brown, G. E., Chivers, D. P. & Smith, R. J. F. (1997). Differential learning rates of chemical versus visual cues of a northern pike by fathead minnows in a natural habitat. *Environmental Biology of Fishes*, **49**, 89–96.

Brown, G. E., Adrian, J. C., Smyth, E., Leet, H. & Brennan, S. (2000). Ostariophysan alarm pheromones: laboratory and field tests of the functional significance of nitrogen-oxides. *Journal of Chemical Ecology*, **26**, 139–154.

Brown, J. L. & Eklund, A. (1994). Kin recognition and the major histocompatibility complex – an integrative review. *American Naturalist*, **143**, 435–461.

Brown, R. E. (1994). *An introduction to neuroendocrinology*. Cambridge: Cambridge University Press.

Brown, R. E., Roser, B. & Singh, P. B. (1989). Class I and class II regions of the major histocompatibility complex both contribute to individual odors in congenic inbred strains of rats. *Behaviour Genetics*, **19**, 659–674.

Browne, K. A., Tamburri, M. N. & Zimmer-Faust, R. K. (1998). Modelling quantitative structure-activity relationships between animal behaviour and environmental signal molecules. *Journal of Experimental Biology*, **201**, 245–258.

Buck, L. B. (2000). The molecular architecture of odor and pheromone sensing in mammals. *Cell*, **100**, 611–18.

Buck, L. [B.] & Axel, R. (1991). A novel multigene family may encode odorant receptors – a molecular-basis for odor recognition. *Cell*, **65**, 175–187.

Burke, R. D. (1984). Pheromonal control of metamorphosis in the Pacific sand dollar, *Dendraster excentricus*. *Science*, **225**, 442–443.

Burkholder, W. E. (1982). Reproductive biology and communication among grain storage and warehouse beetles. *Journal of the Georgia Entomological Society*, **17** (II. suppl.), 1–10.

Bushmann, P. J. & Atema, J. (1997). Shelter sharing and chemical courtship signals in the lobster *Homarus americanus*. *Canadian Journal of Fisheries and Aquatic Sciences*, **54**, 647–654.

Butlin, R. K. (1987). Speciation by reinforcement. *Trends in Ecology & Evolution*, **2**, 8–13.

Butlin, R. K. (1989). Reinforcement of premating isolation. In *Speciation and its consequences*, ed. D. A. Otte & J. A. Endler, pp. 158–179. Sunderland, Mass.: Sinauer Associates.

Butlin, R. K. & Ritchie, M. G. (1989). Genetic coupling in mate recognition systems – what is the evidence. *Biological Journal of the Linnean Society*, **37**, 237–246.

Butlin, R. K. & Ritchie, M. G. (1994). Behaviour and speciation. In *Behaviour and evolution*, ed. P. J. B. Slater & T. R. Halliday, pp. 43–79. Cambridge: Cambridge University Press.

Butlin, R. K. & Tregenza, T. (1997). Evolutionary biology – is speciation no accident? *Nature*, **387**, 551.

Byers, J. A. (1992). Optimal fractionation and bioassay plans for isolation of synergistic chemicals: the subtractive-combination method. *Journal of Chemical Ecology*, **18**, 1603–1621.

Byers, J. A. (1995). Host-tree chemistry affecting colonization of bark beetles. In *Chemical ecology of insects 2*, ed. R. T. Cardé & W. J. Bell, pp. 154–213. London: Chapman and Hall.

Calenbuhr, V. & Deneubourg, J. L. (1992). A model for osmotropotactic orientation. 1. *Journal of Theoretical Biology*, **158**, 359–393.

Calenbuhr, V., Chretien, L., Deneubourg, J. L. & Detrain, C. (1992). A model for osmotropotactic orientation. 2. *Journal of Theoretical Biology*, **158**, 395–407.

Cammaerts, M. C. & Cammaerts, R. (1980). Food recruitment strategies of the ants *Myrmica sabuleti* and *Myrmica ruginodis*. *Behavioural Processes*, **5**, 251–270.

Campan, R. (1997). Tactic components in orientation. In *Orientation and communication in arthropods*, ed. M. Lehrer, pp. 1–40. Basel: Birkhäuser Verlag.

Cardé, R. T. & Baker, T. C. (1984). Sexual communication with pheromones. In *Chemical ecology of insects*, 1st edn, ed. W. J. Bell & R. T. Cardé, pp. 355–386. London: Chapman and Hall.

Cardé, R. T. & Bell, W. J. (ed.) (1995). *Chemical ecology of insects 2*. London: Chapman and Hall.

Cardé, R. T. & Mafra-Neto, A. (1997). Mechanisms of flight of male moths to pheromone. In *Insect pheromone research: new directions*, ed. R. T. Cardé & A. K. Minks, pp. 275–290. New York: Chapman and Hall.

Cardé, R. T. & Minks, A. K. (1995). Control of moth pests by mating disruption – successes and constraints. *Annual Review of Entomology*, **40**, 559–585.

Cardé, R. T. & Minks, A. K. (ed.) (1997). *Insect pheromone research: new directions*. New York: Chapman and Hall.

Cardwell, J. R., Stacey, N. E., Tan, E. S. P., McAdam, D. S. O. & Lang, S. L. C. (1995). Androgen increases olfactory receptor response to a vertebrate sex pheromone. *Journal of Comparative Physiology A-Sensory Neural and Behavioral Physiology*, **176**, 55–61.

Carlin, N. F. & Hölldobler, B. (1987). The kin recognition system of carpenter ants (*Camponotus* spp.): II. Larger colonies. *Behavioral Ecology and Sociobiology*, **20**, 209–218.

Carlson, J. R. (1996). Olfaction in *Drosophila*: From odor to behavior. *Trends in Genetics* **12**, 175–180.

Carroll, J. F., Mills, G. D. & Schmidtmann, E. T. (1996). Field and laboratory responses of adult *Ixodes scapularis* (Acari: Ixodidae) to kairomones produced by white-tailed deer. *Journal of Medical Entomology*, **33**, 640–644.

Carter, C. S. & Getz, L. L. (1993). Monogamy and the prairie vole. *Scientific American*, **268** (6), 100–106.

Carter, C. S. & Roberts, R. L. (1997). The psychobiological basis of cooperative breeding in rodents. In *Cooperative breeding in mammals*, ed. N. G. Solomon & J. A. French, pp. 231–266. Cambridge: Cambridge University Press.

Carter, C. S., Devries, A. C. & Getz, L. L. (1995). Physiological substrates of mammalian monogamy – the prairie vole model. *Neuroscience and Biobehavioral Reviews*, **19**, 303–314.

Chapman, J. W., Howse, P. E., Knapp, J. J. & Goulson, D. (1998). Evaluation of three (Z)-9 tricosene formulations for control of *Musca domestica* (Diptera: Muscidae) in caged-layer poultry units. *Journal of Economic Entomology*, **91**, 915–922.

Chapman, R. F. (1998). *The insects. Structure and function*, 4th edn. Cambridge: Cambridge University Press.

Chapman, T., Liddell, L. F., Kalb, J. M., Wolfner, M. F. & Partridge, L. (1995). Cost of mating in *Drosophila melanogaster* females is mediated by male accessory gland products. *Nature*, **373**, 241–244.

Chivers, D. P., Brown, G. E. & Smith, R. J. F. (1996). The evolution of chemical alarm signals – attracting predators benefits alarm signal senders. *American Naturalist*, **148**, 649–659.

Christensen, T. A. & White, J. (2000). Representation of olfactory information in the brain. In *The neurobiology of taste and smell*, 2nd edn, ed. T. E. Finger, W. L. Silver & D. Restrepo, pp. 201–232. New York: Wiley-Liss.

Christy, J. H. (1987). Competitive mating, mate choice and mating associations of brachyuran crabs. *Bulletin of Marine Science*, **41**, 177–191.

Clare, A. S. (1995). Chemical signals in barnacles: old problems, new approaches. In *New frontiers in barnacle evolution*, ed. F. R. Schram & J. T. Høeg, pp. 49–67. Rotterdam: A. A. Balkema.

Clare, A. S. (1997). Eicosanoids and egg-hatching synchrony in barnacles: evidence against a dietary precursor to egg-hatching pheromone. *Journal of Chemical Ecology*, **23**, 2299–2312.

Clare, A. S. (1998). Towards nontoxic antifouling. *Journal of Marine Biotechnology*, **6**, 3–6.

Clare, A. S. & Matsumura, K. (2000). Nature and perception of barnacle settlement patterns. *Biofouling*, **15**, 57–71.

Clarke, G. M. (1998). Developmental stability and fitness: the evidence is not quite so clear. *American Naturalist*, **152**, 762–766.

Classen, C., Honer, D. & Synott, A. (1994). *Aroma. The cultural history of smell*. London: Routledge.

Claus, R., Hoppen, H. O. & Karg, H. (1981). The secret of truffles: a steroidal pheromone? *Experientia*, **37**, 1178–1179.

Clegg, J. M. & Barlow, C. A. (1982). Escape behavior of the pea aphid *Acyrthosiphon pisum* (Harris) in response to alarm pheromone and vibration. *Canadian Journal of Zoology-Journal Canadien de Zoologie*, **60**, 2245–2252.

Clément, J.-L. & Bagnères, A.-G. (1998). Nestmate recognition in termites. In *Pheromone communication in social insects: ants, wasps, bees, and termites*, ed. R. K. Vander Meer, M. D. Breed, M. L. Winston & K. Espelie, pp. 126–158. Boulder, Colo.: Westview Press.

Clutton-Brock, T. H. & Parker, G. A. (1992). Potential reproductive rates and the operation of sexual selection. *Quarterly Review of Biology*, **67**, 437–456.

Clutton-Brock, T. H. & Vincent, A. C. J. (1991). Sexual selection and the potential reproductive rates of males and females. *Nature*, **351**, 58–60.

Clyne, P. J., Warr, C. G., Freeman, M. R., Lessing, D., Kim, J. H. & Carlson, J. R. (1999). A novel family of divergent seven-transmembrane proteins: candidate odorant receptors in *Drosophila*. *Neuron*, **22**, 327–338.

Coats, J. R. (1994). Risks from natural versus synthetic insecticides. *Annual Review of Entomology*, **39**, 489–515.

Cobb, M. & Ferveur, J. F. (1996). Evolution and genetic-control of mate recognition and stimulation in *Drosophila*. *Behavioural Processes*, **35**, 35–54.

Cockburn, A. (1991). *An introduction to evolutionary ecology*. Oxford: Blackwell Science.

Cognato, A. I., Seybold, S. J. & Sperling, F. A. H. (1999). Incomplete barriers to mitochondrial gene flow between pheromone races of the North American pine engraver, *Ips pini* (Say) (Coleoptera: Scolytidae). *Proceedings of the Royal Society of London Series B-Biological Sciences*, **266**, 1843–1850.

Cohen-Tannoudji, J., Lavenet, C., Locatelli, A., Tillet, Y. & Signoret, J. P. (1989). Non-involvement of the accessory olfactory system in the LH response of anestrous ewes to male odor. *Journal of Reproduction and Fertility* **86**: 135–144.

Cohen-Tannoudji, J., Einhorn, J. & Signoret, J. P. (1994). Ram sexual pheromone – first approach of chemical-identification. *Physiology & Behavior*, **56**, 955–961.

Conner, W. E. & Best, B. A. (1988). Biomechanics of the release of sex pheromone in moths: effects of body posture on local airflow. *Physiological Entomology*, **13**, 15–20.

Conner, W. E., Eisner, T., Vander Meer, R. K., Guerrero, A. & Meinwald, J. (1980). Sex attractant of an arctiid moth (*Utetheisa ornatrix*): a pulsed chemical signal. *Behavioral Ecology and Sociobiology*, **7**, 55–63.

Coopersmith, C. B. & Lenington, S. (1998). Pregnancy block in house mice (*Mus domesticus*) as a function of t-complex genotype: examination of the mate choice and male infanticide hypotheses. *Journal of Comparative Psychology*, **112**, 82–91.

Corbin, A. (1996). *The foul and the fragrant: odour and the social imagination*. London: Papermac.

Cossé, A. A., Campbell, M. G., Glover, T. J., Linn, C. E., Todd, J. L., Baker, T. C. & Roelofs, W. L. (1995). Pheromone behavioral-responses in unusual male European corn-borer hybrid progeny not correlated to electrophysiological phenotypes of their pheromone-specific antennal neurons. *Experientia*, **51**, 809–816.

Courchamp, F., Clutton-Brock, T. & Grenfell, B. (1999). Inverse density dependence and the Allee effect. *Trends in Ecology & Evolution*, **14**, 405–410.

Coyne, J. A. & Charlesworth, B. (1997). Genetics of a pheromonal difference affecting sexual isolation between *Drosophila mauritiana* and *D. sechellia*. *Genetics*, **145**, 1015–1030.

Coyne, J. A. & Orr, H. A. (1997). "Patterns of speciation in *Drosophila*" revisited. *Evolution*, **51**, 295–303.

Crespi, B. J. & Choe, J. C. (1997). Explanation and evolution of social systems. In *The evolution of social behavior in insects and arachnids*, ed. J. C. Choe & B. J. Crespi, pp. 499–524. Cambridge: Cambridge University Press.

Crespi, B. J. & Yanega, D. (1995). The definition of eusociality. *Behavioral Ecology*, **6**, 109–115.

Cutler, W. B., Preti, G., Krieger, A., Huggins, G. R., Garcia, C. R. & Lawley, H. J. (1986). Human axillary secretions influence womens menstrual cycles – the role of donor extract from men. *Hormones and Behavior*, **20**, 463–473.

Dafni, A. & Bernhardt, P. (1990). Pollination of terrestrial orchids of southern Australia and the mediterranean region – systematic, ecological, and evolutionary implications. *Evolutionary Biology*, **24**, 193–252.

Dahl, J., Nilsson, P. A. & Pettersson, L. B. (1998). Against the flow: chemical detection of downstream predators in running waters. *Proceedings of the Royal Society of London Series B-Biological Sciences*, **265**, 1339–1344.

Dalton, P., Doolittle, N. & Breslin, P. A. S. (2002). Gender-specific induction of enhanced sensitivity to odors. *Nature Neuroscience*, **5**, 199–200.

Darwin, C. (1871). *The descent of man and selection in relation to sex*, 1st edn. London: John Murray.

Darwin, C. (1874). *The descent of man and selection in relation to sex*, 2nd edn. London: John Murray.

Dawkins, M. S. (1993). *Through our eyes only?: the search for animal consciousness*. Oxford: W H Freeman.

Dawkins, M. S. (1995). *Unravelling animal behaviour*, 2nd edn. Harlow: Longman Scientific & Technical.

Dawkins, R. (1982). *The extended phenotype*. San Francisco: W. H. Freeman.

Dawkins, R. & Krebs, J. R. (1978). Animal signals: information or manipulation. In *Behavioural ecology: an evolutionary approach*, ed. J. R. Krebs & N. B. Davies, pp. 282–309. Oxford: Blackwell Science.

Dawley, E. M. (1987). Species discrimination between hybridizing and non-hybridizing terrestrial salamanders. *Copeia*, **1987**, 924–931.

Dehnhard, M. & Claus, R. (1996). Attempts to purify and characterize the estrus-signalling pheromone from cow urine. *Theriogenology*, **46**, 13–22.

Deneubourg, J. L., Goss, S., Franks, N. & Pasteels, J. M. (1989). The blind leading the blind – modeling chemically mediated army ant raid patterns. *Journal of Insect Behavior*, **2**, 719–725.

Denny, M. W. (1993). *Air and water*. Princeton, N. J.: Princeton University Press.

Dent, D. (1991). *Insect pest management*. Wallingford, UK: CAB International.

Dethier, V. G. (1987). Sniff, flick, and pulse: an appreciation of intermittency. *Proceedings of the American Philosophical Society*, **131**, 159–176.

Detrain, C. & Deneubourg, J. L. (1997). Scavenging by *Pheidole pallidula*: a key for understanding decision-making systems in ants. *Animal Behaviour*, **53**, 537–547.

Detrain, C., Deneubourg, J. L. & Pasteels, J. M. (ed.) (1999). *Information processing in social insects*. Basel: Birkhäuser.

Dettner, K. & Liepert, C. (1994). Chemical mimicry and camouflage. *Annual Review of Entomology*, **39**, 129–154.

Deutsch, J. C. & Nefdt, R. J. C. (1992). Olfactory cues influence female choice in 2 lek-breeding antelopes. *Nature*, **356**, 596–598.

Diamond, J. M. (1997). Aaaaaaaaaaargh, no! *Nature*, **385**, 295–296.

Dicke, M. & Sabelis, M. W. (1992). Costs and benefits of chemical information conveyance: proximate and ultimate factors. In *Insect chemical ecology. An evolutionary approach*, ed. B. D. Roitberg & M. B. Isman, pp. 122–155. New York: Chapman and Hall.

Dill, L. M., Fraser, A. H. G. & Roitberg, B. D. (1990). The economics of escape behavior in the pea aphid, *Acyrthosiphon pisum*. *Oecologia*, **83**, 473–478.

Dixson, A. F. (1998). *Primate sexuality. Comparative studies of the prosimians, monkeys, apes, and human beings*. Oxford: Oxford University Press.

Doall, M. H., Colin, S. P., Strickler, J. R. & Yen, J. (1998). Locating a mate in 3D: the case of *Temora longicornis*. *Philosophical Transactions of the Royal Society of London Series B-Biological Sciences*, **353**, 681–689.

Donascimento, R. R., Morgan, E. D., Billen, J., Schoeters, E., Dellalucia, T. M. C. & Bento, J. M. S. (1993). Variation with caste of the mandibular gland secretion in the leaf-cutting ant – *Atta sexdens rubropilosa*. *Journal of Chemical Ecology*, **19**, 907–918.

Dorries, K. M. (1992). Sex differences in olfaction in mammals. In *Science of olfaction*, ed. M. J. Serby & M. L. Chobor, pp. 245–275. New York: Springer-Verlag.

Dorries, K. M. (1997). Olfactory "consciousness". *Science*, **278**, 1550.

Dorries, K. M., Schmidt, H. J., Beauchamp, G. K. & Wysocki, C. J. (1989). Changes in sensitivity to the odor androstenone during adolescence. *Developmental Psychobiology*, **22**, 423–436.

Dorries, K. M., Adkins-Regan, E. & Halpern, B. P. (1995). Olfactory sensitivity to the pheromone, androstenone, is sexually dimorphic in the pig. *Physiology & Behavior*, **57**, 255–259.

Dorries, K. M., Adkins-Regan, E. & Halpern, B. P. (1997). Sensitivity and behavioral responses to the pheromone androstenone are not mediated by the vomeronasal organ in domestic pigs. *Brain Behavior and Evolution*, **49**, 53–62.

Doty, R. L. (1985). The primates III: humans. In *Social odours in mammals*, Vol. 2, ed. R. E. Brown & D. W. Macdonald, pp. 804–832. Oxford: Oxford University Press.

Doty, R. L. (2001). Olfaction. *Annual Review of Psychology*, **52**, 423–452.

Doty, R. L., Ford, M., Preti, G. & Huggins, G. R. (1975). Changes in the intensity and pleasantness of human vaginal odors during the menstrual cycle. *Science*, **190**, 1316–1318.

Doty, R. L., Hall, J. W., Flickinger, G. L. & Sondheimer, S. J. (1982). Cyclical changes in olfactory and auditory sensitivity during the menstrual cycle: no attenuation by oral contraceptive medication. In *Olfaction and endocrine regulation*, ed. W. Breipohl, pp. 35–42. London: IRL Press.

Doumbia, M., Hemptinne, J. L. & Dixon, A. F. G. (1998). Assessment of patch quality by ladybirds: role of larval tracks. *Oecologia*, **113**, 197–202.

Døving, K. B. & Trotier, D. (1998). Structure and function of the vomeronasal organ. *Journal of Experimental Biology*, **201**, 2913–2925.

Drickamer, L. C. (1992). Estrous female house mice discriminate dominant from subordinate males and sons of dominant from sons of subordinate males by odor cues. *Animal Behaviour*, **43**, 868–870.

Drickamer, L. C. (1995). Rates of urine excretion by house mouse (*Mus domesticus*) – differences by age, sex, social-status, and reproductive condition. *Journal of Chemical Ecology*, **21**, 1481–1493.

Drickamer, L. C. (1999). Sexual attractants. In *Encyclopedia of reproduction*, Vol. 4, ed. E. Knobil & J. D. Neill, pp. 444–448. New York: Academic Press.

Drickamer, L. C. & Mikesic, D. G. (1990). Urinary chemosignals, reproduction, and population-size for house mice (*Mus domesticus*) living in field enclosures. *Journal of Chemical Ecology*, **16**, 2955–2968.

Drumont, A., Gonzalez, R., Dewindt, N., Gregoire, J. C., Deproft, M. & Seutin, E. (1992). Semiochemicals and the integrated management of *Ips typographus* (L) (Col., Scolytidae) in Belgium. *Journal of Applied Entomology-Zeitschrift Fur Angewandte Entomologie*, **114**, 333–337.

Dryer, L. & Berghard, A. (1999). Odorant receptors: a plethora of G-protein-coupled receptors. *Trends in Pharmacological Sciences*, **20**, 413–417.

Duchamp-Viret, P., Chaput, M. A. & Duchamp, A. (1999). Odor response properties of rat olfactory receptor neurons. *Science*, **284**, 2171–2174.

Dulac, C. (2000). Sensory coding of pheromone signals in mammals. *Current Opinion in Neurobiology*, **10**, 511–518.

Dulka, J. G. (1993). Sex pheromone systems in goldfish: comparisons to vomeronasal systems in tetrapods. *Brain Behavior and Evolution*, **42**, 265–280.

Dusenbery, D. B. (1992). *Sensory ecology. How organisms aquire and respond to information*. New York: W. H. Freeman and Company.

Dusenbery, D. B. & Snell, T. W. (1995). A critical body size for use of pheromones in mate location. *Journal of Chemical Ecology*, **21**, 427–438.

Dussourd, D. E., Harvis, C. A., Meinwald, J. & Eisner, T. (1991). Pheromonal advertisement of a nuptial gift by a male moth (*Utetheisa ornatrix*). *Proceedings of the National Academy of Sciences of the United States of America*, **88**, 9224–9227.

Eberhard, W. G. (1996). *Female control: sexual selection by cryptic female choice*. Princeton, N. J.: Princeton University Press.

Eberhard, W. G. (1997). Sexual selection by cryptic female choice in insects and arachnids. In *The evolution of mating systems in insects and arachnids*, ed. J. C. Choe & B. J. Crespi, pp. 32–57. Cambridge: Cambridge University Press.

Eberhard, W. G. & Cordero, C. (1995). Sexual selection by cryptic female choice on male seminal products – a new bridge between sexual selection and reproductive physiology. *Trends in Ecology & Evolution*, **10**, 493–496.

Edwards, S. V. & Hedrick, P. W. (1998). Evolution and ecology of MHC molecules: from genomics to sexual selection. *Trends in Ecology & Evolution*, **13**, 305–311.

Eggert, A. K. & Müller, J. K. (1997). Biparental care and social evolution in burying beetles: lessons from the larder. In *The evolution of social behavior in insects and arachnids*, ed. J. C. Choe & B. J. Crespi, pp. 216–236. Cambridge: Cambridge University Press.

Eggert, A. K. & Sakaluk, S. K. (1995). Female-coerced monogamy in burying beetles. *Behavioral Ecology and Sociobiology*, **37**, 147–153.

Eggert, F., Holler, C., Luszyk, D., Muller-Ruchholtz, W. & Ferstl, R. (1996). MHC-associated and MHC-independent urinary chemosignals in mice. *Physiology & Behavior*, **59**, 57–62.

Eggert, F., Muller-Ruchholtz, W. & Ferstl, R. (1999). Olfactory cues associated with the major histocompatibility complex. *Genetica*, **104**, 191–197.

Egid, K. & Brown, J. L. (1989). The major histocompatibility complex and female mating preferences in mice. *Animal Behaviour*, **38**, 548–550.

Eisner, T. & Meinwald, J. (1995). Defense-mechanisms of arthropods. 129. The chemistry of sexual selection. *Proceedings of the National Academy of Sciences of the United States of America*, **92**, 50–55.

Eisner, T., Hicks, K., Eisner, M. & Robson, D. S. (1978). "Wolf-in-sheep's-clothing" strategy of a predaceous insect larva. *Science*, **199**, 790–794.

Eisthen, H. L. (1992). Phylogeny of the vomeronasal system and of receptor cell-types in the olfactory and vomeronasal epithelia of vertebrates. *Microscopy Research and Technique*, **23**, 1–21.

Eisthen, H. L. (1997). Evolution of vertebrate olfactory systems. *Brain Behavior and Evolution*, **50**, 222–233.

Elliott, J. K., Mariscal, R. N. & Roux, K. H. (1994). Do anemonefishes use molecular mimicry to avoid being stung by host anemones. *Journal of Experimental Marine Biology and Ecology*, **179**, 99–113.

Elliott, J. K., Lougheed, S. C., Bateman, B., McPhee, L. K. & Boag, P. T. (1999). Molecular phylogenetic evidence for the evolution of specialization in anemonefishes. *Proceedings of the Royal Society of London Series B-Biological Sciences*, **266**, 677–685.

Elmes, G. W., Thomas, J. A. & Wardlaw, J. C. (1991). Larvae of *Maculinea rebeli*, a large-blue butterfly, and their *Myrmica* host ants – wild adoption and behavior in ant-nests. *Journal of Zoology*, **223**, 447–460.

Elmes, G. W., Barr, B., Thomas, J. A. & Clarke, R. T. (1999). Extreme host specificity by *Microdon mutabilis* (Diptera: Syrphidae), a social parasite of ants. *Proceedings of the Royal Society of London Series B-Biological Sciences*, **266**, 447–453.

Emlen, S. T. (1997). Predicting family dynamics in social vertebrates. In *Behavioural ecology*, 4th edn, ed. J. R. Krebs & N. B. Davies, pp. 228–253. Oxford: Blackwell Science.

Endler, J. A. (1992). Signals, signal conditions, and the direction of evolution. *American Naturalist Special*, **139**, 125–153.

Endler, J. A. & Basolo, A. L. (1998). Sensory ecology, receiver biases and sexual selection. *Trends in Ecology & Evolution*, **13**, 415–420.

Engen, T. & Ross, B. M. (1973). Long-term memory of odors with and without verbal descriptions. *Journal of Experimental Psychology*, **100**, 221–227.

Epple, G. & Katz, Y. (1984). Social influences on estrogen excretion and ovarian cyclicity in saddle back tamarins (*Saguinus fuscicollis*). *American Journal of Primatology*, **6**, 215–227.

Epple, G., Belcher, A. M., Kuderling, I., Zeller, U., Scolnick, L., Greenfield, K. L. & Smith, A. B. (1993). Making sense out of scents – species-differences in scent glands, scent marking behavior and scent mark composition in the Callitrichidae. In *Marmosets and tamarins. Systematics, behaviour, and ecology*, ed. A. B. Rylands, pp. 123–151. Oxford: Oxford Science Publications, Oxford University Press.

Espelie, K. E., Gamboa, G. J., Grudzien, T. A. & Bura, E. A. (1994). Cuticular hydrocarbons of the paper wasp, *Polistes fuscatus* – a search for recognition pheromones. *Journal of Chemical Ecology*, **20**, 1677–1687.

Fabré, J. H. (1911). *Social life in the insect world*. Harmondsworth, UK: Penguin.

Farbman, A. I. (1992). *Cell biology of olfaction*. Cambridge: Cambridge University Press.

Faulkes, C. G. & Abbott, D. H. (1993). Evidence that primer pheromones do not cause social suppression of reproduction in male and female naked mole-rats (*Heterocephalus glaber*). *Journal of Reproduction and Fertility*, **99**, 225–230.

Fautin, D. G. (1991). The anemonefish symbiosis – what is known and what is not. *Symbiosis*, **10**, 23–46.

Feener, D. H., Jacobs, L. F. & Schmidt, J. O. (1996). Specialized parasitoid attracted to a pheromone of ants. *Animal Behaviour*, **51**, 61–66.

Ferkin, M. H., Sorokin, E. S., Renfroe, M. W. & Johnston, R. E. (1994). Attractiveness of male odors to females varies directly with plasma testosterone concentration in meadow voles. *Physiology & Behavior*, **55**, 347–353.

Ferkin, M. H., Sorokin, E. S., Johnston, R. E. & Lee, C. J. (1997). Attractiveness of scents varies with protein content of the diet in meadow voles. *Animal Behaviour*, **53**, 133–141.

Ferstl, R., Eggert, F., Westphal, E., Zavazava, N. & Muller-Ruchholtz, W. (1992). MHC-related odors in humans. In *Chemical signals in vertebrates*, Vol. VI, ed. R. L. Doty & D. Müller-Schwarze, pp. 206–211. New York: Plenum Press.

Ferveur, J. F., Cobb, M., Boukella, H. & Jallon, J. M. (1996). World-wide variation in *Drosophila melanogaster* sex-pheromone-behavioral-effects, genetic bases and potential evolutionary consequences. *Genetica*, **97**, 73–80.

Fiedler, K., Hölldobler, B. & Seufert, P. (1996). Butterflies and ants – the communicative domain. *Experientia*, **52**, 14–24.

Finelli, C. M., Pentcheff, N. D., Zimmer-Faust, R. K. & Wethey, D. S. (1999). Odor transport in turbulent flows: constraints on animal navigation. *Limnology and Oceanography*, **44**, 1056–1071.

Finelli, C. M., Pentcheff, N. D., Zimmer, R. K. & Wethey, D. S. (2000). Physical constraints on ecological processes: a field test of odor-mediated foraging. *Ecology*, **81**, 784–797.

Finger, T. E. (1997). Evolution of taste and solitary chemoreceptor cell systems. *Brain Behavior and Evolution*, **50**, 234–243.

Finger, T. E., Silver, W. L. & Restrepo, D. (ed.) (2000). *The neurobiology of taste and smell*, 2nd edn. New York: Wiley-Liss.

Firestein, S. (2001). How the olfactory system makes sense of scents. *Nature*, **413**, 211–218.

Fitzgerald, T. D. (1993). Sociality in caterpillars. In *Caterpillars: ecological and evolutionary constraints on foraging*, 1st edn, ed. N. E. Stamp & T. M. Casey, pp. 372–403. New York: Chapman and Hall.

Fitzgerald, T. D. & Edgerly, J. S. (1982). Site of secretion of the trail marker of the eastern tent caterpillar. *Journal of Chemical Ecology*, **8**, 31–39.

Fitzgerald, T. D. & Gallagher, E. M. (1976). A chemical trail factor from the silk of the eastern tent caterpillar *Malacosoma americanum* (Lepidoptera: Lasiocamidae). *Journal of Chemical Ecology*, **2**, 564–574.

Fitzgerald, T. D. & Peterson, S. C. (1983). Elective recruitment by the eastern tent caterpillar (*Malacosoma americanum*). *Animal Behaviour*, **31**, 417–423.

Fleming, A. S., Steiner, M. & Corter, C. (1997). Cortisol, hedonics, and maternal responsiveness in human mothers. *Hormones and Behavior*, **32**, 85–98.

Flood, P. (1985). Sources of significant smells: the skin and other organs. In *Social odours in mammals*, Vol. 1, ed. R. E. Brown & D. W. Macdonald, pp. 19–36. Oxford: Oxford University Press.

Ford, N. B. (1986). The role of pheromone trails in the sociobiology of snakes. In *Chemical signals in vertebrates 4*, ed. D. Duvall, D. Müller-Schwarze & R. M. Silverstein, pp. 261–278. New York: Plenum Press.

Ford, N. B. & Low, J. R. (1984). Sex pheromone source location by garter snakes: a mechanism for detection of direction in non-volatile trails. *Journal of Chemical Ecology*, **10**, 1193–1199.

Forward, R. B. (1987). Larval release rhythms of decapod crustaceans – an overview. *Bulletin of Marine Science*, **41**, 165–176.

Fraenkel, G. S. & Gunn, D. L. (1940). *The orientation of animals. Kineses, taxes and compass reactions*. Oxford: Oxford University Press.

Frank, D. F., Erb, H. N. & Houpt, K. A. (1999). Urine spraying in cats: presence of concurrent disease and effects of a pheromone treatment. *Applied Animal Behaviour Science*, **61**, 263–272.

Franks, N. R., Gomez, N., Goss, S. & Deneubourg, J. L. (1991). The blind leading the blind in army ant raid patterns – testing a model of self-organization (Hymenoptera, Formicidae). *Journal of Insect Behavior*, **4**, 583–607.

Free, J. B. (1987). *Pheromones of social bees*. London: Chapman and Hall.

French, J. A. (1997). Proximate regulation of singular breeding in callitrichid primates. In *Cooperative breeding in mammals*, ed. N. G. Solomon & J. A. French, pp. 34–75. Cambridge: Cambridge University Press.

Gadagkar, R. (1997). The evolution of communication and the communication of evolution: the case of the honey bee queen pheromone. In *Orientation and communication in arthropods*, ed. M. Lehrer, pp. 375–395. Basel: Birkhäuser Verlag.

Gadenne, C., Renou, M. & Sreng, L. (1993). Hormonal-control of pheromone responsiveness in the male black cutworm *Agrotis ipsilon. Experientia*, **49**, 721–724.

Galef, B. G. & Buckley, L. L. (1996). Use of foraging trails by Norway rats. *Animal Behaviour*, **51**, 765–771.

Galizia, C. G. & Menzel, R. (2000). Odour perception in honeybees: coding information in glomerula patterns. *Current Opinion in Neurobiology*, **10**, 504–510.

Galizia, C. G., Sachse, S., Rappert, A. & Menzel, R. (1999). The glomerular code for odor representation is species specific in the honeybee *Apis mellifera. Nature Neuroscience*, **2**, 473–478.

Gamboa, G. J. (1996). Kin recognition in social wasps. In *Natural history and evolution of paper wasps*, ed. S. Turillazzi & M. J. West-Eberhard, pp. 161–173. Oxford: Oxford University Press.

Gamboa, G. J., Grudzien, T. A., Espelie, K. E. & Bura, E. A. (1996). Kin recognition pheromones in social wasps: combining chemical and behavioural evidence. *Animal Behaviour*, **51**, 625–629.

Gangestad, S. W. & Thornhill, R. (1998). Menstrual cycle variation in women's preferences for the scent of symmetrical men. *Proceedings of the Royal Society of London Series B-Biological Sciences*, **265**, 927–933.

Gautier, J. P. & Gautier, A. (1977). Communication in Old World monkeys. In *How animals communicate*, ed. T. Seboek, pp. 890–964. Bloomington, Ind.: Indiana University Press.

Getz, W. (1991). The honey bee as a model kin recognition system. In *Kin recognition*, ed. P. G. Hepper, pp. 358–412. Cambridge: Cambridge University Press.

Gibson, R. W. & Pickett, J. A. (1983). Wild potato repels aphids by release of aphid alarm pheromone. *Nature*, **302**, 608–609.

Gilbert, A. N., Yamazaki, K., Beauchamp, G. K. & Thomas, L. (1986). Olfactory discrimination of mouse strains (*Mus musculus*) and major histocompatibility types by humans (*Homo sapiens*). *Journal of Comparative Psychology*, **100**, 262–265.

Giurfa, M. & Nunez, J. A. (1992). Honeybees mark with scent and reject recently visited flowers. *Oecologia*, **89**, 113–117.

Godfray, H. C. J. (1994). *Parasitoids: behavioral and evolutionary ecology*. Princeton, N. J.: Princeton University Press.

Goldfoot, D. A. (1981). Olfaction, sexual-behavior, and the pheromone hypothesis in rhesus monkeys – a critique. *American Zoologist*, **21**, 153–164.

Goldfoot, D. A., Kravetz, M. A., Goy, R. W. & Freeman, S. K. (1976). Lack of effect of vaginal lavages and aliphatic acids on ejaculatory responses in rhesus monkeys. *Hormones and Behavior*, **7**, 1–27.

Gonzalez, A., Rossini, C., Eisner, M. & Eisner, T. (1999). Sexually transmitted chemical defense in a moth (*Utetheisa ornatrix*). *Proceedings of the National Academy of Sciences of the United States of America*, **96**, 5570–5574.

Gorman, M. L. (1976). A mechanism for individual recognition by odour in *Herpestes auropunctatus*. *Animal Behaviour*, **24**, 141–146.

Gorman, M. L. & Mills, M. G. L. (1984). Scent marking strategies in hyaenas (Mammalia). *Journal of Zoology*, **202**, 535–547.

Gorman, M. L. & Stone, R. D. (1990). Mutual avoidance by European moles *Talpa europaea*. In *Chemical signals in vertebrates*, Vol. 5, ed. D. W. Macdonald, D. Müller-Schwarze & S. E. Natynczuk, pp. 367–377. Oxford: Oxford Science Publications, Oxford University Press.

Gosling, L. M. (1981). Demarcation in a gerenuk territory – an economic-approach. *Zeitschrift für Tierpsychologie-Journal of Comparative Ethology*, **56**, 305–322.

Gosling, L. M. (1982). A reassessment of the function of scent marking in territories. *Zeitschrift für Tierpsychologie – Journal of Comparative Ethology*, **60**, 89–118.

Gosling, L. M. (1990). Scent marking by resource holders: alternative mechanisms for advertising the cost of competition. In *Chemical signals in vertebrates,* Vol. 5, ed. D. W. Macdonald, D. Müller-Schwarze & S. E. Natynczuk, pp. 315–328. Oxford: Oxford Science Publications, Oxford University Press.

Gosling, L. M. & McKay, H. V. (1990). Competitor assessment by scent matching – an experimental test. *Behavioral Ecology and Sociobiology*, **26**, 415–420.

Gosling, L. M. & Roberts, S. C. (2001). Scent-marking by male mammals: cheat-proof signals to competitors and mates. *Advances in the Study of Behavior*, **30**, 169–217.

Gosling, L. M., Atkinson, N. W., Dunn, S. & Collins, S. A. (1996). The response of subordinate male mice to scent marks varies in relation to their own competitive ability. *Animal Behaviour*, **52**, 1185–1191.

Goulson, D, Hawson, S. A. & Stout, J. C. (1998). How bumblebees avoid flowers already visited by conspecifics or by other bumblebee species. *Animal Behaviour*, **55**, 199–206.

Gower, D. B. (1990). Quantification of odorous 16-androstene steroids in vertebrates. In *Chemical signals in vertebrates*, Vol. 5, ed. D. W. Macdonald, D. Müller-Schwarze & S. E. Natynczuk, pp. 34–47. Oxford: Oxford Science Publications, Oxford University Press.

Gower, D. B. & Ruparelia, B. A. (1993). Olfaction in humans with special reference to odorous 16-androstenes – their occurrence, perception and possible social, psychological and sexual impact. *Journal of Endocrinology*, **137**, 167–187.

Grafen, A. (1990a). Do animals really recognize kin? *Animal Behaviour*, **39**, 42–54.

Grafen, A. (1990b). Sexual selection unhandicapped by the Fisher process. *Journal of Theoretical Biology*, **144**, 473–516.

Grassé P. P. (1959). La reconstruction du nid et les coordinations inter-individuelles chez *Bellicoitermes natalenis* et *Cubitermes* sp. La théorie de la stigmergie: essai d'interprétation des termites constructeurs. *Insectes Sociaux*, **6**, 41–83.

Grasso, F. W., Consi, T., Mountain, D. & Atema, J. (1996). Locating odor sources in turbulence with a lobster inspired robot. In *From animals to animats 4: Proceedings 4th International Conference on Simulation of Adaptive Behavior*, ed. P. Maes, M. Mataric, J.-A. Meyer, J. Pollack & S. Wilson, pp. 104–112. Cambridge, Mass.: MIT Press.

Gray, S. & Hurst, J. L. (1995). The effects of cage cleaning on aggression within groups of male laboratory mice. *Animal Behaviour*, **49**, 821–826.

Greenberg, L. (1979). Genetic component of bee odor in kin recognition. *Science*, **206**, 1095–1097.

Greenfield, M. D. (1981). Moth sex pheromones: an evolutionary perspective. *Florida Entomologist*, **64**, 4–17.

Greenfield, M. D. & Karandinos, M. G. (1979). Resource partitioning of the sex communication channel in clearwing moths (Lepidoptera: Sesiidae) of Wisconsin. *Ecological Monographs*, **49**, 403–426.

Greenspan, R. J. & Ferveur, J. F. (2000). Courtship in *Drosophila*. *Annual Review of Genetics*, **34**, 205–232.

Gross, M. R. (1996). Alternative reproductive strategies and tactics: diversity within sexes. *Trends in Ecology & Evolution*, **11**, 92–98.

Guilford, T. (1995). Animal signals – all honesty and light. *Trends in Ecology & Evolution*, **10**, 100–101.

Guilford, T., Nicol, C., Rothschild, M. & Moore, B. P. (1987). The biological roles of pyrazines – evidence for a warning odor function. *Biological Journal of the Linnean Society*, **31**, 113–128.

Guillamon, A. & Segovia, S. (1997). Sex differences in the vomeronasal system. *Brain Research Bulletin*, **44**, 377–382.

Gullan, P. J. & Cranston, P. S. (1994). *The insects: an outline of entomology*. London: Chapman and Hall.

Guo, C. C., Hwang, J. S. & Fautin, D. G. (1996). Host selection by shrimps symbiotic with sea anemones: a field survey and experimental laboratory analysis. *Journal of Experimental Marine Biology and Ecology*, **202**, 165–176.

Gut, L. J. & Brunner, J. F. (1998). Pheromone-based management of codling moth (Lepidoptera: Tortricidae) in Washington apple orchards. *Journal of Agricultural Entomology*, **15**, 387–406.

Haig, D. (1997). The social gene. In *Behavioural ecology*, 4th edn, ed. J. R. Krebs & N. B. Davies, pp. 284–304. Oxford: Blackwell Science.

Halpin, Z. T. (1986). Individual odors among mammals – origins and functions. *Advances in the Study of Behavior*, **16**, 39–70.

Hamilton, J. G. C. (1992). The role of pheromones in tick biology. *Parasitology Today*, **8**, 130–132.

Hamilton, W. D. (1964). The genetical evolution of social behaviour. I and II. *Journal of Theoretical Biology*, **7**, 1–32.

Hamilton, W. D. (1971). Geometry for the selfish herd. *Journal of Theoretical Biology*, **31**, 295–311.

Hamilton, W. D. (1987). Kinship, recognition, disease, and intelligence: constraints of social evolution. In *Animal societies: theories and facts*, ed. Y. Itô, J. L. Brown & J. Kikkawa, pp. 88–102. Tokyo: Japan Science Society Press.

Hangartner, W. (1967). Spezifität und Inaktivierung des Spurpheromons von *Lasius fuliginosus* Latr. und Orientierung der Arbeiterinnen in Duftfeld. *Zeitschrift für Vergleichende Physiologie*, **57**, 103–136.

Hansen, K. (1983). Reception of bark beetle pheromone in the predaceous clerid beetle, *Thanasimus formicarius* (Coleoptera, Cleridae). *Journal of Comparative Physiology*, **150**, 371–378.

Hansson, B. S. (1997). Antennal lobe projection patterns of pheromone-specific olfactory receptor neurons in moths. In *Insect pheromone research: new directions,* ed. R. T. Cardé & A. K. Minks, pp. 164–183. New York: Chapman and Hall.

Hara, T. J. (1994). The diversity of chemical-stimulation in fish olfaction and gustation. *Reviews in Fish Biology and Fisheries*, **4**, 1–35.

Hardee, D. D. & Mitchell, E. B. (1997). Boll weevil, *Anthonomus grandis* Boheman (Coleoptera: Curculionidae): a summary of research on behavior as affected by chemical communication. *Southwestern Entomologist*, **22**, 466–491.

Hardege, J. D., Müller, C. T., Beckmann, M., Hardege, D. B. & Bentley, M. G. (1998). Timing of reproduction in marine polychaetes: the role of sex pheromones. *Ecoscience*, **5**, 395–404.

Hardie, J. & Minks, A. K. (ed.) (1999). *Pheromones of non-lepidopteran insects associated with agricultural plants*. Wallingford, UK: CAB International.

Hardie, J., Nottingham, S. F., Powell, W. & Wadhams, L. J. (1991). Synthetic aphid sex-pheromone lures female parasitoids. *Entomologia Experimentalis et Applicata*, **61**, 97–99.

Hardie, J., Pickett, J. A., Pow, E. M. & Smiley, D. W. M. (1999). Aphids. In *Pheromones of non-lepidopteran insects associated with agricultural plants*, ed. J. Hardie & A. K. Minks, pp. 227–250. Wallingford, UK: CAB International.

Harris, M. O. & Foster, S. P. (1995). Behavior and integration. In *Chemical ecology of insects 2*, ed. R. T. Cardé & W. J. Bell, pp. 3–46. London: Chapman and Hall.

Harter, J. (ed.) (1979). *Animals: 1419 copyright-free illustrations of mammals, birds, fish, insects, etc.* New York: Dover Publications.

Hassanali, A. & Torto, B. (1999). Grasshoppers and locusts. In *Pheromones of non-lepidopteran insects associated with agricultural plants*, ed. J. Hardie & A. K. Minks, pp. 305–328. Wallingford, UK: CAB International.

Hassanali, A., Nyandat, E., Obenchain, F. A., Otieno, D. A. & Galun, R. (1989). Humidity effects on response of *Argas persicus* (Oken) to guanine, an assembly pheromone of ticks. *Journal of Chemical Ecology*, **15**, 791–793.

Hasson, O. (1991). Pursuit-deterrent signals: communication between prey and predator. *Trends in Ecology & Evolution*, **6**, 325–329.

Hatt, H. & Bauer, U. (1980). Single unit analysis of mechano- and chemosensitive neurones in the crayfish claw. *Neuroscience Letters*, **17**, 203–207.

Haynes, K. F. (1997). Genetics of pheromone communication in the cabbage looper moth, *Trichoplusia ni*. In *Insect pheromone research: new directions*, ed. R. T. Cardé & A. K. Minks, pp. 525–534. New York: Chapman and Hall.

Haynes, K. F. & Millar, J. G. (ed.) (1998). *Methods in chemical ecology, Vol. 2, Bioassay methods.* London: Chapman and Hall.

Haynes, K. F. & Yeargan, K. V. (1999). Exploitation of intraspecific communication systems: illicit signalers and receivers. *Annals of the Entomological Society of America*, **92**, 960–970.

Haynes, K. F., Gaston, L. K., Pope, M. M. & Baker, T. C. (1984). Potential for evolution of resistance to pheromones – interindividual and interpopulational variation in chemical communication-system of pink-bollworm moth. *Journal of Chemical Ecology*, **10**, 1551–1565.

Hecker, E. & Butenandt, A. (1984). Bombykol revisited – reflections on a pioneering period and on some of its consequences. In *Techniques in pheromone research*, ed. H. E. Hummel & T. A. Miller, pp. 1–44. New York: Springer-Verlag.

Hedin, P. A., Hardee, D. D., Thompson, A. C. & Gueldner, R. C. (1974). An assessment of the lifetime biosynthesis potential of the male boll weevil. *Journal of Insect Physiology*, **20**, 1707–1712.

Hedrick, P. W. (1999). Balancing selection and MHC. *Genetica*, **104**, 207–214.

Hedrick, P. W. & Black, F. L. (1997). HLA and mate selection: no evidence in South Amerindians. *American Journal of Human Genetics*, **61**, 505–511.

Hedrick, P. W. & Loeschcke, V. (1996). MHC and mate selection in humans? *Trends in Ecology & Evolution*, **11**, 24.

Hefetz, A., Bergstrom, G. & Tengo, J. (1986). Species, individual and kin specific blends in Dufour's gland secretions of halictine bees – chemical evidence. *Journal of Chemical Ecology*, **12**, 197–208.

Heifetz, Y., Miloslavski, I., Aizenshtat, Z. & Applebaum, S. W. (1998). Cuticular surface hydrocarbons of desert locust nymphs, *Schistocerca gregaria*, and their effect on phase behavior. *Journal of Chemical Ecology*, **24**, 1033–1047.

Heller, S. B. & Halpern, M. (1982). Laboratory observations of aggregative behavior of garter snakes, *Thamnophis sirtalis* – roles of the visual, olfactory, and vomeronasal senses. *Journal of Comparative and Physiological Psychology*, **96**, 984–999.

Henderson, G. (1998). Primer pheromones and the possible caste influence on the evolution of sociality in lower termites. In *Pheromone communication in social insects: ants, wasps, bees, and termites*, ed. R. K. Vander Meer, M. D. Breed, M. L. Winston & K. Espelie, pp. 314–330. Boulder, Colo: Westview Press.

Henderson, P. A., Irving, P. W. & Magurran, A. E. (1997). Fish pheromones and evolutionary enigmas: a reply to Smith. *Proceedings of the Royal Society of London Series B-Biological Sciences*, **264**, 451–453.

Hendrichs, J., Katsoyannos, B. I., Wornoayporn, V. & Hendrichs, M. A. (1994). Odor-mediated foraging by yellowjacket wasps (Hymenoptera, Vespidae) – predation on leks of pheromone-calling Mediterranean fruit-fly males (Diptera, Tephritidae). *Oecologia*, **99**, 88–94.

Hendrichs, M. A. & Hendrichs, J. (1998). Perfumed to be killed: interception of Mediterranean fruit fly (Diptera: Tephritidae) sexual signaling by predatory foraging wasps (Hymenoptera: Vespidae). *Annals of the Entomological Society of America*, **91**, 228–234.

Hendrikse, A. (1986). Intra- and interspecific sex-pheromone communication in the genus *Yponomeuta*. *Physiological Entomology*, **11**, 159–169.

Henning, S. F. (1983). Chemical communciation between lycaenid larvae (Lepidoptera: Lycaenidae) and ants (Hymenoptera: Formicidae). *Journal of the Entomological Society of Southern Africa*, **46**, 341–366.

Herms, D. A., Haack, R. A. & Ayres, B. D. (1991). Variation in semiochemical-mediated prey-predator interaction – *Ips pini* (Scolytidae) and *Thanasimus dubius* (Cleridae). *Journal of Chemical Ecology*, **17**, 515–524.

Herz, R. S. (1998). Are odors the best cues to memory? A cross-modal comparison of associative memory stimuli. *Annals of the New York Academy of Sciences*, **855**, 670–674.

Herz, R. S. & Engen, T. (1996). Odor memory: review and analysis. *Psychonomic Bulletin and Review*, **3**, 300–313.

Hews, D. K. (1988). Alarm response in larval western toads, *Bufo boreas* – release of larval chemicals by a natural predator and its effect on predator capture efficiency. *Animal Behaviour*, **36**, 125–133.

Hick, A. J., Pickett, J. A., Smiley, D. W. M., Wadhams, L. J. & Woodcock, C. M. (1997). Higher plants as a clean source of semiochemicals and genes for their biotechnological production. In *Phytochemical diversity: a source of new industrial products*, ed. S. Wrigley, M. Hayes, R. Thomas & E. Chrystal, pp. 220–236. Cambridge: The Royal Society of Chemistry.

Hildebrand, J. G. (1995). Analysis of chemical signals by nervous systems. *Proceedings of the National Academy of Sciences of the United States of America*, **92**, 67–74.

Hildebrand, J. G. & Shepherd, G. M. (1997). Mechanisms of olfactory discrimination: converging evidence for common principles across phyla. *Annual Review of Neuroscience*, **20**, 595–631.

Hodges, R. J., Birkinshaw, L. A. & Smith, R. H. (2000). Host selection or mate selection? Lessons from *Prostephanus truncatus*, a pest poorly adapted to stored products. In *Stored Product Protection, Proceedings of the 7th International Working Conference on Stored-Product Protection, 14–19 October 1998, Beijing, China*, Vol. 2, pp. 1788–1794.

Högland, J. & Alatalo, R. V. (1995). *Leks*. Princeton, N. J.: Princeton University Press.

Hölldobler, B. (1970). Orientierungsmechanismen des Ameisengastes *Atemeles* (Coleoptera, Staphylinidae) bei der Wirtssuche. In: *Verhandlungen der Zoologischen Gesellschaft (Würzburg, 1969) (Zoologischer Anzeiger Supplement)* Vol. 33, ed. W. Herre. pp. 580–585. Leipzig: Akademische Verlagsgesellschaft, Geest & Portig K.-G.

Hölldobler, B. (1995). The chemistry of social regulation – multicomponent signals in ant societies. *Proceedings of the National Academy of Sciences of the United States of America*, **92**, 19–22.

Hölldobler, B. & Carlin, N. F. (1987). Anonymity and specificity in the chemical communication signals of social insects. *Journal of Comparative Physiology A-Sensory Neural and Behavioral Physiology*, **161**, 567–581.

Hölldobler, B. & Wilson, E. O. (1977). Weaver ants. *Scientific American*, **237** (6), 146–154.

Hölldobler, B. & Wilson, E. O. (1978). The multiple recruitment systems of the African weaver ant *Oecophylla longinoda* (Latreille) (Hymenoptera: Formicidae). *Behavioral Ecology and Sociobiology*, **3**, 19–60.

Hölldobler, B. & Wilson, E. O. (1990). *The ants*. Berlin: Springer-Verlag.

Hölldobler, B. & Wilson, E. O. (1994). *Journey to the ants. A story of scientific exploration*. Cambridge, Mass.: Harvard University Press.

Hölldobler, B., Stanton, R. C. & Markl, H. (1978). Recruitment and food-retrieving behavior in *Novomessor* (Formicidae, Hymenoptera). I. Chemical signals. *Behavioral Ecology and Sociobiology*, **4**, 163–181.

Hoffmeister, T. S. & Roitberg, B. D. (1997). To mark the host or the patch: decisions of a parasitoid searching for concealed host larvae. *Evolutionary Ecology*, **11**, 145–168.

Hold, B. & Schleidt, M. (1977). The importance of human odour in non-verbal communication. *Zeitschrift für Tierpsychologie-Journal of Comparative Ethology*, **43**, 225–238.

Holmes, W. G. (1986). Identification of paternal half-siblings by captive Belding ground-squirrels. *Animal Behaviour*, **34**, 321–327.

Holmes, W. G. & Sherman, P. W. (1982). The ontogeny of kin recognition in 2 species of ground-squirrels. *American Zoologist*, **22**, 491–517.

Houck, L. D. & Reagan, N. L. (1990). Male courtship pheromones increase female receptivity in a plethodontid salamander. *Animal Behaviour*, **39**, 729–734.

Houpt, K. A. (1998). *Domestic animal behavior for veterinarians and animal scientists*, 3rd edn. London: Manson Publishing/The Veterinary Press.

Houpt, K. A., Rivera, W. & Glickstein, L. (1989). The flehmen response of bulls and cows. *Theriogenology*, **32**, 343–350.

Howard, L. O. & Fiske, W. F. (1911). The importation into the United States of the parasites of the gypsy moth and the brown-tail moth. *Bulletin* 91. Washington, D. C.; US Department of Agriculture, Bureau of Entomology.

Howard, R. W. & Akre, R. D. (1995). Propaganda, crypsis, and slave-making. In *Chemical ecology of insects 2*, ed. R. T. Cardé & W. J. Bell, pp. 364–424. London: Chapman and Hall.

Howard, R. W., McDaniel, C. A. & Blomquist, G. J. (1980). Chemical mimicry as an integrating mechanism: cuticular hydrocarbons of a termitophile and its host. *Science*, **210**, 431–433.

Howard, R. W., Akre, R. D. & Garnett, W. B. (1990a). Chemical mimicry in an obligate predator of carpenter ants (Hymenoptera, Formicidae). *Annals of the Entomological Society of America*, **83**, 607–616.

Howard, R. W., Stanley-Samuelson, D. W. & Akre, R. D. (1990b). Biosynthesis and chemical mimicry of cuticular hydrocarbons from the obligate predator, *Microdon albicomatus* Novak (Diptera, Syrphidae) and its ant prey, *Myrmica incompleta provancher* (Hymenoptera, Formicidae). *Journal of the Kansas Entomological Society*, **63**, 437–443.

Howe, N. R. & Harris, L. G. (1978). Transfer of the sea anemone pheromone, anthopleurine, by the nudibranch *Aedidia papillosa*. *Journal of Chemical Ecology*, **4**, 551–561.

Howe, N. R. & Sheik, Y. M. (1975). Anthopleurine: a sea anemone alarm pheromone. *Science*, **189**, 386–388.

Howse, P. E., Stevens, I. D. R. & Jones, O. T. (1998). *Insect pheromones and their use in pest management*. London: Chapman and Hall.

Hrdy, S. B. & Whitten, P. L. (1987). Patterning of sexual activity. In *Primate societies*, ed. B. Smuts, D. Cheney, R. Seyfarth, R. Wrangham & T. Struhsaker, pp. 370–384. Chicago: Chicago University Press.

Hubbard, S. F., Harvey, I. F. & Fletcher, J. P. (1999). Avoidance of superparasitism: a matter of learning? *Animal Behaviour*, **57**, 1193–1197.

Hudson, R. (1993). Olfactory imprinting. *Current Opinion in Neurobiology*, **3**, 548–552.

Hudson, R. & Distel, H. (1995). On the nature and action of the rabbit nipple-search pheromone: a review. In *Chemical signals in vertebrates VII. Advances in the Biosciences*, Vol. 93, ed. R. Apfelbach, Müller-Schwarze, K. Reutter & E. Weiler, pp. 223–232. Oxford: Pergamon Press.

Hughes, M. (1996). The function of concurrent signals: visual and chemical communication in snapping shrimp. *Animal Behaviour*, **52**, 247–257.

Humphries, R. E., Robertson, D. H. L., Beynon, R. J. & Hurst, J. L. (1999). Unravelling the chemical basis of competitive scent marking in house mice. *Animal Behaviour*, **58**, 1177–1190.

Hunter, J. R. & Hasler, A. D. (1965). Spawning association of the redfin shiner *Notropis umbratilis* and the green sunfish *Lepomis cyanellus*. *Copeia*, **1965** (3), 265–281.

Hurst, J. L. (1993). The priming effects of urine substrate marks on interactions between male house mice, *Mus musculus domesticus* Schwarz and Schwarz. *Animal Behaviour*, **45**, 55–81.

Hurst, J. L. & Rich, T. J. (1999). Scent marks as competitive signals of mate quality. In *Advances in chemical signals in vertebrates*, ed. R. E. Johnston, D. Müller-Schwarze & P. W. Sorenson, pp. 209–226. New York: Kluwer Academic Publishers/Plenum Press.

Hurst, J. L., Robertson, D. H. L., Tolladay, U. & Beynon, R. J. (1998). Proteins in urine scent marks of male house mice extend the longevity of olfactory signals. *Animal Behaviour*, **55**, 1289–1297.

Hurst, J. L., Payne, C. E., Nevison, C. M., Marie, A. D., Humphries, R. E., Robertson, D. H. L., Cavaggioni, A. & Beynon, R. J. (2001). Individual recognition in mice mediated by major urinary proteins. *Nature*, **414**, 631–634.

Hutchison, L. V. & Wenzel, B. M. (1980). Olfactory guidance in procellariiforms. *Condor*, **82**, 314–319.

Irving, P. W. & Magurran, A. E. (1997). Contest-dependent fright reactions in captive European minnows: the importance of naturalness in laboratory experiments. *Animal Behaviour*, **53**, 1193–1201.

Iyengar, V. K. & Eisner, T. (1999a). Heritability of body mass, a sexually selected trait, in an arctiid moth (*Utetheisa ornatrix*). *Proceedings of the National Academy of Sciences of the United States of America*, **96**, 9169–9171.

Iyengar, V. K. & Eisner, T. (1999b). Female choice increases offspring fitness in an arctiid moth (*Utetheisa ornatrix*). *Proceedings of the National Academy of Sciences of the United States of America*, **96**, 15013–15016.

Izard, M. K. (1983). Pheromones and reproduction in domestic animals. In *Pheromones and reproduction in mammals*, ed. J. G. Vandenbergh, pp. 253–285. New York: Academic Press.

Jacob, J., Balthazart, J. & Schoffeniels, E. (1979). Sex differences in the chemical composition of uropygial gland waxes in domestic ducks. *Biochemical Systematics and Ecology*, **7**, 149–153.

Jacob, S. & McClintock, M. K. (2000). Psychological state and mood effects of steroidal chemosignals in women and men. *Hormones and Behavior*, **37**, 57–78.

Jacob, S., McClintock, M. K., Zelano, B. & Ober, C. (2002). Paternally inherited HLA alleles are associated with women's choice of male odor. *Nature Genetics*, **30**, 175–179.

Jaeger, R. G. & Gabor, C. R. (1993). Intraspecific chemical communication by a territorial salamander via the postcloacal gland. *Copeia*, **1993**, 1171–1174.

Jaeger, R. G., Schwarz, J. & Wise, S. E. (1995). Territorial male salamanders have foraging tactics attractive to gravid females. *Animal Behaviour*, **49**, 633–639.

Jefferis, G. S. X. E., Marin, E. C., Stocker, R. F. & Luo, L. Q. (2001). Target neuron prespecification in the olfactory map of *Drosophila*. *Nature*, **414**, 204–208.

Jemiolo, B., Xie, T. M., Andreolini, F., Baker, A. E. M. & Novotny, M. (1991). The t-complex of the mouse – chemical characterization by urinary volatile profiles. *Journal of Chemical Ecology*, **17**, 353–367.

Johnsen, P. B. (1981). A behavioral control model for homestream selection in migratory salmonids. In *Proceedings of the salmon and trout migratory behavior symposium*, ed. E. L. Brannon & E. O. Salo, pp. 266–273. Seattle: School of Fisheries, University of Washington.

Johnston, C. E. (1994). Nest association in fishes – evidence for mutualism. *Behavioral Ecology and Sociobiology*, **35**, 379–383.

Johnston, R. E. (1998). Pheromones, the vomeronasal system, and communication – from hormonal responses to individual recognition. *Annals of the New York Academy of Sciences*, **855**, 333–348.

Johnston, R. E. (2000). Chemical communication and pheromones: the types of chemical signals and the role of the vomeronasal system. In *The neurobiology of taste and smell*, 2nd edn, ed. T. E. Finger, W. L. Silver & D. Restrepo, pp. 99–125. New York: Wiley-Liss.

Johnston, R. E. & Jernigan, P. (1994). Golden hamsters recognize individuals, not just individual scents. *Animal Behaviour*, **48**, 129–136.

Johnston, R. E. & Rasmussen, K. (1984). Individual recognition of female hamsters by males: role of chemical cues and of the olfactory and vomeronasal systems. *Physiology & Behavior*, **33**, 95–104.

Johnston, R. E., Derzie, A., Chiang, G., Jernigan, P. & Lee, H. C. (1993). Individual scent signatures in golden hamsters: evidence for specialization of function. *Animal Behaviour*, **45**, 1061–1070.

Johnston, R. E., Chiang, G. & Tung, C. (1994). The information in scent over-marks of golden hamsters. *Animal Behaviour*, **48**, 323–330.

Johnston, R. E., Müller-Schwarze, D. & Sorensen, P. W. (ed.) (1999). *Advances in chemical signals in vertebrates*. New York: Kluwer Academic Publishers/Plenum Press.

Johnstone, R. A. (1995). Sexual selection, honest advertisement and the handicap principle – reviewing the evidence. *Biological Reviews of the Cambridge Philosophical Society*, **70**, 1–65.

Johnstone, R. A. & Norris, K. (1993). Badges of status and the cost of aggression. *Behavioral Ecology and Sociobiology*, **32**, 127–134.

Jolly, A. (1966). *Lemur behavior. A Madagascar field study*. Chicago: Chicago University Press.

Jones, O. T. (1994). The current and future prospects for semiochemicals in the integrated management of insect pests. *Proceedings of Brighton Crop Protection Conference – Pests and diseases 1994*, **3**, 1213–1222.

Jones, T. M., Quinnell, R. J. & Balmford, A. (1998). Fisherian flies: benefits of female choice in a lekking sandfly. *Proceedings of the Royal Society of London Series B-Biological Sciences*, **265**, 1651–1657.

Judd, T. M. & Sherman, P. W. (1996). Naked mole-rats recruit colony mates to food sources. *Animal Behaviour*, **52**, 957–969.

Jumper, G. Y. & Baird, R. C. (1991). Location by olfaction – a model and application to the mating problem in the deep-sea hatchetfish *Argyropelecus hemigymnus*. *American Naturalist*, **138**, 1431–1458.

Jutsum, A. R. & Gordon, R. F. S. (1989). Epilogue. In *Insect pheromones in plant protection*, ed. A. R. Jutsum & R. F. S. Gordon, pp. 353–355. Chichester: John Wiley.

Jütte, A. (1995). Weibliche Pheromone – Wirkung und Rolle von synthetischen 'Kopulinen' bei der versteckten Ovulation des Menschen. Diplomarbeit Thesis, Universität Wien, Vienna (Unpublished).

Kaib, M. (1999). Termites. In *Pheromones of non-lepidopteran insects associated with agricultural plants*, ed. J. Hardie & A. K. Minks, pp. 329–353. Wallingford, UK: CAB International.

Kaib, M., Bruinsma, O. & Leuthold, R. H. (1982). Trail-following in termites – evidence for a multicomponent system. *Journal of Chemical Ecology*, **8**, 1193–1205.

Kaib, M., Husseneder, C., Epplen, C., Epplen, J. T. & Brandl, R. (1996). Kin-biased foraging in a termite. *Proceedings of the Royal Society of London Series B-Biological Sciences*, **263**, 1527–1532.

Kainoh, Y. (1999). Parasitoids. In *Pheromones of non-lepidopteran insects associated with agricultural plants,* ed. J. Hardie & A. K. Minks, pp. 383–404. Wallingford, UK: CAB International.

Kaissling, K.-E. (1987). *R. H. Wright lectures on insect olfaction,* ed. K. Colbow. Burnaby, B. C.: Simon Fraser University.

Kaissling, K.-E. (1998a). Olfactory transduction in moths: I. Generation of receptor potentials and nerve impulses. In *From structure to information in sensory systems,* ed. C. Taddei-Ferretti & C. Musio, pp. 93–112. Singapore: World Scientific.

Kaissling, K.-E. (1998b). Olfactory transduction in moths: II. Extracellular transport, deactivation and degradation of stimulus molecules. In *From structure to information in sensory systems,* ed. C. Taddei-Ferretti & C. Musio, pp. 113–137. Singapore: World Scientific.

Kaissling, K.-E. (2001). Olfactory perireceptor and receptor events in moths: a kinetic model. *Chemical Senses,* **26,** 125–150.

Kaissling, K.-E. & Kramer, E. (1990). Sensory basis of pheromone-mediated orientation in moths. *Verhandlunger der Deutschen Zoologischen Gesellschaft,* **83,** 109–131.

Kaitz, M., Good, A., Rokem, A. M. & Eidelman, A. I. (1987). Mothers' recognition of their newborns by olfactory cues. *Developmental Psychobiology,* **20,** 587–591.

Kanaujia, S. & Kaissling, K. E. (1985). Interactions of pheromone with moth antennae – adsorption, desorption and transport. *Journal of Insect Physiology,* **31,** 71–81.

Kaneshiro, K. Y. (1989). The dynamics of sexual selection and founder effects in species formation. In *Genetics, speciation, and the founder principle,* ed. L. V. Giddings, K. Y. Kaneshiro & W. W. Anderson, pp. 279–296. Oxford: Oxford University Press.

Kaneshiro, K. Y. & Boake, C. R. B. (1987). Sexual selection and speciation – issues raised by Hawaiian *Drosophila. Trends in Ecology & Evolution,* **2,** 207–212.

Karavanich, C. & Atema, J. (1998a). Individual recognition and memory in lobster dominance. *Animal Behaviour,* **56,** 1553–1560.

Karavanich, C. & Atema, J. (1998b). Olfactory recognition of urine signals in dominance fights between male lobster, *Homarus americanus. Behaviour,* **135,** 719–730.

Karlson, P. & Lüscher, M. (1959). 'Pheromones': a new term for a class of biologically active substances. *Nature,* **183,** 155–156.

Katsoyannos, B. I. & Boller, E. F. (1980). Second field application of oviposition-deterring pheromone of the European cherry fruit fly, *Rhagoletis ceresi* L. (Diptera: Tephritidae). *Zeitschrift für Angewandte Entomologie-Journal of Applied Entomology,* **89,** 278–281.

Kauer, J. S. & White, J. (2001). Imaging and coding in the olfactory system. *Annual Review of Neuroscience,* **24,** 963–979.

Kavaliers, M. & Colwell, D. D. (1995). Discrimination by female mice between the odors of parasitized and non-parasitized males. *Proceedings of the Royal Society of London Series B-Biological Sciences,* **261,** 31–35.

Keller, L. & Chapuisat, M. (1999). Conflict and cooperation in social insects. *Bioscience* : **49,** 899–909.

Keller, L. & Nonacs, P. (1993). The role of queen pheromones in social insects – queen control or queen signal. *Animal Behaviour,* **45,** 787–794.

Keller, L. & Reeve, H. K. (1994). Partitioning of reproduction in animal societies. *Trends in Ecology & Evolution,* **9,** 98–102.

Keller, L. & Reeve, H. K. (1999). Dynamics of conflicts within insect societies. In *Levels of selection in evolution,* ed. L. Keller, Princeton, N. J.: Princeton University Press.

Keller, L. & Ross, K. G. (1998). Selfish genes: a green beard in the red fire ant. *Nature,* **394,** 573–575.

Kelliher, K. R., Chang, Y. M., Wersinger, S. R. & Baum, M. J. (1998). Sex difference and testosterone modulation of pheromone-induced neuronal *Fos* in the ferret's main olfactory bulb and hypothalamus. *Biology of Reproduction,* **59,** 1454–1463.

Kelly, D. R. (1996). When is a butterfly like an elephant? *Chemistry & Biology*, **3**, 595–602.

Kelly, D. W. & Dye, C. (1997). Pheromones, kairomones and the aggregation dynamics of the sandfly *Lutzomyia longipalpis*. *Animal Behaviour*, **53**, 721–731.

Kendrick, K. M., Da Costa, A. P. C., Broad, K. D., Ohkura, S., Guevara, R., Lévy, F. & Keverne, E. B. (1997). Neural control of maternal behaviour and olfactory recognition of off-spring. *Brain Research Bulletin*, **44**, 383–395.

Kennedy, J. S. (1986). Some current issues in orientation to odour sources. In *Mechanisms in insect olfaction*, ed. T. L. Payne, M. C. Birch & C. E. J. Kennedy, pp. 1–25. Oxford: Oxford Scientific Publications, Oxford University Press.

Kennedy, J. S. (1992). *The new anthropomorphism*. Cambridge: Cambridge University Press.

Kennedy, J. S., Ludlow, A. R. & Sanders, C. J. (1981). Guidance of flying male moths by wind-borne sex-pheromone. *Physiological Entomology*, **6**, 395–412.

Keverne, E. B. (1983). Chemical communication in primate reproduction. In *Pheromones and reproduction in mammals*, ed. J. G. Vandenbergh, pp. 79–92. New York: Academic Press.

Keverne, E. B. (1998). Vomeronasal/accessory olfactory system and pheromonal recognition. *Chemical Senses*, **23**, 491–494.

Keverne, E. B. (1999). The vomeronasal organ. *Science*, **286**, 716–720.

Keverne, E. B. & Brennan, P. A. (1996). Olfactory recognition memory. *Journal of Physiology-Paris*, **90**, 399–401.

Keverne, E. B. & Rosser, A. E. (1986). The evolutionary significance of the olfactory block to pregnancy. In *Chemical signals in vertebrates 4*, ed. D. Duvall, D. Müller-Schwarze & R. M. Silverstein, pp. 433–439. New York: Plenum Press.

Kikuyama, S. & Toyoda, F. (1999). Sodefrin: a novel sex pheromone in a newt. *Reviews of Reproduction*, **4**, 1–4.

Kikuyama, S., Toyoda, F., Ohmiya, Y., Matsuda, K., Tanaka, S. & Hayashi, H. (1995). Sodefrin: a female-attracting peptide pheromone in newt cloacal glands. *Science*, **267**, 1643–1645.

Kiltie, R. A. (1982). On the significance of menstrual synchrony in closely associated women. *American Naturalist*, **119**, 414–419.

Kirkendall, L. R., Kent, D. S. & Raffa, K. A. (1997). Interactions among males, females and offspring in bark and ambrosia beetles: the significance of living in tunnels for the evolution of social behavior. In *The evolution of social behavior in insects and arachnids*, ed. J. C. Choe & B. J. Crespi, pp. 181–215. Cambridge: Cambridge University Press.

Klun, K. A. & Huettel, M. D. (1988). Genetic regulation of sex pheromone production and response: interaction of sympatric pheromonal types of the European corn borer, *Ostrinia nubilalis* (Lepidoptera: Pyralidae). *Journal of Chemical Ecology*, **14**, 2047–2061.

Koehl, M. A. R. (1995). Fluid flow through hair-bearing appendages: feeding, smelling and swimming at low and intermediate Reynolds numbers. In *Biological fluid dynamics. Symposia of the Society of Experimental Biology 27*, ed. T. Pedley & C. Ellington, pp. 157–182. Cambridge: Company of Biologists.

Koivula, M. & Viitala, J. (1999). Rough-legged buzzards use vole scent marks to assess hunting areas. *Journal of Avian Biology*, **30**, 329–332.

Kouros-Mehr, H., Pintchovski, S., Melnyk, J., Chen, Y. J., Friedman, C., Trask, B. & Shizuya, H. (2001). Identification of non-functional human VNO receptor genes provides evidence for vestigiality of the human VNO. *Chemical Senses*, **26**, 1167–1174.

Krause, J. (1993a). The effect of 'Schreckstoff' on the shoaling behaviour of the minnow – a test of Hamilton's selfish herd theory. *Animal Behaviour*, **45**, 1019–1024.

Krause, J. (1993b). Transmission of fright reaction between different species of fish. *Behaviour*, **127**, 37–48.

Krause, J. (1994). Differential fitness returns in relation to spatial position in groups. *Biological Reviews of the Cambridge Philosophical Society*, **69**, 187–206.

Krebs, J. R. & Davies, N. B. (1993). *An introduction to behavioural ecology*, 3rd edn. Oxford: Blackwell Scientific Publications.

Krebs, J. R. & Davies, N. B. (ed.) (1997). *Behavioural ecology: an evolutionary approach*, 4th edn. Oxford: Blackwell Science.

Kruczek, M. (1997). Male rank and female choice in the bank vole, *Clethrionomys glareolus*. *Behavioural Processes*, **40**, 171–176.

Kruuk, H. (1972). *The spotted hyena. A study of predation and social behavior*. Chicago: Chicago University Press.

Kruuk, H. (1989). *The social badger. Ecology and behaviour of a group-living carnivore* (Meles meles). Oxford: Oxford University Press.

Kruuk, H., Gorman, M. & Leitch, A. (1984). Scent-marking with the subcaudal gland by the European badger, *Meles meles* L. *Animal Behaviour*, **32**, 899–907.

Labov, J. B. (1981). Pregnancy blocking in rodents: adaptive advantages for females. *American Naturalist*, **118**, 361–371.

Labows, J. N. & Preti, G. (1992). Human semiochemicals. In *Fragrance: the psychology and biology of perfume,* ed. S. Van Toller & G. H. Dodd, pp. 69–90. London: Elsevier Science.

Lacey, E. A. & Sherman, P. W. (1997). Cooperative breeding in naked mole-rats: implications for vertebrate and invertebrate sociality. In *Cooperative breeding in mammals*, ed. N. G. Solomon & J. A. French, pp. 267–301. Cambridge: Cambridge University Press.

Lacey, R. C. & Sherman, P. W. (1983). Kin recognition by phenotype matching. *American Naturalist*, **121**, 489–512.

Lamunyon, C. W. & Eisner, T. (1993). Postcopulatory sexual selection in an arctiid moth (*Utetheisa ornatrix*). *Proceedings of the National Academy of Sciences of the United States of America*, **90**, 4689–4692.

Lamunyon, C. W. & Eisner, T. (1994). Spermatophore size as determinant of paternity in an arctiid moth (*Utetheisa ornatrix*). *Proceedings of the National Academy of Sciences of the United States of America*, **91**, 7081–4.

Lancet. D. (1991). The strong scent of success. *Nature*, **351**, 275–276.

Lande, R. (1982). Rapid origin of sexual isolation and character divergence in a cline. *Evolution*, **36**, 213–223.

Landolt, P. J. (1997). Sex attractant and aggregation pheromones of male phytophagous insects. *American Entomologist*, **43**, 12–22.

Landolt, P. J. & Averill, A. L. (1999). Fruit flies. In *Pheromones of non-lepidopteran insects associated with agricultural plants*, ed. J. Hardie & A. K. Minks, pp. 3–25. Wallingford, UK: CAB International.

Landolt, P. J., Reed, H. C. & Heath, R. R. (1992). Attraction of female papaya fruit-fly (Diptera, Tephritidae) to male pheromone and host fruit. *Environmental Entomology*, **21**, 1154–1159.

Landolt, P. J., Molina, O. H., Heath, R. R., Ward, K., Dueben, B. D. & Millar, J. G. (1996). Starvation of cabbage looper moths (Lepidoptera: Noctuidae) increases attraction to male pheromone. *Annals of the Entomological Society of America*, **89**, 459–465.

Laurent, G. Stopfer, M., Friedrich, R. W., Rabinovich, M.I., Volkovskii, A., and Abarbanel, H.D.I. (2001). Odor encoding as an active, dynamical process: experiments, computation, and theory. *Annual Review of Neuroscience*, **24**, 263–297.

Law, R. H. & Regnier, F. E. (1971). Pheromones. *Annual Review of Biochemistry*, **40**, 533–548.

Lawrence, B. J. & Smith, R. J. F. (1989). Behavioral-response of solitary fathead minnows, *Pimephales promelas*, to alarm substance. *Journal of Chemical Ecology*, **15**, 209–219.

Leal, W. S. (1999). Scarab beetles. In *Pheromones of non-lepidopteran insects associated with agricultural plants*, ed. J. Hardie & A. K. Minks, pp. 51–68. Wallingford, UK: CAB International.

Leal, W. S., Hasegawa, M. & Sawada, M. (1992). Identification of *Anomala schonfeldti* sex-pheromone by high-resolution GC-behavior bioassay. *Naturwissenschaften*, **79**, 518–519.

Le Conte, Y., Sreng, L. & Trouiller, J. (1994). The recognition of larvae by worker honeybees. *Naturwissenschaften*, **81**, 462–465.

Leinders-Zufall, T., Lane, A. P., Puche, A. C., Ma, W. D., Novotny, M. V., Shipley, M. T. & Zufall, F. (2000). Ultrasensitive pheromone detection by mammalian vomeronasal neurons. *Nature*, **405**, 792–796.

Lenington, S., Coopersmith, C. B. & Erhart, M. (1994). Female preference and variability among t-haplotypes in wild house mice. *American Naturalist*, **143**, 766–784.

Lenoir, A., Malosse, C. & Yamaoka, R. (1997). Chemical mimicry between parasitic ants of the genus *Formicoxenus* and their host *Myrmica* (Hymenoptera, Formicidae). *Biochemical Systematics and Ecology*, **25**, 379–389.

Lenoir, A., D'Ettorre, P., Errard, C. & Hefetz, A. (2001). Chemical ecology and social parasitism in ants. *Annual Review of Entomology*, **46**, 573–599.

Levesley, P. B. & Magurran, A. E. (1988). Population differences in the reaction of minnows to alarm substance. *Journal of Fish Biology*, **32**, 699–706.

Levinson, A. & Levinson, H. (1995). Reflections on structure and function of pheromone glands in storage insect species. *Anzeiger für Schadlingskunde Pflanzenschutz Umweltschutz*, **68**, 99–118.

Levitan, D. R. (1996). Effects of gamete traits on fertilization in the sea and the evolution of sexual dimorphism. *Nature*, **382**, 153–155.

Levitan, D. R. & Petersen, C. (1995). Sperm limitation in the sea. *Trends in Ecology & Evolution*, **10**, 228–231.

Lévy, F., Locatelli, A., Piketty, V., Tillet, Y. & Poindron, P. (1995). Involvement of the main but not the accessory olfactory system in maternal-behavior of primiparous and multiparous ewes. *Physiology & Behavior*, **57**, 97–104.

Lévy, F., Porter, R. H., Kendrick, K. M., Keverne, E. B. & Romeyer, A. (1996). Physiological, sensory, and experiential factors of parental care in sheep. *Advances in the Study of Behavior*, **25**, 385–422.

Lewis, S. E. & Pusey, A. E. (1997). Factors influencing the occurance of communal care in plural breeding mammals. In *Cooperative breeding in mammals*, ed. N. G. Solomon & J. A. French, pp. 335–363. Cambridge: Cambridge University Press.

Lewis, S. M. & Austad, S. N. (1994). Sexual selection in flour beetle. *Behavioral Ecology*, **5**, 219–224.

Li, W., Sorensen, P. W. & Gallaher, D. D. (1995). The olfactory system of migratory adult sea lamprey (*Petromyzon marinus*) is specifically and acutely sensitive to unique bile acids released by conspecific larvae. *Journal of General Physiology*, **105**, 569–587.

Licht, G. & Meredith, M. (1987). Convergence of main and accessory olfactory pathways onto single neurons in the hamster amygdala. *Experimental Brain Research*, **69**, 7–18.

Lindauer, M. (1961). *Communication among social bees*. Cambridge, Mass.: Harvard University Press.

Linn, C. E. & Roelofs, W. L. (1989). Response specificity of male moths to multicomponent pheromones. *Chemical Senses*, **14**, 421–437.

Linn, C. E. & Roelofs, W. L. (1995). Pheromone communication in moths and its role in the speciation process. In *Speciation and the recognition concept. Theory and application*, ed. D. M. Lambert & H. G. Spencer, pp. 263–300. Baltimore: Johns Hopkins University Press.

Linn, C. E., Campbell, M. G. & Roelofs, W. L. (1987). Pheromone components and active spaces: what do moths smell and where do they smell it? *Science*, **237**, 650–652.

Linn, J. E. (1997). Neuroendocrine factors in the photoperiodic control of male moth responsiveness to pheromone. In *Insect pheromone research: new directions*, ed. R. T. Cardé & A. K. Minks, pp. 184–193. New York: Chapman and Hall.

Liu, Y. B. & Haynes, K. F. (1992). Filamentous nature of pheromone plumes protects integrity of signal from background chemical noise in cabbage-looper moth, *Trichoplusia ni. Journal of Chemical Ecology*, **18**, 299–307.

Lloyd, J. E. (1979). Sexual selection in bioluminsecent beetles. In *Sexual selection and reproductive competition in insects,* ed. M. Blum & N. A. Blum, pp. 293–342. London: Academic Press.

Löfstedt C. (1986). Sexual feromoner och reproductive isolering hos natfjärilar. *Entomologisk Tidskoift*, **107**, 125–137.

Löfstedt, C. (1990). Population variation and genetic control of pheromone communication systems in moths. *Entomologia Experimentalis et Applicata*, **54**, 199–218.

Löfstedt, C. (1991). Evolution of moth pheromones. In *Insect chemical ecology, Proceedings as a conference on insect chemical ecology,* ed. I. Hrdy, pp. 57–73. The Hague: Academia Prague and SPB Academic Publishing.

Löfstedt, C. (1993). Moth pheromone genetics and evolution. *Philosophical Transactions of the Royal Society of London Series B-Biological Sciences*, **340**, 167–177.

Löfstedt, C., Vickers, N. J., Roelofs, W. L. & Baker, T. C. (1989). Diet related courtship success in the oriental fruit moth, *Grapholita molesta* (Tortricidae). *Oikos*, **55**, 402–408.

Löfstedt, C., Herrebout, W. M. & Menken, S. B. J. (1991). Sex pheromones and their potential role in the evolution of reproductive isolation in small ermine moths (Yponomeutidae). *Chemoecology*, **2**, 20–28.

Lopez, F., Acosta, F. J. & Serrano, J. M. (1994). Guerrilla vs. phalanx strategies of resource capture – growth and structural plasticity in the trunk trail system of the harvester ant *Messor barbarus. Journal of Animal Ecology*, **63**, 127–138.

Lorig, T. S. (1999). On the similarity of odor and language perception. *Neuroscience and Biobehavioral Reviews*, **23**, 391–398.

Lorig, T. S., Huffman, E., DeMartino, A. & DeMarco, J. (1991). The effects of low concentration odors on EEG activity and behavior. *Journal of Psychophysiology*, **5**, 69–77.

Lu, X. C. M. & Slotnick, B. M. (1998). Olfaction in rats with extensive lesions of the olfactory bulbs: implications for odor coding. *Neuroscience*, **84**, 849–866.

Lüscher, M. (1961). Air-conditioned termite nests. *Scientific American*, **205** (1), 138–145.

Ma, W. D., Miao, Z. S. & Novotny, M. V. (1998). Role of the adrenal gland and adrenal-mediated chemosignals in suppression of estrus in the house mouse: the Lee-Boot effect revisited. *Biology of Reproduction*, **59**, 1317–1320.

Ma, W. [D.], Miao, Z. [S.] & Novotny, M. V. (1999). Induction of estrus in grouped female mice (*Mus domesticus*) by synthetic analogues of preputial gland constituants. *Chemical Senses*, **24**, 289–293.

Macdonald, D. W. (1985a). The carnivores: order Carnivora. In *Social odours in mammals,* Vol. 2, ed. R. E. Brown & D. W. Macdonald, pp. 619–722. Oxford: Oxford University Press.

Macdonald, D. W. (1985b). The rodents IV: suborder Hystricomorpha. In *Social odours in mammals,* Vol. 1, ed. R. E. Brown & D. W. Macdonald, pp. 480–506. Oxford: Oxford University Press.

Madsen, T., Shine, R., Loman, J. & Håkansson, T. (1992). Why do female adders copulate so frequently? *Nature*, **355**, 440–441.

Mafra-Neto, A. & Baker, T. C. (1996). Timed, metered sprays of pheromone disrupt mating of *Cadra cautella* (Lepidoptera: Pyralidae). *Journal of Agricultural Entomology*, **13**, 149–168.

Magurran, A. E., Irving, P. W. & Henderson, P. A. (1996). Is there a fish alarm pheromone? A wild study and critique. *Proceedings of the Royal Society of London Series B-Biological Sciences*, **263**, 1551–1556.

Malnic, B., Hirono, J., Sato, T. & Buck, L. B. (1999). Combinatorial receptor codes for odors. *Cell*, **96**, 713–723.

Manning, A. & Dawkins, M. S. (1998). *An introduction to animal behaviour*. 5th edn. Cambridge: Cambridge University Press.

Manning, C. J., Wakeland, E. K. & Potts, W. K. (1992). Communal nesting patterns in mice implicate MHC genes in kin recognition. *Nature*, **360**, 581–583.

Marchlewska-Koj, A., Lepri, J. J. & Müller-Schwarze, D. (ed.) (2001). *Chemical signals in vertebrates 9*. Dordrecht: Kluwer Academic Publishers/Plenum Press.

Marsden, H. M. & Bronson, F. H. (1964). Estrous synchrony in mice: alteration by exposure to male urine. *Science*, **144**, 1469.

Marshall, D. A., Blumer, L. & Moulton, D. G. (1981). Odor detection curves for *n*-pentanoic acid in dogs and humans. *Chemical Senses*, **6**, 445–453.

Martin, G. N. (1996). Olfactory remediation: current evidence and possible applications. *Social Science & Medicine*, **43**, 63–70.

Mason, R. T. (1993). Chemical ecology of the red-sided garter snake, *Thamnophis sirtalis parietalis*. *Brain Behavior and Evolution*, **41**, 261–268.

Mason, R. T. (1994). Hormonal and pheromonal correlates of reproductive behavior in garter snakes. In *Perspectives in comparative endocrinology*, ed. K. G. Davey, R. E. Peter & S. S. Tobe, pp. 427–432. Ottawa: National Research Council.

Mason, R. T. & Crews, D. (1985). Female mimicry in garter snakes. *Nature*, **316**, 59–60.

Mateo, J. M. & Johnston, R. E. (2000). Kin recognition and the 'armpit effect': evidence of self-referent phenotype matching. *Proceedings of the Royal Society of London Series B-Biological Sciences*, **267**, 695–700.

Mathis, A. (1990a). Territorial salamanders assess sexual and competitive information using chemical signals. *Animal Behaviour*, **40**, 953–962.

Mathis, A. (1990b). Territoriality in a terrestrial salamander: the influence of resource quality and body size. *Behaviour*, **112**, 162–175.

Mathis, A. & Smith, R. J. F. (1993). Chemical alarm signals increase the survival-time of fathead minnows (*Pimephales promelas*) during encounters with northern pike (*Esox lucius*). *Behavioral Ecology*, **4**, 260–265.

Mathis, A., Chivers, D. P. & Smith, R. J. F. (1995). Chemical alarm signals – predator deterrents or predator attractants. *American Naturalist*, **145**, 994–1005.

Matsumura, K., Mori, S., Nagano, M. & Fusetani, N. (1998a). Lentil lectin inhibits adult extract-induced settlement of the barnacle, *Balanus amphitrite*. *Journal of Experimental Zoology*, **280**, 213–219.

Matsumura, K., Nagano, M. & Fusetani, N. (1998b). Purification of a larval settlement-inducing protein complex (SIPC) of the barnacle, *Balanus amphitrite*. *Journal of Experimental Zoology*, **281**, 12–20.

McAllister, M. K. & Roitberg, B. D. (1987). Adaptive suicidal behaviour in pea aphids. *Nature*, **328**, 797–799.

McCaffery, A. R., Simpson, S. J., Islam, M. S. & Roessingh, P. (1998). A gregarizing factor present in the egg pod foam of the desert locust *Schistocerca gregaria*. *Journal of Experimental Biology*, **201**, 347–363.

McCall, P. J. & Cameron, M. M. (1995). Oviposition pheromones in insect vectors. *Parasitology Today*, **11**, 352–55.

McClintock, M. K. (1971). Menstrual synchrony and suppression. *Nature*, **229**, 244–245.

McClintock, M. K. (1983). Modulation of the estrous-cycle by pheromones from pregnant and lactating rats. *Biology of Reproduction*, **28**, 823–829.

McClintock, M. K. (1998). On the nature of mammalian and human pheromones. *Annals of the New York Academy of Sciences*, **855**, 390–392.

McClintock, M. K. (2000). Human pheromones: primers, releasers, signalers, or modulators? In *Reproduction in context*, ed. K. Wallen & J. Schneider, pp. 355–420. Cambridge, Mass.: MIT Press.

McGlone, J. J. & Morrow, J. L. (1988). Reduction of pig agonistic behavior by androstenone. *Journal of Animal Science*, **66**, 880–884.

McNeil, J. N. (1991). Behavioral ecology of pheromone-mediated communication in moths and its importance in the use of pheromone traps. *Annual Review of Entomology*, **36**, 407–430.

McNeil, J. N. (1992). Evolutionary perspectives and insect pest control: an attractive blend for the deployment of semiochemicals in management systems. In *Insect chemical ecology. An evolutionary approach*, ed. B. D. Roitberg & M. B. Isman, pp. 334–352. New York: Chapman and Hall.

Mead, K. S., Koehl, M. A. R. & O'Donnell, M. J. (1999). Stomatopod sniffing: the scaling of chemosensory sensillae and flicking behavior with body size. *Journal of Experimental Marine Biology and Ecology*, **241**, 235–261.

Menzel, R., Hammer, M., Müller, U. & Rosenboom, H. (1996). Behavioral, neural and cellular components underlying olfactory learning in the honeybee. *Journal of Physiology-Paris*, **90**, 395–398.

Meredith, M. (1994). Chronic recording of vomeronasal pump activation in awake behaving hamsters. *Physiology & Behavior*, **56**, 345–354.

Meredith, M. (1998). Vomeronasal, olfactory, hormonal convergence in the brain – cooperation or coincidence? *Annals of the New York Academy of Sciences*, **855**, 349–361.

Meredith, M. (1999). Vomeronasal organ. In *Encyclopedia of reproduction*, Vol. 4, ed. E. Knobil & J. D. Neill, pp. 1004–1014. New York: Academic Press.

Meredith, M. (2001). Human vomeronasal organ function: a critical review of best and worst cases. *Chemical Senses*, **26**, 433–445.

Merlin, J., Lemaitre, O. & Grégoire, J. C. (1996). Chemical cues produced by conspecific larvae deter oviposition by the coccidophagous ladybird beetle, *Cryptolaemus montrouzieri*. *Entomologia Experimentalis et Applicata*, **79**, 147–151.

Meyer, S. L. F., Johnson, G., Dimock, M., Fahey, J. W. & Huettel, R. N. (1997). Field efficacy of *Verticillium lecanii*, sex pheromone, and pheromone analogs as potential management agents for soybean cyst nematode. *Journal of Nematology*, **29**, 282–288.

Michael, R. P. & Keverne, E. B. (1970). Primate sex pheromones of vaginal origin. *Nature*, **225**, 84–85.

Milinski, M. & Wedekind, C. (2001). Evidence for MHC-correlated perfume preferences in humans. *Behavioral Ecology*, **12**, 140–149.

Millar, J. G. & Haynes, K. F. (ed.) (1998). *Methods in chemical ecology, Vol. 1, Chemical methods*. London: Chapman and Hall.

Mills, M. G. L., Gorman, M. L. & Mills, M. E. J. (1980). The scent marking behaviour of the brown hyaena, *Hyaena brunea*. *South African Journal of Zoology*, **15**, 240–248.

Minguez, E. (1997). Olfactory nest recognition by British storm-petrel chicks. *Animal Behaviour*, **53**, 701–707.

Minks, A. K. & Cardé, R. T. (1988). Disruption of pheromone communication in moths – is the natural blend really most efficacious. *Entomologia Experimentalis et Applicata*, **49**, 25–36.

Minks, A. K. & Kirsch, P. A. (1998). Application of pheromones: toxicological aspects, effects on beneficials and state of registration. In *Ecotoxicology: pesticides and beneficial organisms*, ed. P. T. Haskell & P. McEwen, pp. 337–347. London: Chapman and Hall.

Møller, A. P. & Thornhill, R. (1998). Bilateral symmetry and sexual selection: a meta-analysis. *American Naturalist*, **151**, 174–192.

Mombaerts, P. (1999). Odorant receptor genes in humans. *Current Opinion in Genetics & Development*, **9**, 315–320.

Monaco, E. L., Tallamy, D. W. & Johnson, R. K. (1998). Chemical mediation of egg dumping in the lace bug *Gargaphia solani* Heidemann (Heteroptera: Tingidae). *Animal Behaviour*, **56**, 1491–1495.

Monnin, T., Malosse, C. & Peeters, C. (1998). Solid-phase microextraction and cuticular hydrocarbon differences related to reproductive activity in queenless ant *Dinoponera quadriceps*. *Journal of Chemical Ecology*, **24**, 473–490.

Montagna, W. & Parakkal, P. F. (1974). *The structure and function of skin*. New York: Academic Press.

Montague, C. E. & Oldroyd, B. P. (1998). The evolution of worker sterility in honey bees: an investigation into a behavioral mutant causing failure of worker policing. *Evolution*, **52**, 1408–1415.

Montgomery, M. E. & Nault, L. R. (1978). Effects of age and wing polymorphism on the sensitivity of *Myzus persicae* to alarm pheromone. *Annals of the Entomological Society of America*, **71**, 788–790.

Monti-Bloch, L., Jennings-White, C. & Berliner, D. L. (1998). The human vomeronasal system – a review. *Annals of the New York Academy of Sciences*, **855**, 373–389.

Moore, A. J. & Moore, P. J. (1999). Balancing sexual selection through opposing mate choice and male competition. *Proceedings of the Royal Society of London Series B-Biological Sciences*, **266**, 711–716.

Moore, A. J., Reagan, N. L. & Haynes, K. F. (1995). Conditional signaling strategies – effects of ontogeny, social experience and social-status on the pheromonal signal of male cockroaches. *Animal Behaviour*, **50**, 191–202.

Moore, B. P., Brown, W. V. & Rothschild, M. (1990). Methylalkylpyrazines in aposematic insects, their host plants and mimics. *Chemoecology*, **1**, 43–51.

Moore, P. J., Reagan-Wallin, N. L., Haynes, K. F. & Moore, A. J. (1997). Odour conveys status on cockroaches. *Nature*, **389**, 25.

Mori, I. (1999). Genetics of chemotaxis and thermotaxis in the nematode *Caenorhabditis elegans*. *Annual Review of Genetics*, **33**, 399–422.

Mori, K. (1996). Molecular asymmetry and pheromone science. *Bioscience Biotechnology and Biochemistry*, **60**, 1925–1932.

Mori, K., Nagao, H. & Yoshihara, Y. (1999). The olfactory bulb: coding and processing of odor molecule information. *Science*, **286**, 711–715.

Morris, N. M. & Udry, R. J. (1978). Pheromonal influences on human sexual behavior: an experimental search. *Journal of Biosocial Science*, **10**, 147–157.

Mucignat-Caretta, C. & Caretta, A. (1999). Urinary chemical cues affect light avoidance behaviour in male laboratory mice, *Mus musculus*. *Animal Behaviour*, **57**, 765–769.

Mucignat-Caretta, C., Caretta, A. & Cavaggioni, A. (1995). Acceleration of puberty onset in female mice by male urinary proteins. *Journal of Physiology-London*, **486**, 517–522.

Mucignat-Caretta, C., Caretta, A. & Baldini, E. (1998). Protein-bound male urinary pheromones: differential responses according to age and gender. *Chemical Senses*, **23**, 67–70.

Müller, C. T., Beckmann, M. & Hardege, J. D. (1999). Sex pheromones in *Nereis succinea*. *Invertebrate Reproduction & Development*, **36**, 183–186.

Müller-Schwarze, D. (1971). Pheromones in black-tailed deer. *Animal Behaviour*, **19**, 141–152.

Müller-Schwarze, D. (1990). Leading them by their noses: animal and plant odours for managing vertebrates. In *Chemical signals in vertebrates*, Vol. 5, ed. D. W. Macdonald, D. Müller-Schwarze & S. E. Natynczuk, pp. 586–598. Oxford: Oxford Science Publications, Oxford University Press.

Müller-Schwarze, D. (1999). Signal specialization and evolution in mammals. In *Advances in chemical signals in vertebrates*, ed. R. E. Johnston, D. Müller-Schwarze & P. W. Sorenson, pp. 1–14. New York: Kluwer Academic Publishers/Plenum Press.

Müller-Schwarze, D. & Müller-Schwarze, C. (1971). Olfactory imprinting in a precocial mammal. *Nature*, **229**, 55–56.

Müller-Schwarze, D. & Müller-Schwarze, C. (1972). Social scents in hand-reared proghorn. *Zoologica Africana*, **7**, 251–327.

Müller-Schwarze, D., Altieri, R. & Porter, N. (1984). Alert odor from skin gland in deer. *Journal of Chemical Ecology*, **10**, 1707–1729.

Murlis, J. (1997). Odor plumes and the signal they provide. In *Insect pheromone research: new directions*, ed. R. T. Cardé & A. K. Minks, pp. 221–231. New York: Chapman and Hall.

Murlis, J. & Jones, C. D. (1981). Fine-scale structure of odor plumes in relation to insect orientation to distant pheromone and other attractant sources. *Physiological Entomology*, **6**, 71–86.

Murlis, J., Elkinton, J. S. & Cardé, R. T. (1992). Odor plumes and how insects use them. *Annual Review of Entomology*, **37**, 505–532.

Naish, K. A., Carvalho, G. R. & Pitcher, T. J. (1993). The genetic-structure and microdistribution of shoals of *Phoxinus phoxinus*, the European minnow. *Journal of Fish Biology*, **43** (Supplement A), 75–89.

Nault, L. R. (1973). Alarm pheromones help aphids escape predators. *Ohio Report*, **58**, 16–17.

Nault, L. R. & Phelan, P. L. (1984). Alarm pheromones and sociality in presocial insects. In *Chemical ecology of insects*, 1st edn, ed. W. J. Bell & R. T. Cardé, pp. 237–256. London: Chapman and Hall.

Nault, L. R., Montgomery, M. E. & Bowers, W. S. (1976). Ant – aphid association: role of aphid alarm pheromone. *Science*, **192**, 1349–1351.

Naumann, K., Winston, M. L., Slessor, K. N., Prestwich, G. D. & Latli, B. (1992). Intra-nest transmission of aromatic honey-bee queen mandibular gland pheromone components – movement as a unit. *Canadian Entomologist*, **124**, 917–934.

Nelson, R. J. (1995). *An introduction to behavioral endocrinology*. Sunderland, Mass.: Sinauer Associates.

Nevitt, G. [A.] (1999a). Foraging by seabirds on an olfactory landscape. *American Scientist*, **87**, 46–53.

Nevitt, G. [A.] (1999b). Olfactory foraging in Antarctic seabirds: a species-specific attraction to krill odors. *Marine Ecology-Progress Series*, **177**, 235–241.

Nevitt, G. A. (2000). Olfactory foraging by Antarctic procellariiform seabirds: life at high Reynolds numbers. *Biological Bulletin*, **198**, 245–253.

Newcomer, S. D., Zeh, J. A. & Zeh, D. W. (1999). Genetic benefits enhance the reproductive success of polyandrous females. *Proceedings of the National Academy of Sciences of the United States of America*, **96**, 10236–10241.

Nicolis, S. C. & Deneubourg, J. L. (1999). Emerging patterns and food recruitment in ants: an analytical study. *Journal of Theoretical Biology*, **198**, 575–592.

Nijhout, H. F. (1994). *Insect hormones*. Princeton, N. J.: Princeton University Press.

Noldus, L. P. J. J., Potting, R. P. J. & Barendregt, H. E. (1991). Moth sex pheromone adsorption to leaf surface: bridge in time for chemical spies. *Physiological Entomology*, **16**, 329–344.

Nolte, D. L., Mason, J. R., Epple, G., Aronov, E. & Campbell, D. L. (1994). Why are predator urines aversive to prey. *Journal of Chemical Ecology*, **20**, 1505–1516.

Nordlund, D. A. (1981). Semiochemicals: a review of the terminology. In *Semiochemicals: their role in pest control*, ed. D. A. Nordlund, R. L. Jones & W. J. Lewis, pp. 13–28. New York: John Wiley.

North, R. D., Jackson, C. W. & Howse, P. E. (1997). Evolutionary aspects of ant – fungus interactions in leaf-cutting ants. *Trends in Ecology & Evolution*, **12**, 386–389.

North, R. D., Jackson, C. W. & Howse, P. E. (1999). Communication between the fungus garden and workers of the leaf-cutting ant, *Atta sexdens rubropilosa*, regarding choice of substrate for the fungus. *Physiological Entomology*, **24**, 127–133.

Norval, R. A. I., Sonenshine, D. E., Allan, S. A. & Burridge, M. J. (1996). Efficacy of pheromone-acaricide-impregnated tail-tag decoys for controlling the bont tick, *Amblyomma hebraeum* (Acari: Ixodidae) on cattle in Zimbabwe. *Experimental & Applied Acarology*, **20**, 31–46.

Novotny, M. V., Harvey, S., Jemiolo, B. & Alberts, J. R. (1985). Synthetic pheromones that promote inter-male aggression in mice. *Proceedings of the National Academy of Sciences of the United States of America*, **82**, 2059-2061.

Novotny, M. V., Xie, T. M., Harvey, S., Wiesler, D., Jemiolo, B. & Carmack, M. (1995). Stereoselectivity in mammalian chemical communication – male-mouse pheromones. *Experientia*, **51**, 738–743.

Novotny, M. V., Ma, W. D., Wiesler, D. & Zidek, L. (1999a). Positive identification of the puberty-accelerating pheromone of the house mouse: the volatile ligands associating with the major urinary protein. *Proceedings of the Royal Society of London Series B-Biological Sciences*, **266**, 2017–2022.

Novotny, M. V., Ma, W., Zidek, L. & Daev, E. (1999b). Recent biochemical insights into puberty acceleration, estrus induction and puberty delay in the house mouse. In *Advances in chemical signals in vertebrates*, ed. R. E. Johnston, D. Müller-Schwarze & P. W. Sorenson, pp. 99–116. New York: Kluwer Academic Publishers/Plenum Press.

Ober, C. (1999). Studies of HLA, fertility and mate choice in a human isolate. *Human Reproduction Update*, **5**, 103–107.

Ober, C., Weitkamp, L. R., Cox, N., Dytch, H., Kostyu, D. & Elias, S. (1997). HLA and mate choice in humans. *American Journal of Human Genetics*, **61**, 497–504.

Obin, M. S. & Vander Meer, R. K. (1989). Nestmate recognition in fire ants (*Solenopsis invicta* Buren) – do queens label workers. *Ethology*, **80**, 255–264.

Ode, P. R. & Wissinger, S. A. (1993). Interaction between chemical and tactile cues in mayfly detection of stoneflies. *Freshwater Biology*, **30**, 351–357.

O'Hara, R. K. & Blaustein, A. R. (1982). Kin preference behavior in *Bufo boreas* tadpoles. *Behavioral Ecology and Sociobiology*, **11**, 43–49.

Okano, K. & Fusetani, N. (1997). Larval settlement in barnacles [in Japanese]. *Seikagaku*, **69**, 1348–1360.

Olberg, R. M. & Willis, M. A. (1990). Pheromone-modulated optomotor response in male gypsy moths, *Lymantria dispar* L. – directionally selective visual interneurons in the ventral nerve cord. *Journal of Comparative Physiology A-Sensory Neural and Behavioral Physiology*, **167**, 707–714.

Oldroyd, B. P., Clifton, M. J., Wongsiri, S., Rinderer, T. E., Sylvester, H. A. & Crozier, R. H. (1997). Polyandry in the genus *Apis*, particularly *Apis andreniformis*. *Behavioral Ecology and Sociobiology*, **40**, 17–26.

Oliver, S. J., Grasso, F. W. & Atema, J. (1996). Filament tracking and casting in American elvers (*Anguilla rostrata*). *Biological Bulletin*, **191**, 314–315.

Olsen, H. (1999). Present knowledge of kin discrimination in salmonids. *Genetica*, **104**, 295–299.

Olsen, K. H., Grahn, M., Lohm, J. & Langefors, A. (1998). MHC and kin discrimination in juvenile Arctic charr, *Salvelinus alpinus* (L.). *Animal Behaviour*, **56**, 319–327.

Olsson, M. & Shine, R. (1998). Chemosensory mate recognition may facilitate prolonged mate guarding by male snow skinks, *Niveoscincus microlepidotus*. *Behavioral Ecology and Sociobiology*, **43**, 359–363.

O'Neill, P. L. (1978). Hydrodynamic analysis of feeding in sand dollars. *Oecologia*, **34**, 157–174.

Ono, M., Igarashi, T., Ohno, E. & Sasaki, M. (1995). Unusual thermal defence by a honey-bee against mass attack by hornets. *Nature*, **377**, 334–336.

O'Riain, M. J. & Jarvis, J. U. M. (1997). Colony member recognition and xenophobia in the naked mole-rat. *Animal Behaviour*, **53**, 487–498.

Orr, M. R. & Smith, T. B. (1998). Ecology and speciation. *Trends in Ecology & Evolution*, **13**, 502–506.

Otieno, D. A., Hassanali, A., Obenchain, F. D., Sternberg, A. & Galun, R. (1985). Identification of guanine as an assembly pheromone of ticks. *Insect Science and Its Application*, **6**, 667–670.

Owens, I. P. F., Rowe, C. & Thomas, A. L. R. (1999). Sexual selection, speciation and imprinting: separating the sheep from the goats. *Trends in Ecology & Evolution*, **14**, 131–132.

Painter, S. D., Clough, B., Garden, R. W., Sweedler, J. V. & Nagle, G. T. (1998). Characterization of *Aplysia* attractin, the first water-borne peptide pheromone in invertebrates. *Biological Bulletin*, **194**, 120–131.

Pankiw, T., Winston, M. L. & Slessor, K. N. (1995). Queen attendance behavior of worker honey-bees (*Apis mellifera* L) that are high and low responding to queen mandibular pheromone. *Insectes Sociaux*, **42**, 371–378.

Papaj, D. R. & Lewis, A. C. (ed.) (1993). *Insect learning. Ecological and evolutionary perspectives*. New York: Chapman and Hall.

Pasteels, J. M. & Bordereau, C. (1998). Releaser pheromones in termites. In *Pheromone communication in social insects: ants, wasps, bees, and termites*, ed. R. K. Vander Meer, M. D. Breed, M. L. Winston & K. Espelie, pp. 193–215. Boulder, Colo.: Westview Press.

Pasteels, J. M., Deneubourg, J. L. & Goss, S. (1987). Self-organization mechanisms in ant societies (I): trail recruitment to newly discovered food sources. *Experimentia Supplementum*, **54**, 155–175.

Paterson, H. E. H. (1993). Variation and the specific-mate recognition system. In *Perspectives in Ethology, Vol. 10, Behavior and evolution*, ed. P. P. G. Bateson, P. H. Klopfer & N. S. Thompson, pp. 209–27. New York: Plenum Press.

Paterson, S. & Pemberton, J. M. (1997). No evidence for major histocompatibility complex-dependent mating patterns in a free-living ruminant population. *Proceedings of the Royal Society of London Series B-Biological Sciences*, **264**, 1813–1819.

Patiño, R. (1997). Manipulations of the reproductive system of fishes by means of exogenous chemicals. *Progressive Fish-Culturist*, **59**, 118–128.

Pawliszyn, J. (1997). *Solid phase microextraction. Theory and practice*. New York: John Wiley.

Pawson, M. G. (1977). The responses of cod *Gadus morhua* (L.) to chemical attractants in moving water. *Journal du Conseil, Conseil Internationale pour l' Exploration de la Mer*, **37**, 316–318.

Peakall, R. (1990). Responses of male *Zaspilothynnus trilobatus* Turner wasps to females and the sexually deceptive orchid it pollinates. *Functional Ecology*, **4**, 159–168.

Peeters, C. (1997). Morphologically 'primitive' ants: comparative review of social characters, and the importance of queen-worker dimorphism. In *The evolution of social behavior in insects and arachnids*, ed. J. C. Choe & B. J. Crespi, pp. 372–391. Cambridge: Cambridge University Press.

Peeters, C., Monnin, T. & Malosse, C. (1999). Cuticular hydrocarbons correlated with reproductive status in a queenless ant. *Proceedings of the Royal Society of London Series B-Biological Sciences*, **266**, 1323–1327.

Pelosi, P. (1996). Perireceptor events in olfaction. *Journal of Neurobiology*, **30**, 3–19.

Pelosi, P. (1998). Odorant-binding proteins: Structural aspects. *Annals of the New York Academy of Sciences*, **855**, 281–293.

Penn, D. [J.] & Potts, W. K. (1998a). Chemical signals and parasite-mediated sexual selection. *Trends in Ecology & Evolution*, **13**, 391–396.

Penn, D. [J.] & Potts, W. K. (1998b). How do major histocompatibility complex genes influence odor and mating preferences? *Advances in Immunology*, **69**, 411–436.

Penn, D. [J.] & Potts, W. (1998c). MHC-disassortative mating preferences reversed by cross-fostering. *Proceedings of the Royal Society of London Series B-Biological Sciences*, **265**, 1299–1306.

Penn, D. [J.] & Potts, W. K. (1998d). Untrained mice discriminate MHC-determined odors. *Physiology & Behavior*, **64**, 235–243.

Penn, D. J. & Potts, W. K. (1999). The evolution of mating preferences and major histocompatibility complex genes. *American Naturalist*, **153**, 145–164.

Penn, D. [J.], Schneider, G., White, K., Slev, P. & Potts, W. (1998). Influenza infection neutralizes the attractiveness of male odor to female mice (*Mus musculus*). *Ethology*, **104**, 685–694.

Penn, D. J. (2002). The scent of genetic compatibility: Sexual selection and the major histocompatibility complex. *Ethology*, **108**, 1–21.

Perret, M. (1992). Environmental and social determinants of sexual function in the male lesser mouse lemur (*Microcebus murinus*). *Folia Primatologica*, **59**, 1–25.

Perret, M. (1995). Chemocommunication in the reproductive function of mouse lemurs. In *Creatures of the dark: the nocturnal prosimians*, ed. L. Alterman, G. A. Doyle & M. K. Izard, pp. 377–391. New York: Plenum Press.

Perret, M. (1996). Manipulation of sex ratio at birth by urinary cues in a prosimian primate. *Behavioral Ecology and Sociobiology*, **38**, 259–266.

Pes, D., Robertson, D. H. L., Hurst, J. L., Gaskell, S. J. & Beynon, R. J. (1999). How many major urinary proteins are produced by the house mouse *Mus domesticus*. In *Advances in chemical signals in vertebrates*, ed. R. E. Johnston, D. Müller-Schwarze & P. W. Sorenson, pp. 149–162. New York: Kluwer Academic Publishers/Plenum Press.

Peschke, K. (1990). Chemical traits in sexual selection of the rove beetle, *Aleochara curtula* (Coleoptera, Staphylinidae). *Entomologia Generalis*, **15**, 127–132.

Pettis, J., Pankiw, T. & Plettner, E. (1999). Bees. In *Pheromones of non-lepidopteran insects associated with agricultural plants*, ed. J. Hardie & A. K. Minks, pp. 429–450. Wallingford, UK: CAB International.

Pettis, R. J., Erickson, B. W., Forward, R. B. & Rittschof, D. (1993). Superpotent synthetic tripeptide mimics of the mud-crab pumping pheromone. *International Journal of Peptide and Protein Research*, **42**, 312–319.

Pfeiffer, W. (1974). Pheromones in fish and amphibia. In *Pheromones,* ed. M. C. Birch, pp. 269–296. Amsterdam: North-Holland.

Pfennig, D. W. & Sherman, P. W. (1995). Kin recognition. *Scientific American*, **272** (6), 98–103.

Phelan, P. L. (1992). Evolution of sex pheromones and the role of asymmetric tracking. In *Insect chemical ecology. An evolutionary approach*, ed. B. D. Roitberg & M. B. Isman, pp. 245–264. New York: Chapman and Hall.

Phelan, P. L. (1997). Evolution of mate-signalling in moths: phylogenetic considerations and predictions from the asymmetric tracking hypothesis. In *The evolution of mating systems in insects and arachnids*, ed. J. C. Choe & B. J. Crespi, pp. 240–256. Cambridge: Cambridge University Press.

Phelan, P. L. & Baker, T. C. (1986). Male-size-related courtship success and intersexual selection in the tobacco moth, *Ephestia elutella. Experientia*, **42**, 1291–1293.

Phelan, P. L. & Baker, T. C. (1987). Evolution of male pheromones in moths – reproductive isolation through sexual selection. *Science*, **235**, 205–207.

Phillips, T. W. (1997). Semiochemicals of stored-product insects: Research and applications. *Journal of Stored Products Research*, **33**, 17–30.

Pickett, J. A. (1985). Production of behaviour-controlling chemicals by crop plants. *Philosophical Transactions of the Royal Society of London Series B-Biological Sciences*, **310**, 235–239.

Pickett, J. A. (1992). Potential of novel chemical aproaches for overcoming insecticide resistance. In *Resistance '91: Achievements and developments in combating pesticide resistance,* ed. I. Denholm, A. L. Devonshire & D. W. Hollomon, pp. 354–356. London: Elsevier Applied Science.

Pickett, J. A. (ed.) (1999). *Insect-plant interactions and induced plant defence.* Chichester: Wiley-Interscience.

Pickett, J. A. & Woodcock, C. M. (1996). The role of mosquito olfaction in oviposition site location and in the avoidance of unsuitable hosts. In *Olfaction in mosquito-host interactions,* CIBA Foundation Symposium No. 200, pp. 109–123. Chichester: Wiley.

Pickett, J. A., Wadhams, L. J., Woodcock, C. M. & Hardie, J. (1992). The chemical ecology of aphids. *Annual Review of Entomology,* **37,** 67–90.

Pickett, J. A., Wadhams, L. J. & Woodcock, C. M. (1997). Developing sustainable pest control from chemical ecology. *Agriculture Ecosystems & Environment,* **64,** 149–156.

Pierce, N. E. (1989). Butterfly-ant mutualisms. In *Toward a more exact ecology,* ed. P. J. Grubb & J. B. Whittaker, pp. 299–324. Oxford: Blackwell Science.

Pike, A. W. & Wadsworth, S. L. (1999). Sealice on salmonids: their biology and control. *Advances in Parasitology,* **44,** 234–337.

Pilpel, Y. & Lancet, D. (1999). Good reception in fruitfly antennae. *Nature,* **398,** 285–287.

Plarre, R. (1998). Pheromones and other semiochemicals of store product insects – a historical review, current application, and perspective needs. In *100 Jahre Pflanzenschutzforschung (One hundred years research in plant protection),* ed. C. Reichmuth, pp. 13–84. Berlin: Parey.

Plarre, R. & Vanderwell, D. C. (1999). Stored-product beetles. In *Pheromones of non-lepidopteran insects associated with agricultural plants,* ed. J. Hardie & A. K. Minks, pp. 149–198. Wallingford, UK: CAB International.

Porter, R. H. & Blaustein, A. R. (1989). Mechanisms and ecological correlates of kin recognition. *Science Progress,* **73,** 53–66.

Porter, R. H. & Winberg, J. (1999). Unique salience of maternal breast odors for newborn infants. *Neuroscience and Biobehavioral Reviews,* **23,** 439–449.

Porter, R. H., Tepper, V. J. & White, D. M. (1981). Experiential influences on the development of huddling preferences and sibling recognition in spiny mice. *Developmental Psychobiology,* **14,** 375–382.

Porter, R. H., Cernoch, J. M. & McLaughlin, F. J. (1983). Maternal recognition of neonates through olfactory cues. *Physiology & Behavior,* **30,** 151–154.

Porter, R. H., Balogh, R. D., Cernoch, J. M. & Franchi, C. (1986). Recognition of kin through characteristic body odors. *Chemical Senses,* **11,** 389–395.

Porter, R. H., McFadyenketchum, S. A. & King, G. A. (1989). Underlying bases of recognition signatures in spiny mice, *Acomys cahirinus. Animal Behaviour,* **37,** 638–644.

Potts, W. K. & Wakeland, E. K. (1993). Evolution of MHC genetic diversity – a tale of incest, pestilence and sexual preference. *Trends in Genetics,* **9,** 408–412.

Potts, W. K., Manning, C. J. & Wakeland, E. K. (1991). Mating patterns in seminatural populations of mice influenced by MHC genotype. *Nature,* **352,** 619–621.

Powell, W. (1998). Semiochemicals to increase parasitism in aphid pest control. *Pesticide Science,* **54,** 291–293.

Prestwich, K. N. (1994). The energetics of acoustic signalling in anurans and insects. *American Naturalist,* **34,** 625–643.

Preti, G. & Wysocki, C. J. (1999). Human pheromones: releasers or primers, fact or myth. In *Advances in chemical signals in vertebrates,* ed. R. E. Johnston, D. Müller-Schwarze & P. W. Sorenson, pp. 315–332. New York: Kluwer Academic Publishers/Plenum Press.

Preti, G., Cutler, W. B., Garcia, C. R., Huggins, G. R. & Lawley, H. J. (1986). Human axillary secretions influence womens menstrual cycles – the role of donor extract of females. *Hormones and Behavior,* **20,** 474–482.

Quinet, Y. & Pasteels, J. M. (1996). Spatial specialization of the foragers and foraging strategy in *Lasius fuliginosus* (Latreille) (Hymenoptera, Formicidae). *Insectes Sociaux,* **43,** 333–346.

Qvarnström, A. & Forsgren, E. (1998). Should females prefer dominant males? *Trends in Ecology & Evolution*, **13**, 498–501.

Raffa, K. F. & Berryman, A. A. (1983). The role of host plant-resistance in the colonization behavior and ecology of bark beetles (Coleoptera, Scolytidae). *Ecological Monographs*, **53**, 27–49.

Raffa, K. F. & Dahlsten, D. L. (1995). Differential responses among natural enemies and prey to bark beetle pheromones. *Oecologia*, **102**, 17–23.

Raffa, K. F. & Klepzig, K. D. (1989). Chiral escape of bark beetles from predators responding to a bark beetle pheromone. *Oecologia*, **80**, 566–569.

Raffa, K. F., Phillips, T. W. & Salom, S. M. (1993). Strategies and mechanisms of host colonization by bark beetles. In *Beetle-pathogens interactions in conifer forests*, ed. T. D. Schowalter & G. M. Filip, pp. 103–128. London: Academic Press.

Rai, M. M., Hassanali, A., Saini, R. K., Odongo, H. & Kahoro, H. (1997). Identification of components of the oviposition aggregation pheromone of the gregarious desert locust, *Schistocerca gregaria* (Forskal). *Journal of Insect Physiology*, **43**, 83–87.

Rasa, O. A. E. (1973). Marking behaviour and its social significance in the African dwarf mongoose, *Helogale undulata rufula*. *Zeitschrift für Tierpsychologie-Journal of Comparative Ethology*, **32**, 293–318.

Rasmussen, L. E. L. (1998). Chemical communication: an integral part of functional Asian elephant (*Elephas maximus*) society. *Ecoscience*, **5**, 410–426.

Rasmussen, L. E. L. (1999). Evolution of chemical signals in the Asian elephant, *Elephas maximas*: behavioural and ecological influences. *Journal of Biosciences*, **24**, 241–251.

Rasmussen, L. E. L. & Schulte, B. A. (1998). Chemical signals in the reproduction of Asian (*Elephas maximus*) and African (*Loxodonta africana*) elephants. *Animal Reproduction Science*, **53**, 19–34.

Rasmussen, L. E. L., Lee, T. D., Roelofs, W. L., Zhang, A. J. & Daves, G. D. (1996). Insect pheromone in elephants. *Nature*, **379**, 684.

Rasmussen, L. E. L., Lee, T. D., Zhang, A. J., Roelofs, W. L. & Daves, G. D. (1997). Purification, identification, concentration and bioactivity of (Z)-7-dodecen-1-yl acetate: sex pheromone of the female Asian elephant, *Elephas maximus*. *Chemical Senses*, **22**, 417–437.

Rasmussen, L. E. L., Riddle. H. S. & Krishnamurthy, V. (2002). Mellifluous matures to malodorous in musth. *Nature*, **415**, 975–976.

Ratnieks, F. L. W. (1993). Egg-laying, egg-removal, and ovary development by workers in queenright honey-bee colonies. *Behavioral Ecology and Sociobiology*, **32**, 191–198.

Ratnieks, F. L. W. (1995). Evidence for a queen-produced egg-marking pheromone and its use in worker policing in the honey-bee. *Journal of Apicultural Research*, **34**, 31–37.

Ratnieks, F. L. W. & Visscher, P. K. (1989). Worker policing in the honeybee. *Nature*, **342**, 796–797.

Rechav, Y., Norval, R. A. I., Tannock, J. & Colborne, J. (1978). Attraction of the tick *Ixodes neitzi* to twigs marked by the klipspringer antelope. *Nature*, **275**, 310–311.

Redd, W. H., Manne, S. L., Peters, B., Jacobsen, P. B. & Schmidt, H. (1994). Fragrance administration to reduce anxiety during MR-imaging. *Journal of Magnetic Resonance Imaging*, **4**, 623–626.

Reeve, H. K. & Sherman, P. W. (1993). Adaptation and the goals of evolutionary research. *Quarterly Review of Biology*, **68**, 1–32.

Regnier, F. E. & Wilson, E. O. (1969). The alarm-defense system of the ant *Lasius alienus*. *Journal of Insect Physiology*, **15**, 893–898.

Regnier, F. E. & Wilson, E. O. (1971). Chemical communication and 'propaganda' in slave-maker ants. *Science*, **172**, 267–269.

Reilly, J. T., Vowels, B. R., Leyden, J. J., Sondheimer, S. & Preti, G. (1996). Quantitative comparison of female axillary secretions as a function of the menstrual cycle. *Chemical Senses*, **21**, 661.

Reinhard, J. & Kaib, M. (1995). Interaction of pheromones during food exploitation by the termite *Schedorhinotermes lamanianus*. *Physiological Entomology*, **20**, 266–272.

Reusch, T. B. H., Haberli, M. A., Aeschlimann, P. B. & Milinski, M. (2001). Female sticklebacks count alleles in a strategy of sexual selection explaining MHC polymorphism. *Nature*, **414**, 300–302.

Reynolds, J. D. (1996). Animal breeding systems. *Trends in Ecology & Evolution*, **11**, 68–72.

Rice, W. R. (1996). Sexually antagonistic male adaptation triggered by experimental arrest of female evolution. *Nature*, **381**, 232–234.

Rich, T. J. & Hurst, J. L. (1999). The competing countermarks hypothesis: reliable assessment of competitive ability by potential mates. *Animal Behaviour*, **58**, 1027–1037.

Ridley, M. (1983). *The explanation of organic diversity*. Oxford: Clarendon Press.

Roberts, R. L., Zullo, A., Gustafson, E. A. & Carter, C. S. (1996). Perinatal steroid treatments alter alloparental and affiliative behavior in prairie voles. *Hormones and Behavior*, **30**, 576–582.

Roberts, R. L., Williams, J. R., Wang, A. K. & Carter, C. S. (1998). Cooperative breeding and monogamy in prairie voles: Influence of the sire and geographical variation. *Animal Behaviour*, **55**, 1131–1140.

Roberts, S. C. & Lowen, C. (1997). Optimal patterns of scent marks in klipspringer (*Oreotragus oreotragus*) territories. *Journal of Zoology*, **243**, 565–578.

Robinson, G. E. & Huang, Z. Y. (1998). Colony integration in honey bees: genetic, endocrine and social control of division of labor. *Apidologie*, **29**, 159–170.

Robinson, S. W., Jutsum, A. R., Cherrett, J. M. & Quinlan, R. J. (1982). Field evaluation of methyl 4-methylpyrrole-2-carboxylate, an ant trail pheromone, as a component of baits for leaf-cutting ant (Hymenoptera, Formicidae) control. *Bulletin of Entomological Research*, **72**, 345–356.

Rodriguez, I., Greer, C. A., Mok, M Y. & Mombaerts, P. (2000). A putative pheromone receptor gene expressed in human olfactory mucosa. *Nature Genetics*, **26**, 18–19.

Rodriguez-Manzo, G. (1999). Blockade of the establishment of the sexual inhibition resulting from sexual exhaustion by the Coolidge effect. *Behavioural Brain Research*, **100**, 245–254.

Roelofs, W. L. (1995). The chemistry of sex attraction. In *Chemical ecology: the chemistry of biotic interaction*, ed. T. Eisner & J. Meinwald, pp. 103–117. Washington, D.C.: National Academy of Sciences.

Roelofs, W. L., Glover, T., Tang, X. H., Sreng, I., Robbins, P., Eckenrode, C., Löfstedt, C., Hansson, B. S. & Bengtsson, B. O. (1987). Sex-pheromone production and perception in European corn-borer moths is determined by both autosomal and sex-linked genes. *Proceedings of the National Academy of Sciences of the United States of America*, **84**, 7585–7589.

Roessingh, P., Peterson, S. C. & Fitzgerald, T. D. (1988). The sensory basis of trail following in some lepidopterous larvae-contact chemoreception. *Physiological Entomology*, **13**, 219–224.

Roitberg, B. D. & Mangel, M. (1988). On the evolutionary ecology of marking pheromones. *Evolutionary Ecology*, **2**, 289–315.

Roitberg, B. D. & Prokopy, R. J. (1981). Experience required for pheromone recognition by the apple maggot fly. *Nature*, **292**, 540–541.

Roitberg, B. D. & Prokopy, R. J. (1987). Insects that mark host plants. *Bioscience*, **37**, 400–406.

Rollmann, S. M., Houck, L. D. & Feldhoff, R. C. (1999). Proteinaceous pheromone affecting female receptivity in a terrestrial salamander. *Science*, **285**, 1907–1909.

Romeyer, A., Porter, R. H., Poindron, P., Orgeur, P., Chesné, P. & Poulin, N. (1993). Recognition of dizygotic and monozygotic twin lambs by ewes. *Behaviour*, **127**, 119–139.

Roper, T. J. (1999). Olfaction in birds. In *Advances in the study of behavior*, Vol. 28, ed. P. J. B. Slater, J. S. Rosenblatt, C. T. Snowdon & T. J. Roper, pp. 247–332. New York: Academic Press.

Roper, T. J., Conradt, L., Butler, J., Christian, S. E., Ostler, J. & Schmid, T. K. (1993). Territorial marking with feces in badgers (*Meles meles*) – a comparison of boundary and hinterland latrine use. *Behaviour*, **127**, 289–307.

Ross, K. G. & Mathews, R. W. (ed.) (1991). *The social biology of wasps*. Ithaca, N.Y.: Cornell University Press.

Rothschild, G. H. L. & Vickers, R. A. (1991). Biology, ecology and control of the oriental fruit moth. In *Tortricid moths, their biology, natural enemies and control*, ed. L. P. S. van der Geest & H. H. Evenhuis, pp. 389–412. Amsterdam: Elsevier Science.

Roubik, D. W. (1989). *Ecology and natural history of tropical bees*. Cambridge: Cambridge University Press.

Rouquier, S., Friedman, C., Delettre, C., van den Engh, G., Blancher, A., Crouau-Roy, B., Trask, B. J. & Giorgi, D. (1998a). A gene recently inactivated in human defines a new olfactory receptor family in mammals. *Human Molecular Genetics*, **7**, 1337–1345.

Rouquier, S., Taviaux, S., Trask, B. J., BrandArpon, V., van den Engh, G., Demaille, J. & Giorgi, D. (1998b). Distribution of olfactory receptor genes in the human genome. *Nature Genetics*, **18**, 243–250.

Russell, E. M. (1985). The metatherians: order Marsupialia. In *Social odours in mammals*, Vol. 1, ed. R. E. Brown & D. W. Macdonald, pp. 45–104. Oxford: Oxford University Press.

Rust, M. K., Owens, J. M. & Reierson, D. A. (1995). *Understanding and controlling the German cockroach*. Oxford: Oxford University Press.

Ryan, M. J. (1997). Sexual selection and mate choice. In *Behavioural ecology*, 4th edn ed. J. R. Krebs & N. B. Davies, pp. 179–202. Oxford: Blackwell Science.

Ryan, M. J. (1998). Sexual selection, receiver biases, and the evolution of sex differences. *Science*, **281**, 1999–2003.

Ryan, M. J. & Keddy-Hector, A. (1992). Directional patterns of female mate choice and the role of sensory biases. *American Naturalist*, **139**, S4–S35.

Sachs, B. D. (1999). Airborne aphrodisiac odor from estrous rats: implication for pheromonal classification. In *Advances in chemical signals in vertebrates*, ed. R. E. Johnston, D. Müller-Schwarze & P. W. Sorenson, pp. 333–342. New York: Kluwer Academic Publishers Plenum Press.

Sakagami, S. F., Roubik, D. W. & Zucchi, R. (1993). Ethology of the robber stingless bee, *Lestrimelitta limao* (Hymenoptera, Apidae). *Sociobiology*, **21**, 237–277.

Salom, S. M., Billings, R. F., Upton, W. W., Dalusky, M. J., Grosman, D. M., Payne, T. L., Berisford, C. W. & Shaver, T. N. (1992). Effect of verbenone enantiomers and racemic *endo*-brevicomin on response of *Dendroctonus frontalis* (Coleoptera: Scolytidae) to attractant-baited traps. *Canadian Journal of Forest Research-Journal Canadien de la Recherche Forestiere*, **22**, 925–931.

Sam, M., Vora, S., Malnic, B., Ma, W. D., Novotny, M. V. & Buck, L. B. (2001). Odorants may arouse instinctive behaviours. *Nature*, **412**, 142.

Sanders, C. J. (1997). Mechanisms of mating disruption in moths. In *Insect pheromone research: new directions*, ed. R. T. Cardé & A. K. Minks, pp. 333–346. New York: Chapman and Hall.

Savarit, F., Sureau, G., Cobb, M. & Ferveur, J. F. (1999). Genetic elimination of known pheromones reveals the fundamental chemical bases of mating and isolation in *Drosophila*. *Proceedings of the National Academy of Sciences of the United States of America*, **96**, 9015–9020.

Schaal, B. (1988). Olfaction in infants and children – developmental and functional perspectives. *Chemical Senses*, **13**, 145–190.

Schaal, B. & Porter, R. H. (1991). Microsmatic humans revisited – the generation and perception of chemical signals. *Advances in the Study of Behavior*, **20**, 135–99.

Schaal, B., Montagner, H., Hertling, E., Bolzoni, D., Moyse, A. & Quichon, R. (1980). Les stimulations olfactives dans les relations entre l'enfant et la mère. *Reproduction Nutrition Developpement*, **20**, 843–858.

Schackwitz, W. S., Inoue, T. & Thomas, J. H. (1996). Chemosensory neurons function in parallel to mediate a pheromone response in *C. elegans*. *Neuron*, **17**, 719–728.

Schellinck, H. M. & Brown, R. E. (1994). Methodological questions in the study of the rat's ability to discriminate between the odours of individual conspecifics. *Advances in the Biosciences*, **93**, 427–436.

Schellinck, H. M., Rooney, E. & Brown, R. E. (1995). Odors of individuality of germ-free mice are not discriminated by rats in a habituation-dishabituation procedure. *Physiology & Behavior*, **57**, 1005–1008.

Schemske, D. W. & Lande, R. (1984). Fragrance collection and territorial display by male orchid bees. *Animal Behaviour*, **32**, 935–937.

Schiestl, F. P., Ayasse, M., Paulus, H. F., Löfstedt, C., Hansson, B. S., Ibarra, F. & Francke, W. (1999). Orchid pollination by sexual swindle. *Nature*, **399**, 421–422.

Schleidt, M. (1980). Personal odor and nonverbal communication. *Ethology and Sociobiology*, **1**, 225–231.

Schlyter, F. & Birgersson, G. A. (1999). Forest beetles. In *Pheromones of non-lepidopteran insects associated with agricultural plants*, ed. J. Hardie & A. K. Minks, pp. 113–148. Wallingford, UK: CAB International.

Schmidt, J. O. (1998). Mass action in honey bees: alarm, swarming and role of releaser pheromones. In *Pheromone communication in social insects: ants, wasps, bees, and termites*, ed. R. K. Vander Meer, M. D. Breed, M. L. Winston & K. Espelie, pp. 257–292. Boulder, Colo.: Westview Press.

Schmidt, J. O. & Thoenes, S. C. (1987). Swarm traps for survey and control of Africanized honey bees. *Bulletin of the Entomological Society of America*, **33**, 155–158.

Schmidt, J. O., Thoenes, S. C. & Hurley, R. (1989). Swarm traps. *American Bee Journal*, **129**, 468–471.

Schneider, M. A. & Hendrix, L. (2000). Olfactory sexual inhibition and the Westermarck effect. *Human Nature*, **11**, 65–91.

Schneider, R. W. S. & Moore, P. A. (2000). The physics of chemoreception revisited: how the environment influences chemically mediated behavior. In *Biomechanics in animal behaviour*, ed. P. Domenici & R. W. Blake, pp. 159–176. Oxford: BIOS Scientific.

Schneider, R. W. S., Lanzen, J. & Moore, P. A. (1998). Boundary-layer effect on chemical signal movement near the antennae of the sphinx moth, *Manduca sexta*, temporal filters for olfaction. *Journal of Comparative Physiology A-Sensory Neural and Behavioral Physiology*, **182**, 287–298.

Schöne, H. (1984). *Spatial orientation. The spatial control of behaviour in animals and man*. Princeton, N.J. Princeton University Press.

Schoon, G. A. A. (1997). Scent identifications by dogs (*Canis familiaris*): a new experimental design. *Behaviour*, **134**, 531–550.

Schulte, B. A. (1998). Scent marking and responses to male castor fluid by beavers. *Journal of Mammalogy*, **79**, 191–203.

Schulte, B. A. & Rasmussen, L. E. L. (1999). Musth, sexual selection, testosterone, and metabolites. In *Advances in chemical signals in vertebrates*, ed. R. E. Johnston, D. Müller-Schwarze & P. W. Sorenson, pp. 383–398. New York: Kluwer Academic Publishers/Plenum Press.

Schwagmeyer, P. L. (1979). The Bruce effect: an evaluation of male/female advantages. *American Naturalist*, **114**, 932–938.

Schwenk, K. (1994). Why snakes have forked tongues. *Science*, **263**, 1573–1577.

Seeley, T. D. (1979). Queen substance dispersal by messenger workers in honey bee colonies. *Behavioral Ecology and Sociobiology*, **5**, 391–415.

Seeley, T. D. (1985). *Honeybee ecology: a study of adaptation in social life*. Princeton, N. J. Princeton University Press.

Seeley, T. D. (1995). *The wisdom of the hive. The social physiology of honey bee colonies*. Cambridge, Mass. Harvard University Press.

Settle, R. H., Sommerville, B. A., McCormick, J. & Broom, D. M. (1994). Human scent matching using specially trained dogs. *Animal Behaviour*, **48**, 1443–1448.

Seybold, S. J. (1993). Role of chirality in olfactory-directed behavior – aggregation of pine engraver beetles in the genus *Ips* (Coleoptera, Scolytidae). *Journal of Chemical Ecology*, **19**, 1809–1831.

Shelley, W. B., Hurley, H. J. & Nichols, A. C. (1953). Axillary odour: experimental study of the role of bacteria, apocrine sweat, and deodorants. *Archives of Dermatology and Syphiology*, **68**, 430–446.

Shellman-Reeve, J. S. (1997). The spectrum of eusociality in termites. In *The evolution of social behavior in insects and arachnids*, ed. J. C. Choe & B. J. Crespi, pp. 52–93. Cambridge: Cambridge University Press.

Shelly, T. W. & Whittier, T. S. (1997). Lek behavior of insects. In *The evolution of mating systems in insects and arachnids*, ed. J. C. Choe & B. J. Crespi, pp. 273–293. Cambridge: Cambridge University Press.

Sherman, P. W., Jarvis, J. U. M. & Alexander, R. D. (ed.) (1991). *The biology of the naked mole-rat*. Princeton, N. J.: Princeton University Press.

Sherman, P. W., Lacey, E. A., Reeve, H. K. & Keller, L. (1995). The eusociality continuum. *Behavioral Ecology*, **6**, 102–108.

Sherman, P. W., Reeve, H. K. & Pfennig, D. W. (1997). Recognition systems. In *Behavioural ecology*, 4th edn, ed. J. R. Krebs & N. B. Davies, pp. 69–96. Oxford: Blackwell Science.

Shine, R., Harlow, P., Lemaster, M. P., Moore, I. T. & Mason, R. T. (2000). The transvestite serpent: why do male garter snakes court (some) other males? *Animal Behaviour*, **59**, 349–359.

Shorey, H. H. & Gerber, R. G. (1996). Use of puffers for disruption of sex pheromone communication of codling moths (Lepidoptera: Tortricidae) in walnut orchards. *Environmental Entomology*, **25**, 1398–1400.

Sillén-Tullberg, B. & Møller, A. P. (1993). The relationship between concealed ovulation and mating systems in anthropoid primates: a phylogenetic analysis. *American Naturalist*, **141**, 1–25.

Sillero-Zubiri, C. & Macdonald, D. W. (1998). Scent-marking and territorial behaviour of Ethiopian wolves *Canis simensis*. *Journal of Zoology*, **245**, 351–361.

Silverstein, R. M. (1977). Complexity, diversity, and specificity of behavior-modifying chemicals: examples mainly from Coleoptera and Hymenoptera. In *Chemical control of insect behaviour. Theory and application*, ed. H. H. Shorey & J. J. J. McKelvey, pp. 231–251. New York: Plenum Press.

Silverstein, R. M. (1979). Enantiomeric composition and bioactivity of chiral semiochemicals in insects. In *Chemical ecology: odour communication in animals*, ed. F. J. Ritter, P. E. Howse & G. Le Masne, pp. 133–146. Amsterdam: Elsevier North-Holland.

Silverstein, R. M. (1990). Practical use of pheromones and other behavior-modifying compounds: overview. In *Behavior-modifying chemicals for insect management*, ed. R. L. Ridgeway, R. M. Silverstein & M. N. Inscoe, pp. 1–8. New York: Marcel Dekker.

Simmons, L. W. (1989). Kin recognition and its influence on mating preferences of the field cricket, *Gryllus bimaculatus* (Degeer). *Animal Behaviour*, **38**, 68–77.

Simons, R. R., Jaeger, R. G. & Felgenhauer, B. E. (1997). Competitor assessment and area defense by territorial salamanders. *Copeia*, **1997**, 70–76.

Simpson, S. J., McCaffery, A. R. & Hägele, B. F. (1999). A behavioural analysis of phase change in the desert locust. *Biological Reviews of the Cambridge Philosophical Society*, **74**, 461–480.

Singer, A. G., Beauchamp, G. K. & Yamazaki, K. (1997). Volatile signals of the major histocompatibility complex in male mouse urine. *Proceedings of the National Academy of Sciences of the United States of America*, **94**, 2210–2214.

Singer, T. L., Espelie, K. E. & Gamboa, G. J. (1998). Nest and nestmate discrimination in independent-founding paper wasps. In *Pheromone communication in social insects: ants, wasps, bees, and termites*, ed. R. K. Vander Meer, M. D. Breed, M. L. Winston & K. Espelie, pp. 104–125. Boulder, Colo. Westview Press.

Singh, D. & Bronstad, P. M. (2001). Female body odour is a potential cue to ovulation. *Proceedings of the Royal Society of London Series B-Biological Sciences*, **268**, 797–801.

Sirugue, D., Bonnard, O., Lequere, J. L., Farine, J. P. & Brossut, R. (1992). 2-Methylthiazolidine and 4-ethylguaiacol, male sex-pheromone components of the cockroach *Nauphoeta cinerea* (Dictyoptera, Blaberidae) – a reinvestigation. *Journal of Chemical Ecology*, **18**, 2261–2276.

Slater, P. J. B. (1994). Kinship and altruism. In *Behaviour and evolution*, ed. P. J. B. Slater & T. R. Halliday, pp. 193–222. Cambridge: Cambridge University Press.

Slessor, K. N., Foster, L. J. & Winston, M. L. (1998). Royal flavours: honey bee queen pheromones. In *Pheromone communication in social insects: ants, wasps, bees, and termites*, ed. R. K. Vander Meer, M. D. Breed, M. L. Winston & K. Espelie, pp. 331–344. Boulder, Colo. Westview Press.

Sliwa, A. & Richardson, P. R. K. (1998). Responses of aardwolves, *Proteles cristatus*, Sparrman 1783, to translocated scent marks. *Animal Behaviour*, **56**, 137–146.

Smith, B. H. & Breed, M. D. (1995). The chemical basis for nest-mate recognition and mate discrimination in social insects. In *Chemical ecology of insects 2*, ed. R. T. Cardé & W. J. Bell, pp. 287–317. London: Chapman and Hall.

Smith, F. V., Barnard, C. J. & Behnke, J. M. (1996). Social odours, hormone modulation and resistance to disease in male laboratory mice, *Mus musculus*. *Animal Behaviour*, **52**, 141–153.

Smith, J. L., Cork, A., Hall, D. R. & Hodges, R. J. (1996). Investigation of the effect of female larger grain borer, *Prostephanus truncatus* (Horn) (Coleoptera: Bostrichidae), and their residues on the production of aggregation pheromone by males. *Journal of Stored Products Research*, **32**, 171-181.

Smith, R. J. F. (1992). Alarm signals in fishes. *Reviews in Fish Biology and Fisheries*, **2**, 33–63.

Smith, R. J. F. (1997). Does one result trump all others? A response to Magurran, Irving and Henderson. *Proceedings of the Royal Society of London Series B-Biological Sciences*, **264**, 445–450.

Smith, R. J. F. (1999). What good is smelly stuff in the skin? Cross function and cross taxa effects in fish 'alarm substances'. In *Advances in chemical signals in vertebrates*, ed. R. E. Johnston, D. Müller-Schwarze & P. W. Sorenson, pp. 475–488. New York: Kluwer Academic Publishers/Plenum Press.

Smith, R. J. F. & Lemly, A. D. (1986). Survival of fathead minnows after injury by predators and its possible role in the evolution of alarm signals. *Environmental Biology of Fishes*, **15**, 147–149.

Smith, T. D., Siegel, M. I., Burrows, A. M., Mooney, M. P., Burdi, A. R., Fabrizio, P. A. & Clemente, F. R. (1998). Searching for the vomeronasal organ of adult humans: preliminary findings on location, structure, and size. *Microscopy Research and Technique*, **41**, 483–491.

Snyder, N. F. R. & Snyder, H. (1970). Alarm response of *Diadema antillarum*. *Science*, **168**, 276–278.

Sobel, N., Prabhakaran, V., Desmond, J. E., Glover, G. H., Goode, R. L., Sullivan, E. V. & Gabrieli, J. D. E. (1998). Sniffing and smelling: separate subsystems in the human olfactory cortex. *Nature*, **392**, 282–286.

Sobel, N., Khan, R. M., Saltman, A., Sullivan, E. V. & Gabrieli, J. D. E. (1999a). The world smells different to each nostril. *Nature*, **402**, 35.

Sobel, N., Prabhakaran, V., Hartley, C. A., Desmond, J. E., Glover, G. H., Sullivan, E. V. & Gabrieli, J. D. E. (1999b). Blind smell: brain activation induced by an undetected air-borne chemical. *Brain*, **122**, 209–217.

Solomon, N. G. (1991). Current indirect fitness benefits associated with philopatry in juvenile prairie voles. *Behavioral Ecology and Sociobiology*, **29**, 277–282.

Solomon, N. G. & French, J. A. (ed.) (1997). *Cooperative breeding in mammals*. Cambridge: Cambridge University Press.

Solomon, N. G. & Getz, L. L. (1997). Examination of alternative hypotheses for cooperative breeding in rodents. In *Cooperative breeding in mammals*, ed. N. G. Solomon & J. A. French, pp. 199–230. Cambridge: Cambridge University Press.

Sonenshine, D. E., Hamilton, J. G. C. & Lusby, W. R. (1992): Use of cholesteryl esters as mounting sex pheromones in combination with 2, 6-dichlorophenol and pesticides to control populations of hard ticks. Patent Application: 5, 149, 526, 22 Sep 1992. Washington, D. C.: U. S. Patent and Trademark office.

Sorensen, P. W. & Stacey, N. E. (1999). Evolution and specialization of fish hormonal pheromones. In *Advances in chemical signals in vertebrates*, ed. R. E. Johnston, D. Müller-Schwarze & P. W. Sorenson, pp. 15–48. New York: Kluwer Academic Publishers/Plenum Press.

Sorensen, P. W., Hara, T. J. & Stacey, N. E. (1987). Extreme olfactory sensitivity of mature and gonadally-regressed goldfish to a potent steroidal pheromone, 17-alpha, 20-beta-dihydroxy-4-pregnen-3-one. *Journal of Comparative Physiology a Sensory Neural and Behavioral Physiology*, **160**, 305–314.

Soroker, V., Fresneau, D. & Hefetz, A. (1998). Formation of colony odor in ponerine ant *Pachycondyla apicalis*. *Journal of Chemical Ecology*, **24**, 1077–1090.

Soussignan, R., Schaal, B., Marlier, L. & Jiang, T. (1997). Facial and autonomic responses to biological and artificial olfactory stimuli in human neonates: Re-examining early hedonic discrimination of odors. *Physiology & Behavior*, **62**, 745–758.

Spielman, A. I., Zeng, X. N., Leyden, J. J. & Preti, G. (1995). Proteinaceous precursors of human axillary odor – isolation of 2 novel odor–binding proteins. *Experientia*, **51**, 40–47.

Sreng, L. (1990). Seducin, male sex-pheromone of the cockroach *Nauphoeta cinerea* – isolation, identification, and bioassay. *Journal of Chemical Ecology*, **16**, 2899–2912.

Stacey, N. E. & Sorensen, P. W. (1986). 17-alpha, 20-beta-Dihydroxy-4-pregnen-3-one: a steroidal primer pheromone increasing milt volume in the goldfish, *Carassius auratus*. *Canadian Journal of Zoology*, **64**, 2412–2417.

Stacey, N. E. & Sorensen, P. W. (1999). Pheromones, fish. In *Encyclopedia of reproduction*, Vol. 3, ed. E. Knobil & J. D. Neill, pp. 748–755. New York: Academic Press.

Stacey, N. E., Zheng, W. & Cardwell, J. R. (1994). Milt production in common carp (*Cyprinus carpio*): stimulation by a goldfish steroid pheromone. *Aquaculture*, **127**, 265–276.

Stamps, J. [A.] (1994). Territorial behavior – testing the assumptions. *Advances in the Study of Behavior*, **23**, 173–232.

Stamps, J. A. & Krishnan, V. V. (1999). A learning-based model of territory establishment. *Quarterly Review of Biology*, **74**, 291–318.

Staten, R. T., El-Lissy, O. & Antilla, L. (1997). Successful area-wide program to control pink bolilworm by mating disruption. In *Insect pheromone research: new directions*, ed. R. T. Cardé & A. K. Minks, pp. 383–396. New York: Chapman and Hall.

Steel, E. & Keverne, E. B. (1985). Effect of female odor on male hamsters mediated by the vomeronasal organ. *Physiology & Behavior*, **35**, 195–200.

Stein, B. E. & Meredith, M. A. (1993). *The merging of the senses*. Cambridge, Mass.: MIT Press.

Steinbrecht, R. A. (1996). Are odorant-binding proteins involved in odorant discrimination? *Chemical Senses*, **21**, 719–727.

Steinbrecht, R. A. (1999). Olfactory receptors. In *Atlas of arthropod sensory receptors. Dynamic morphology in relation to function*, ed. E. Eguchi, Y. Tominaga & H. Ogawa, pp. 155–176. Tokyo: Springer-Verlag.

Stephens, P. A. & Sutherland, W. J. (1999). Consequences of the Allee effect for behaviour, ecology and conservation. *Trends in Ecology & Evolution*, **14**, 401–405.

Stern, D. L. & Foster, W. A. (1996). The evolution of soldiers in aphids. *Biological Reviews of the Cambridge Philosophical Society*, **71**, 27–79.

Stern, K. & McClintock, M. K. (1998). Regulation of ovulation by human pheromones. *Nature*, **392**, 177–179.

Stoddart, D. M. (1980). *The ecology of vertebrate olfaction*. London: Chapman and Hall.

Stoddart, D. M. (1990). *The scented ape. The biology and culture of human odour*. Cambridge: Cambridge University Press.

Stoddart, D. M., Bradley, A. J. & Mallick, J. (1994). Plasma testosterone concentration, body-weight, social-dominance and scent-marking in male marsupial sugar gliders (*Petaurus breviceps* Marsupialia, Petauridae). *Journal of Zoology*, **232**, 595–601.

Storer, A. J., Wainhouse, D. & Speight, M. R. (1997). The effect of larval aggregation behaviour on larval growth of the spruce bark beetle *Dendroctonus micans*. *Ecological Entomology*, **22**, 109–115.

Stowe, M. K. (1988). Chemical mimicry. In *Chemical mediation of coevolution*, ed. K. C. Spencer, pp. 513–580. San Diego: Academic Press.

Stuart, A. M. (1969). Social behavior and communication. In *The biology of termites*, ed. K. Krishna & F. M. Weesner, pp. 193–232. New York: Academic Press.

Suckling, D. M., Shaw, P. W., Khoo, J. G. I. & Cruickshank, V. (1990). Resistance management of light-brown apple moth, *Epiphyas postvittana* (Lepidoptera, Tortricidae) by mating disruption. *New Zealand Journal of Crop and Horticultural Science*, **18**, 89–98.

Sullivan, T. P. & Crump, D. (1984). Influence of mustelid scent gland compounds on suppression of feeding by snowshoe hares (*Lepus americanus*). *Journal of Chemical Ecology*, **10**, 903–919.

Sun, L. X. & Müller-Schwarze, D. (1997). Sibling recognition in the beaver: a field test for phenotype matching. *Animal Behaviour*, **54**, 493–502.

Sun, L. X. & Müller-Schwarze, D. (1998). Beaver response to recurrent alien scents: scent fence or scent match. *Animal Behaviour*, **55**, 1529–1536

Sun, L. X. & Müller-Schwarze, D. (1999). Chemical signals in the beaver: one species, two secretions, many functions? In *Advances in chemical signals in vertebrates*, ed. R. E. Johnston, D. Müller-Schwarze & P. W. Sorenson, pp. 281–288. New York: Kluwer Academic Publishers/Plenum Press.

Surbey, M. K. (1990). Family composition, stress, and the timing of human menarche. In *Socioendocrinology of primate reproduction*, ed. T. E. Zeigler & F. B. Bercovitch, pp. 11–32. New York: Wiley-Liss.

Süskind, P. (1986). *Perfume: the story of a murderer*, trans. J. E. Wood. London: Hamish Hamilton.

Tallamy, D. W. & Denno, R. F. (1982). Maternal care in *Gargaphia solani* (Hemiptera: Tingidae). *Animal Behaviour*, **29**, 771–778.

Tardif, S. D. (1997). The bioenergetics of parental behavior and the evolution of alloparental care in marmosets and tamarins. In *Cooperative breeding in mammals*, ed. N. G. Solomon & J. A. French, pp. 11–33. Cambridge: Cambridge University Press.

Temeles, E. J. (1994). The role of neighbours in territorial systems: when are they 'dear enemies?' *Animal Behaviour*, **47**, 339–350.

Theisen, B., Zeiske, E., Silver, W. L., Marui, T. & Caprio, J. (1991). Morphological and physiological studies on the olfactory organ of the striped eel catfish, *Plotosus lineatus*. *Marine Biology*, **110**, 127–135.

Thesen, A., Steen, J. B. & Døving, K. B. (1993). Behavior of dogs during olfactory tracking. *Journal of Experimental Biology*, **180**, 247–251.

Thomas, J. A. & Elmes, G. W. (1998). Higher productivity at the cost of increased host-specificity when *Maculinea* butterfly larvae exploit ant colonies through trophallaxis rather than by predation. *Ecological Entomology*, **23**, 457–464.

Thomas, J. A., Elmes, G. W., Wardlaw, J. C. & Woyciechowski, M. (1989). Host specificity among *Maculinea* butterflies in *Myrmica* ant nests. *Oecologia*, **79**, 452–457.

Thomas, J. H. (1993). Chemosensory regulation of development in *C. elegans*. *Bioessays*, **15**, 791–797.

Thornhill, R. (1979). Male pair formation pheromones in *Panorpa* scorpionflies (Mecoptera: Panorpidae). *Environmental Entomology*, **8**, 886–889.

Thornhill, R. (1992). Female preference for the pheromone of males with low fluctuating asymmetry in the Japanese scorpionfly (*Panorpa japonica*, Mecoptera). *Behavioral Ecology*, **3**, 277–283.

Thornhill, R. & Alcock, J. (1983). *The evolution of insect mating systems*. Cambridge, Mass.: Harvard University Press.

Thornton-Manning, J. R. & Dahl, A. R. (1997). Metabolic capacity of nasal tissue inter-species comparisons of xenobiotic-metabolizing enzymes. *Mutation Research-Fundamental and Molecular Mechanisms of Mutagenesis*, **380**, 43–59.

Tompkins, L., McRobert, S. P. & Kaneshiro, K. Y. (1993). Chemical communication in Hawaiian *Drosophila*. *Evolution*, **47**, 1407–1419.

Toonen, R. J. & Pawlik, J. R. (1994). Foundations of gregariousness. *Nature*, **370**, 511–512.

Torto, B., Assad, Y. O. H., Njagi, P. G. N. & Hassanali, A. (1999). Evidence for additional pheromonal components mediating oviposition aggregation in *Schistocerca gregaria*. *Journal of Chemical Ecology*, **25**, 835–845.

Traniello, J. F. A. (1983). Social-organization and foraging success in *Lasius neoniger* (Hymenoptera, Formicidae) – behavioral and ecological aspects of recruitment communication. *Oecologia*, **59**, 94–100.

Traniello, J. F. A. (1989). Foraging strategies of ants. *Annual Review of Entomology*, **34**, 191–210.

Traniello, J. F. A. & Robson, S. K. (1995). Trail and territorial communication in insects. In *Chemical ecology of insects 2*, ed. R. T. Cardé & W. J. Bell, pp. 241–286. London: Chapman and Hall.

Trematerra, P. (1997). Integrated pest management of stored-product insects: Practical utilization of pheromones. *Anzeiger für Schadlingskunde Pflanzenschutz Umweltschutz*, **70**, 41–44.

Trouiller, J., Arnold, G., Chappe, B., Le Conte, Y. & Masson, C. (1992). Semiochemical basis of infestation of honey-bee brood by *Varroa jacobsoni*. *Journal of Chemical Ecology*, **18**, 2041–2053.

Trumble, J. T. (1997). Integrating pheromones into vegetable crop production. In *Insect pheromone research: new directions*, ed. R. T. Cardé & A. K. Minks, pp. 397–410. New York: Chapman and Hall.

Tsuda, A. Z. & Miller, C. B. (1998). Mate finding behaviour in *Calanus marshallae* Frost. *Transactions of the Royal Society B–Biological Science*, **353**, 713–720.

Tumlinson, J. H., Silverstein, R. M., Moser, J. C., Brownlee, R. G. & Ruth, J. M. (1971). Identification of the trail pheromone of a leaf-cutting ant, *Atta texana*. *Nature*, **234**, 348–349.

Turlings, T. C. J., Loughrin, J. H., McCall, P. J., Rose, U. S. R., Lewis, W. J. & Tumlinson, J. H. (1995). How caterpillar-damaged plants protect themselves by attracting parasitic wasps. *Proceedings of the National Academy of Sciences of the United States of America*, **92**, 4169–4174.

Vandenbergh, J. G. (1994). Pheromones and mammalian reproduction. In *The physiology of reproduction*, 2nd edn, ed. E. Knobil & J. D. Neill, pp. 343–359. New York: Academic Press.

Vandenbergh, J. G. (1999a). Pheromones, mammals. In *Encyclopedia of reproduction*, Vol. 3, ed. E. Knobil & J. D. Neill, pp. 130–135. New York: Academic Press.

Vandenbergh, J. G. (1999b). Whitten effect. In *Encyclopedia of reproduction*, Vol. 4, ed. E. Knobil & J. D. Neill, pp. 6–10. New York: Academic Press.

Vandenbergh, J. G. (1999c). Puberty acceleration. In *Encyclopedia of reproduction*, Vol. 4, ed. E. Knobil & J. D. Neill, pp. 563–565. New York: Academic Press.

Vander Meer, R. K. & Alonso, L. E. (1998). Pheromone directed behavior in ants. In *Pheromone communication in social insects: ants, wasps, bees, and termites*, ed. R. K. Vander Meer, M. D. Breed, M. L. Winston & K. Espelie, pp. 159–192. Boulder, Colo.: Westview Press.

Vander Meer, R. K. & Morel, L. (1995). Ant queens deposit pheromones and antimicrobial agents on eggs. *Naturwissenschaften*, **82**, 93–95.

Vander Meer, R. K. & Morel, L. (1998). Nestmate recognition in ants. In *Pheromone communication in social insects: ants, wasps, bees, and termites*, ed. R. K. Vander Meer, M. D. Breed, M. L. Winston & K. Espelie, pp. 79–103. Boulder, Col.: Westview Press.

Vander Meer, R. K. & Wojcik, D. P. (1982). Chemical mimicry in the myrmecophilous beetle *Myrmecaphodius excavaticollis*. *Science*, **218**, 806–808.

Vander Meer, R. K., Breed, M. D., Winston, M. L. & Espelie, K. E. (ed.) (1998). *Pheromone communication in social insects: ants, wasps, bees, and termites*. Boulder, Color.: Westview Press.

van der Pers, J. N. C. & Minks, A. K. (1997). Measuring pheromone dispersion in the field with the single sensillum recording technique. In *Insect pheromone research: new directions*, ed. R. T. Cardé & A. K. Minks, pp. 359–371. New York: Chapman and Hall.

van der Pers, J. N. C. & Minks, A. K. (1998). A portable electroantennogram sensor for routine measurements of pheromone concentrations in greenhouses. *Entomologia Experimentalis et Applicata*, **87**, 209–215.

van Djiken, M. J., van Stratum, P. & van Alphen, J. J. M. (1992). Recognition of individual-specific marked parasitized hosts by the solitary parasitoid *Epidinocarsis lopezi*. *Behavioral Ecology and Sociobiology*, **30**, 77–82.

Vane-Wright, R. I. & Boppré, M. (1993). Visual and chemical signaling in butterflies – functional and phylogenetic perspectives. *Philosophical Transactions of the Royal Society of London Series B-Biological Sciences*, **340**, 197–205.

van Lenteren, J. C. (1981). Host discrimination by parasitoids. In *Semiochemicals: their role in pest control*, ed. D. A. Nordlund, R. L. Jones & W. J. Lewis, pp. 153–180. New York: John Wiley.

van Lenteren, J. C. & Woets, J. (1988). Biological and integrated pest-control in greenhouses. *Annual Review of Entomology*, **33**, 239–269.

Van Toller, S. & Dodd, G. H. (ed.) (1992). *Fragrance: the psychology and biology of perfume*. London: Elsevier Science.

Varendi, H., Porter, R. H. & Winberg, J. (1994). Does the newborn baby find the nipple by smell. *Lancet*, **344**, 989–990.

Vargo, E. L. (1992). Mutual pheromonal inhibition among queens in polygyne colonies of the fire ant *Solenopsis invicta*. *Behavioral Ecology and Sociobiology*, **31**, 205–210.

Vargo, E. L. (1998). Primer pheromones in ants. In *Pheromone communication in social insects: ants, wasps, bees, and termites*, ed. R. K. Vander Meer, M. D. Breed, M. L. Winston & K. Espelie, pp. 293–313. Boulder, Colo.: Westview Press.

Vargo, E. L. (1999). Reproductive development and ontogeny of queen pheromone production in the fire ant *Solenopsis invicta*. *Physiological Entomology*, **24**, 1–7.

Venkataraman, A. B., Swarnalatha, V. B., Nair, P. & Gadagkar, R. (1988). The mechanism of nestmate discrimination in the tropical social wasp *Ropalidia marginata* and its implications for the evolution of sociality. *Behavioral Ecology and Sociobiology*, **23**, 271–279.

Verrell, P. A. (1991). Illegitimate exploitation of sexual signaling systems and the origin of species. *Ethology Ecology & Evolution*, **3**, 273–283.

Vickers, N. J. (1999). The effects of chemical and physical features of pheromone plumes upon the behavioral responses of moths. In *Advances in chemical signals in vertebrates*, ed. R. E. Johnston, D. Müller-Schwarze & P. W. Sorenson, pp. 63–76. New York: Kluwer Academic Publishers/Plenum Press.

Vickers, N. J. (2000). Mechanisms of animal navigation in odor plumes. *Biological Bulletin*, **198**, 203–212.

Vickers, N. J. & Baker, T. C. (1991). The effects of unilateral antennectomy on the flight behavior of male *Heliothis virescens* in a pheromone plume. *Physiological Entomology*, **16**, 497–506.

Vickers, N. J. & Baker, T. C. (1994). Reiterative responses to single strands of odor promote sustained upwind flight and odor source location by moths. *Proceedings of the National Academy of Sciences of the United States of America*, **91**, 5756–5760.

Vickers, N. J. & Baker, T. C. (1996). Latencies of behavioral response to interception of filaments of sex pheromone and clean air influence flight track shape in *Heliothis virescens* (F) males. *Journal of Comparative Physiology A-Sensory Neural and Behavioral Physiology*, **178**, 831–847.

Vickers, N. J., Christensen, T. A. & Hildebrand, J. G. (1998). Integrating behavior with neurobiology: odor-mediated moth flight and olfactory discrimination by glomerular arrays. *Integrative Biology*, **1**, 224–230.

Viitala, J., Korpimaki, E., Palokangas, P. & Koivula, M. (1995). Attraction of kestrels to vole scent marks visible in ultraviolet-light. *Nature*, **373**, 425–427.

Vinson, S. B. & Frankie, G. W. (1990). Territorial and mating behavior of *Xylocopa fimbriata* F. and *Xylocopa gualanensis* Cockerell from Costa Rica. *Journal of Insect Behavior*, **3**, 13–32.

Visscher, P. K. (1998). Colony integration and reproductive conflict in honey bees. *Apidologie*, **29**, 23–45.

Visscher, P. K. & Dukas, R. (1995). Honey-bees recognize development of nestmates ovaries. *Animal Behaviour*, **49**, 542–544.

Vité, J. P. & Francke, W. (1976). The aggregation pheromones of bark beetles: progress and problems. *Naturwissenschaften*, **63**, 550–555.

Vogel, S. (1983). How much air flows through a silkmoth's antenna? *Journal of Insect Physiology*, **29**, 597–602.

Vogel, S. (1994). *Life in moving fluids: the physical biology of flow*, 2nd edn. Princeton, N.J.: Princeton University Press.

Vogt, R. G., Riddiford, L. M. & Prestwich, G. D. (1985). Kinetic-properties of a sex pheromone-degrading enzyme – the sensillar esterase of *Antheraea polyphemus*. *Proceedings of the National Academy of Sciences of the United States of America*, **82**, 8827–8831.

Vosshall, L. B., Amrein, H., Morozov, P. S., Rzhetsky, A. & Axel, R. (1999). A spatial map of olfactory receptor expression in the *Drosophila* antenna. *Cell*, **96**, 725–736.

Vroon, P. A., van Amerongen, A. & de Vries, H. (1997). *Smell: the secret seducer*. New York: Farrar, Straus and Giroux.

Wabnitz, P. A., Bowie, J. H., Tyler, M. J., Wallace, J. C. & Smith, B. P. (1999). Aquatic sex pheromone from a male tree frog. *Nature*, **401**, 444–445.

Waldman, B. (1982). Sibling association among schooling toad tadpoles – field evidence and implications. *Animal Behaviour*, **30**, 700–713.

Waldman, B. (1985). Olfactory basis of kin recognition in toad tadpoles. *Journal of Comparative Physiology A-Sensory Neural and Behavioral Physiology*, **156**, 565–577.

Waldman, B., Frumhoff, P. C. & Sherman, P. W. (1988). Problems of kin recognition. *Trends in Ecology & Evolution*, **3**, 8–13.

Wall, C. (1989). Monitoring and spray timing. In *Insect pheromones in plant protection*, ed. A. R. Jutsum & R. F. S. Gordon, pp. 39–66. Chichester: John Wiley.

Wang, H. W., Wysocki, C. J. & Gold, G. H. (1993). Induction of olfactory receptor sensitivity in mice. *Science*, **260**, 998–1000.

Wang, Z. X. & Novak, M. A. (1994). Parental care and litter development in primiparous and multiparous prairie voles (*Microtus ochrogaster*). *Journal of Mammalogy*, **75**, 18–23.

Waters, R. M. (1993). Odorized air current trailing by garter snakes, *Thamnophis sirtalis*. *Brain Behavior and Evolution*, **41**, 219–223.

Watkins, J., Gehlbach, F. R. & Baldridge, R. S. (1967). Ability of the blind snake, *Leptotyphlops dulcis*, to follow pheromone trails of army ants, *Neivamyrmex nigrescens* and *N. opacithorax*. *Southwestern Naturalist*, **12**, 455–462.

Watson, P. J. (1986). Transmission of a female sex pheromone thwarted by males in the spider *Linyphia litogiosa* Keyserling (Linyphiidae). *Science*, **233**, 219–221.

Webb, J. K. & Shine, R. (1992). To find an ant – trail-following in Australian blindsnakes (Typhlopidae). *Animal Behaviour*, **43**, 941–948.

Wedekind, C. & Folstad, I. (1994). Adaptive or nonadaptive immunosuppression by sex-hormones. *American Naturalist*, **143**, 936–938.

Wedekind, C. & Füri, S. (1997). Body odour preferences in men and women: do they aim for specific MHC combinations or simply heterozygosity? *Proceedings of the Royal Society of London Series B-Biological Sciences*, **264**, 1471–1479.

Wedekind, C., Seebeck, T., Bettens, F. & Paepke, A. J. (1995). MHC-dependent mate preferences in humans. *Proceedings of the Royal Society of London Series B-Biological Sciences*, **260**, 245–249.

Wehner, R. (1997). Sensory systems and behaviour. In *Behavioural ecology*, 4th edn, ed. J. R. Krebs & N. B. Davies, pp. 19–41. Oxford: Blackwell Science.

Weissburg, M. J. (1997). Chemo- and mechanosensory orientation by crustaceans in laminar and turbulent flows: from odor trails to vortex sheets. In *Orientation and communication in arthropods*, ed. M. Lehrer, pp. 216–246. Basel: Birkhäuser Verlag.

Weissburg, M. J. (2000). The fluid dynamical context of chemosensory behavior. *Biological Bulletin*, **198**, 188–202.

Weissburg, M. J. & Zimmer-Faust, R. K. (1993). Life and death in moving fluids – hydrodynamic effects on chemosensory-mediated predation. *Ecology*, **74**, 1428–1443.

Weissburg, M. J., Doall, M. H. & Yen, J. (1998). Following the invisible trail: kinematic analysis of mate-tracking in the copepod *Temora longicornis*. *Philosophical Transactions of the Royal Society of London Series B-Biological Sciences*, **353**, 701–712.

Weldon, P. J. (1990). Responses by vertebrates to chemicals from predators. In *Chemical signals in vertebrates*, Vol. 5, ed. D. W. Macdonald, D. Müller-Schwarze & S. E. Natynczuk, pp. 500–521. Oxford: Oxford Science Publications, Oxford University Press.

Weller, A. & Weller, L. (1997). Menstrual synchrony under optimal conditions: Bedouin families. *Journal of Comparative Psychology*, **111**, 143–151.

Weller, L. & Weller, A. (1993). Human menstrual synchrony: a critical assessment. *Neuroscience and Biobehavioral Reviews*, **17**, 427–439.

Wells, M. J. & Buckley, S. K. L. (1972). Snails and trails. *Animal Behaviour*, **20**, 345–355.

Welsh, R. G. & Müller-Schwarze, D. (1989). Experimental habitat scenting inhibits colonization by beaver, *Castor canadensis*. *Journal of Chemical Ecology*, **15**, 887–893.

Wenhold, B. A. & Rasa, O. A. E. (1994). Territorial marking in the yellow mongoose *Cynictis penicillata* – sexual advertisement for subordinates. *Zeitschrift für Saugetierkunde-International Journal of Mammalian Biology*, **59**, 129–138.

West, M. J., King, A. P. & Eastzer, D. H. (1981). The cowbird: reflections on development from an unlikely source. *American Scientist*, **69**, 56–66.

Westneat, D. F. & Birkhead, T. R. (1998). Alternative hypotheses linking the immune system and mate choice for good genes. *Proceedings of the Royal Society of London Series B-Biological Sciences*, **265**, 1065–1073.

Whittier, T. S. & Kaneshiro, K. Y. (1995). Intersexual selection in the Mediterranean fruit-fly – does female choice enhance fitness. *Evolution*, **49**, 990–996.

Whittier, T. S., Nam, F. Y., Shelly, T. E. & Kaneshiro, K. Y. (1994). Male courtship success and female discrimination in the Mediterranean fruit-fly (Diptera, Tephritidae). *Journal of Insect Behavior*, **7**, 159–170.

Wilcox, R. M. & Johnston, R. E. (1995). Scent counter-marks: specialized mechanisms of perception and response to odors in golden hamster (*Mesocricetus auratus*). *Journal of Comparative Psychology*, **109**, 349–356.

Williams, G. C. (1992). *Natural selection: domains, levels, and challenges*. Oxford: Oxford University Press.

Willis, M. A. & Arbas, E. A. (1991). Odor-modulated upwind flight of the sphinx moth, *Manduca sexta* L. *Journal of Comparative Physiology A-Sensory Neural and Behavioral Physiology*, **169**, 427–440.

Wilson, E. O. (1962). Chemical communication among workers of the fire ant *Solenopsis savissima* (Fr. Smith). 1: The organization of mass foraging. *Animal Behaviour*, **10**, 134–147.

Wilson, E. O. (1970). Chemical communication within animal species. In *Chemical ecology*, Vol. 9, ed. E. Sondheimer & J. B. Simeone, pp. 133–155. New York: Academic Press.

Wilson, E. O. (1971). *The insect societies*. Harvard, Mass.: Belknap Press.

Wilson, E. O. (1975). *Sociobiology*. Harvard, Mass.: Belknap Press.

Wilson, E. O. & Bossert, W. H. (1963). Chemical communication among animals. *Recent Progress in Hormone Research*, **19**, 673–716.

Winberg, J. & Porter, R. H. (1998). Olfaction and human neonatal behaviour: clinical implications. *Acta Paediatrica*, **87**, 6–10.

Winston, M. L. (1987). *The biology of the honey bee*. Cambridge, Mass.: Harvard University Press.

Winston, M. L. (1992). Semiochemicals and insect sociality. In *Insect chemical ecology. An evolutionary approach*, ed. B. D. Roitberg & M. B. Isman, pp. 315–333. New York: Chapman and Hall.

Winston, M. L. (1997). *Nature wars: people vs. pests*. Cambridge, Mass.: Harvard University Press.

Winston, M. L. & Slessor, K. N. (1992). The essence of royalty – honey-bee queen pheromone. *American Scientist*, **80**, 374–385.

Winston, M. L. & Slessor, K. N. (1998). Honey bee primer pheromones and colony organization: gaps in our knowledge. *Apidologie*, **29**, 81–95.

Wisenden, B. D. (1999). Alloparental care in fishes. *Reviews in Fish Biology and Fisheries*, **9**, 45–70.

Wittmann, D., Redtke, R., Zeil, J., Luebke, G. & Francke, W. (1990). Robber bees (*Lestrimelitta limao*) and their host: chemical and visual cues in nest defense by *Trigona angustula* (Apidae: Meliponinae). *Journal of Chemical Ecology*, **16**, 631–642.

Wittzell, H., Madsen, T., Westerdahl, H., Shine, R. & vonSchantz, T. (1999). MHC variation in birds and reptiles. *Genetica*, **104**, 301–309.

Wood, D. L. (1982). The role of pheromones, kairomones, and allomones in the host selection and colonization behavior of bark beetles. *Annual Review of Entomology*, **27**, 411–446.

Wood, R. I. (1998). Integration of chemosensory and hormonal input in the male Syrian hamster brain. *Annals of the New York Academy of Sciences*, **855**, 362–72.

Wood, R. I. & Swann, J. M. (2000). Neuronal integration of chemosensory and hormonal signals in the control of male sexual behavior. In *Reproduction in context*, ed. K. Wallen & J. E. Schneider, pp. 423–444. Cambridge, Mass.: MIT Press.

Wood, T. K. (1977). Role of parent females and attendant ants in the maturation of the treehopper, *Entylia bactriana* (Homoptera: Membracidae). *Sociobiology*, **2**, 257–272.

Wright, R. H. (1964). *The science of smell*. London: Allen & Unwin.

Wyatt, T. D. (1997). Putting pheromones to work: paths forward for direct control. In *Insect pheromone research: new directions*, ed. R. T. Cardé & A. K. Minks, pp. 445–459. New York: Chapman and Hall.

Wysocki, C. J. & Beauchamp, G. K. (1984). Ability to smell androstenone is genetically determined. *Proceedings of the National Academy of Sciences of the United States of America*, **81**, 4899–4902.

Wysocki, C. J. & Lepri, J. J. (1991). Consequences of removing the vomeronasal organ. *Journal of Steroid Biochemistry and Molecular Biology*, **39**, 661–669.

Wysocki, C. J., Dorries, K. M. & Beauchamp, G. K. (1989). Ability to perceive androstenone can be acquired by ostensibly anosmic people. *Proceedings of the National Academy of Sciences of the United States of America*, **86**, 7976–7978.

Yamamoto, D., Jallon, J. M. & Komatsu, A. (1997). Genetic dissection of sexual behavior in *Drosophila melanogaster*. *Annual Review of Entomology*, **42**, 551–585.

Yamamoto, K., Kawai, Y., Hayashi, T., Ohe, Y., Hayashi, H., Toyoda, F., Kawahara, G., Iwata, T. & Kikuyama, S. (2000). Silefrin, a sodefrin-like pheromone in the abdominal gland of the sword-tailed newt, *Cynops ensicauda*. *FEBS Letters*, **472**, 267–270.

Yamazaki, K., Beauchamp, G. K., Wysocki, C. J., Bard, J., Thomas, L. & Boyse, E. A. (1983). Recognition of H-2 types in relation to the blocking of pregnancy in mice. *Science*, **221**, 186–188.

Yeargan, K. V. (1994). Biology of bolas spiders. *Annual Review of Entomology*, **39**, 81–99.

Yen, J., Weissburg, M. J. & Doall, M. H. (1998). The fluid physics of signal perception by mate-tracking copepods. *Philosophical Transactions of the Royal Society of London Series B-Biological Sciences*, **353**, 787–804.

Yoder, J. A. & Grojean, N. C. (1997). Group influence on water conservation in the giant Madagascar hissing-cockroach, *Gromphadorhina portentosa* (Dictyoptera: Blaberidae). *Physiological Entomology*, **22**, 79–82.

Yousem, D. M., Maldjian, J. A., Siddiqi, F., Hummel, T., Alsop, D. C., Geckle, R. J., Bilker, W. B. & Doty, R. L. (1999). Gender effects on odor-stimulated functional magnetic resonance imaging. *Brain Research*, **818**, 480–487.

Zahavi, A. (1975). Mate selection: a selection for a handicap. *Journal of Theoretical Biology*, **53**, 205–214.

Zanen, P. O., Sabelis, M. W., Buonaccorsi, J. P. & Carde, R. T. (1994). Search strategies of fruit-flies in steady and shifting winds in the absence of food odors. *Physiological Entomology*, **19**, 335–341.

Zavazava, N. & Eggert, F. (1997). MHC and behavior. *Immunology Today*, **18**, 8–10.

Zeeck, E., Harder, T., Beckmann, M. & Müller, C. T. (1996). Marine gamete-release pheromones. *Nature*, **382**, 214.

Zeeck, E., Müller, C. T., Beckmann, M., Hardege, J. D., Papke, U., Sinnwell, V., Schroder, F. C. & Francke, W. (1998a). Cysteine-glutatione disulfide, the sperm-release pheromone of the marine polychaete *Nereis succinea* (Annelida: Polychaeta). *Chemoecology*, **8**, 33–38.

Zeeck, E., Harder, T. & Beckmann, M. (1998b). Inosine, L-glutamic acid and L-glutamine as components of a sex pheromone complex of the marine polychaete *Nereis succinea* (Annelida: Polychaeta). *Chemoecology*, **8**, 77–84.

Zeng, X. N., Leyden, J. J., Lawley, H. J., Sawano, K., Nohara, I. & Preti, G. (1991). Analysis of characteristic odors from human male axillae. *Journal of Chemical Ecology*, **17**, 1469–1492.

Zeng, X. N., Leyden, J. J., Brand, J. G., Spielman, A. I., McGinley, K. J. & Preti, G. (1992). An investigation of human apocrine gland secretion for axillary odor precursors. *Journal of Chemical Ecology*, **18**, 1039–1055.

Zeng, X. N., Leyden, J. J., Spielman, A. I. & Preti, G. (1996). Analysis of characteristic human female axillary odors – qualitative comparison to males. *Journal of Chemical Ecology*, **22**, 237–257.

Zhang, X. M. & Firestein, S. (2002). The olfactory receptor gene superfamily of the mouse. *Nature Neuroscience,* **5**, 124–133.

Zimmer, R. K. & Butman, C. A. (2000). Chemical signaling processes in the marine environment. *Biological Bulletin*, **198**, 168–187.

Zimmer-Faust, R. K., Tyre, J. E. & Case, J. F. (1985). Chemical attraction causing aggregation in the spiny lobster, *Panulirus interruptus* (Randall), and its probable ecological significance. *Biological Bulletin*, **169**, 106–118.

Zimmer Faust, R. K., Finelli, C. M., Pentcheff, N. D. & Wethey, D. S. (1995). Odor plumes and animal navigation in turbulent water-flow – a field-study. *Biological Bulletin*, **188**, 111–116.

Zou, Z. H., Horowitz, L. F., Montmayeur, J. P., Snapper, S. & Buck, L. B. (2001). Genetic tracing reveals a stereotyped sensory map in the olfactory cortex. *Nature,* **414**, 173–179.

Zuk, M. & Kolluru, G. R. (1998). Exploitation of sexual signals by predators and parasitoids. *Quarterly Review of Biology*, **73**, 415–438.

Zuk, M., Kim, T., Kristan, D. M. & Luong, L. T. (1997). Sex, pain and parasites. *Parasitology Today*, **13**, 332–333.

List of Credits

AAAS (American Association for the Advancement of Science)

Fig. 9.4 Reprinted with permission from Fig. 1, page 712, in Mori, K., Nagao, H. & Yoshihara, Y. (1999). The olfactory bulb: coding and processing of odor molecule information. *Science*, **286**, 711–715. Copyright (1999) American Association for the Advancement of Science.

Fig. 11.8 Reprinted with permission from Fig. 1, page 267 in Regnier, F. E. & Wilson, E. O. (1971). Chemical communication and 'propaganda' in slave-maker ants. *Science*, **172**, 267–269. Copyright (1971) American Association for the Advancement of Science.

Academic Press

Fig. 1.10 Fig. 1, page 674 in Wilson, E. O. & Bossert, W. H. (1963). Chemical communication among animals. *Recent Progress in Hormone Research*, **19**, 673–716.

Box 3.1 Fig. Fig. 1, page 423, in Penn, D. & Potts, W. K. (1998). How do major histocompatibility complex genes influence odor and mating preferences? *Advances in Immunology*, **69**, 411–436.

Box 7.3 Fig. (bottom) Fig. 4, page 205, Stuart, A. M. (1969). Social behavior and communication. In *The biology of termites*, ed. K. Krishna & F. M. Weesner, pp. 193–232. New York: Academic Press.

Fig. 10.2 (ant) Fig. 2, page 137 in Wilson, E. O. (1962). Chemical communication among workers of the fire ant *Solenopsis savissima* (Fr. Smith). 1: The organization of mass foraging. *Animal Behaviour*, **10**, 134–147.

Fig. 10.2 (lower part) Fig. 4, page 689 in Wilson, E. O. & Bossert, W. H. (1963). Chemical communication among animals. *Recent Progress in Hormone Research*, **19**, 673–716.

Fig. 10.13 (whole) Fig. 1, page 162 in Arbas, E. A., Willis, M. A. & Kanzaki, R. (1993). Organization of goal-oriented locomotion: pheromone-modulated flight behavior of moths. In *Biological neural networks in invertebrate neuroethology and robotics*, ed. R. D. Beer, R. E. Ritzmann & T. McKenna, pp. 159–198. Boston, Mass.: Academic Press.

Fig. 10.13(b) Fig. 1, page 317 in Pawson, M. G. (1977). The responses of cod *Gadus morhua* (L.) to chemical attractants in moving water. *Journal du Conseil International pour l'Exploration de la Mer*, **37**, 316–318.

Fig. 13.7 Fig. 1, page 1006 in Meredith, M. (1999). Vomeronasal organ. In *Encyclopedia of reproduction*, ed. E. Knobil & J. D. Neill, pp. 1004–1014. New York: Academic Press.

Fig. 13.8 Fig. 11.1, page 334 in Montagna, W. & Parakkal, P. F. (1974). *The structure and function of skin*. New York: Academic Press.

American Medical Association

Fig. 13.3 Fig. 4, page 437 in Shelley, W. B., Hurley, H. J. & Nichols, A. C. (1953). Axillary odour: experimental study of the role of bacteria, apocrine sweat, and deodorants. *Archives of Dermatology and Syphiology*, **68**, 430–446. Copyrighted (1953), American Medical Association.

Annual Reviews

Fig. 9.2(a) Fig. 1, page 597 in Hildebrand, J. G. & Shepherd, G. M. (1997). Mechanisms of olfactory discrimination: converging evidence for common principles across phyla.

Annual Review of Neuroscience, **20**, 595–631. With permission from the *Annual Review of Neuroscience*, Volume **20**, © 1997 by Annual Reviews www.AnnualReviews.org

Fig. 10.8 Fig. 2, page 512 in Murlis, J., Elkinton, J. S. & Cardé, R. T. (1992). Odor plumes and how insects use them. *Annual Review of Entomology*, **37**, 505–532. With permission from the *Annual Review of Entomology*, Volume 37, © 1992 by Annual Reviews www.Annual Reviews.org

Balaban International

Fig. 3.1 (left) Fig. 2, page 185 in Müller, C. T., Beckmann, M. & Hardege, J. D. (1999). Sex pheromones in *Nereis succinea. Invertebrate Reproduction & Development*, **36**, 183–186.

Begell House

Fig. 3.8 Fig. 6, page 211 in Apanius, V., Penn, D., Slev, P. R., Ruff, L. R. & Potts, W. K. (1997). The nature of selection on the major histocompatibility complex. *Critical Reviews in Immunology*, **17**, 179–224.

Birkhäuser Verlag

Fig. 3.12 Fig. 2, page 22 in Löfstedt, C., Herrebout, W. M. & Menken, S. B. J. (1991). Sex pheromones and their potential role in the evolution of reproductive isolation in small ermine moths (Yponomeutidae). *Chemoecology*, **2**, 20–28.

Fig. 10.6 Fig. 1, page 219 in Weissburg, M. J. (1997). Chemo- and mechanosensory orientation by crustaceans in laminar and turbulent flows: from odor trails to vortex sheets. In *Orientation and communication in arthropods*, ed. M. Lehrer, pp. 216–246.

Fig. 10.11 Fig. 5, page 228 in Weissburg, M. J. (1997). Chemo- and mechanosensory orientation by crustaceans in laminar and turbulent flows: from odor trails to vortex sheets. In *Orientation and communication in arthropods*, ed. M. Lehrer, pp. 216–246.

Blackwell Science

Fig. 3.6 (whole) Fig. 4.6, page 91 in Sherman, P. W., Reeve, H. K. & Pfennig, D. W. (1997). Recognition systems. In *Behavioural ecology*, 4th edn, ed. J. R. Krebs & N. B. Davies, pp. 69–96.

Fig. 5.1 Fig. 2, page 308 and Fig. 5, pages 313, 308 in Gosling, L. M. (1981). Demarcation in a gerenuk territory – an economic approach. *Zeitschrift für Tierpsychologie-Journal of Comparative Ethology* **56**, 305–322.

Fig. 5.3 Fig. 2, page 98 and Fig 3a, page 100 in Gosling, L. M. (1982). A reassessment of the function of scent marking in territories. *Zeitschrift für Tierpsychologie-Journal of Comparative Ethology*, **60**, 89–118.

Fig. 5.5 (badger) Fig. 9.7, page 267 in Cockburn, A. (1991). *An introduction to evolutionary ecology.*

Box 6.1 Fig. Fig. 1, page 59 in Porter, R. H. & Blaustein, A. R. (1989). Mechanisms and ecological correlates of kin recognition. *Science Progress,* **73**, 53–66.

Fig. 7.3 Fig. 6, page 270 in Reinhard, J. & Kaib, M. (1995). Interaction of pheromones during food exploitation by the termite *Schedorhinotermes lamanianus. Physiological Entomology*, **20**, 266–272.

Fig. 7.5 (original) Fig. 1, page 133 in Burton, J. L. & Franks, N. R. (1985). The foraging ecology of the army ant *Eciton rapax* – an ergonomic enigma. *Ecological Entomology*, **10**, 131–141.

Fig. 9.6 (original) Fig. 5A, page 3976 and Fig. 7B, page 3979 in Sachse, S., Rappert, A. & Galizia, C. G. (1999). The spatial representation of chemical structures in the antennal lobe of honeybees: steps towards the olfactory code. *European Journal of Neuroscience*, **11**, 3970–3982.

Fig. 10.9 Fig. 8, page 83 in Murlis, J. & Jones, C. D. (1981). Fine-scale structure of odor plumes in relation to insect orientation to distant pheromone and other attractant sources. *Physiological Entomology*, **6**, 71–86.

Fig. 11.5 Fig. 1, page 161 in Peakall, R. (1990). Responses of male *Zaspilothynnus trilobatus* Turner wasps to females and the sexually deceptive orchid it pollinates. *Functional Ecology*, **4**, 159–168.

Brill

Fig. 5.5 (main) Fig. 1a, page 294 in Roper, T. J., Conradt, L., Butler, J., Christian, S. E., Ostler, J. & Schmid, T. K. (1993). Territorial marking with feces in badgers (*Meles meles*) – a comparison of boundary and hinterland latrine use. *Behaviour*, **127**, 289–307.

CAB International (CABI Publishing)

Box 4.1 Fig. Fig. 12.1, page 306 in Hassanali, A. & Torto, B. (1999). Grasshoppers and locusts. In *Pheromones of non-lepidopteran insects associated with agricultural plants*, ed. J. Hardie & A. K. Minks, pp. 305–328.

Fig. 7.2 (right) Fig. 13.4, page 339 in Kaib, M. (1999). Termites. In *Pheromones of non-lepidopteran insects associated with agricultural plants*, ed. J. Hardie & A. K. Minks, pp. 329–353.

Fig. 11.2 Fig. 14.2, page 366 in Aldrich, J. R. (1999). Predators. In *Pheromones of non-lepidopteran insects associated with agricultural plants*, ed. J. Hardie & A. K. Minks, pp. 357–381.

Cambridge University Press

Fig. 1.7 Fig. 27.4, page 708 in Chapman, R. F. (1998). *The insects. Structure and function*, 4th edn.

Table 4.1 Table 27.3, page 719 in Chapman, R. F. (1998). *The insects. Structure and function*, 4th edn.

Box 6.2 Fig. (top) Fig. 27.15, page 725 in Chapman, R. F. (1998). *The insects. Structure and function*, 4th edn.

Box 6.2 Fig. (middle) Fig. 27.14, page 724 in Chapman, R. F. (1998). *The insects. Structure and function*, 4th edn.

Fig. 6.2 Fig. 6.11, page 343 in Manning, A. & Dawkins, M. S. (1998). *An introduction to animal behaviour*, 5th edn.

Table 7.1 Table 27.4, page 720 in Chapman, R. F. (1998). *The insects. Structure and function*, 4th edn.

Table 8.1 Table 27.5, page 722 in Chapman, R. F. (1998). *The insects. Structure and function*, 4th edn.

Fig. 8.3 (left) Fig. 6b, page 44 in Stern, D. L. & Foster, W. A. (1996). The evolution of soldiers in aphids. *Biological Reviews of the Cambridge Philosophical Society*, **71**, 27–79.

Fig. 9.2(b) (right) Fig. 2.17, page 58 in Farbman, A. I. (1992). *Cell biology of olfaction*.

Fig. 10.1 Fig. 27.13, page 723 in Chapman, R. F. (1998). *The insects. Structure and function*, 4th edn.

Fig. 13.1 Fig. 4.4, page 88 in Stoddart, D. M. (1990). *The scented ape. The biology and culture of human odour*.

Box 13.1 Fig. Fig. 3.1, page 50 in Stoddart, D. M. (1990). *The scented ape. The biology and culture of human odour*.

Company of Biologists Ltd.

Fig. 10.5 Fig. 1, page 249 in Thesen, A., Steen, J. B. & Døving, K. B. (1993). Behavior of dogs during olfactory tracking. *Journal of Experimental Biology*, **180**, 247–251.

Cooper Ornithological Society

Fig. 10.13 (a) Fig. 3, page 318 in Hutchison, L. V. & Wenzel, B. M. (1980). Olfactory guidance in procellariiforms. *Condor*, **82**, 314–319 (copyright The Cooper Ornithological Society, as published in *The Condor*).

Copyright Council of the Academic Sciences, Japan

Fig. 4.5 Fig. 1A, page 1348 in Okano, K. & Fusetani, N. (1997). Larval settlement in barnacles. *Seikagaku*, **69**, 1348–1360.

Ecological Society of America

Fig. 4.2 Fig. 11a, page 39 in Raffa, K. F. & Berryman, A. A. (1983). The role of host plant-resistance in the colonization behavior and ecology of bark beetles (Coleoptera, Scolytidae). *Ecological Monographs*, **53**, 27–49.

Fig. 10.7 Fig. 1D, page 1433 in Weissburg, M. J. & Zimmer-Faust, R. K. (1993). Life and death in moving fluids – hydrodynamic effects on chemosensory-mediated predation. *Ecology*, **74**, 1428–1443.

Elsevier Science

Fig. 2.4 (top) Fig. 1, page 428. Reprinted from *Advances in the Biosciences*, **93**, Schellinck, H. M. & Brown, R. E. Methodological questions in the study of the rat's ability to discriminate between the odours of individual conspecifics, 427–436, Copyright 1994, with permission from Elsevier Science.

Fig. 3.15 Fig. 3, page 45. Reprinted from *Behavioural Processes*, **35**, Cobb, M. & Ferveur, J. F. Evolution and genetic-control of mate recognition and stimulation in *Drosophila*. 35–54, Copyright 1996, with permission from Elsevier Science.

Box 3.1 Fig. Fig. 1, page 9. Reprinted from *Immunology Today*, **18**, Zavazava, N. & Eggert, F. MHC and behavior. 8–10, Copyright 1997, with permission from Elsevier Science.

Fig. 6.1 Fig. 1, page 263. Reprinted from *Comparative Biochemistry and Physiology C-Pharmacology Toxicology & Endocrinology*, **119**, Abbott, D. H., Saltzman, W., Schultz-Darken, N. J. & Tannenbaum, P. L. Adaptations to subordinate status in female marmoset monkeys. 261–274, Copyright 1998, with permission from Elsevier Science.

Fig. 7.4 Fig. 3, page 266. Reprinted from *Behavioural Processes*, **5**, Cammaerts, M. C. & Cammaerts, R. Food recruitment strategies of the ant *Myrmica sabuleti* and *Myrmica ruginodis*. 251–270, Copyright 1980, with permission from Elsevier Science.

Fig. 9.6 Fig. 3, page 508. Reprinted from *Current Opinion in Neurobiology*, **10**, Galizia, C. G. & Menzel, R. Odour perception in honeybees: coding information in glomerula patterns. 504–510, Copyright 2000, with permission from Elsevier Science.

Fig. 9.10 Fig. 5, page 392. Reprinted from *Brain Research Bulletin*, **44**, Kendrick, K. M., DaCosta, A. P. C., Broad, K. D., Ohkura, S., Guevara, R., Lévy, F. & Keverne, E. B. Neural control of maternal behaviour and olfactory recognition of offspring. 383–395, Copyright 1997, with permission from Elsevier Science.

Fig. 10.12 Fig. 1, page 204. Reprinted from *Neuroscience Letters*, **17**, Hatt, H. & Bauer, U. Single unit analysis of mechano- and chemosensitive neurones in the crayfish claw. 203–207, Copyright 1980, with permission from Elsevier Science.

Fig. 12.1 Fig. 1, page 346. Reprinted from *Theriogenology*, **32**, Houpt, K. A., Rivera, W. & Glickstein, L. The flehmen response of bulls and cows. 343–350, Copyright 1989, with permission from Elsevier Science.

Table 13.5 Table 1, page 393. Reprinted from *Trends in Ecology & Evolution*, **13**, Penn, D. & Potts, W. K. Chemical signals and parasite-mediated sexual selection. 391–396, Copyright 1998, with permission from Elsevier Science.

Florida Entomological Society

Table 11.1 Table 1, page 19 in Alcock, J. (1982). Natural selection and communication in bark beetles. *Florida Entomologist*, **65**, 17–32.

Gordon & Breach

Box 4.2 Fig. Fig. 1, page 58 in Clare, A. S. & Matsumura, K. (2000). Nature and perception of barnacle settlement patterns. *Biofouling*, **15**, 57–71. Reprinted with permission from Gordon & Breach – a Member of Taylor & Francis.

Indiana University Press

Fig. 13.1 Fig. 35, page 941 in Gautier, J. P. & Gautier, A. (1977). Communication in Old World monkeys. In *How animals communicate*, ed. T. Seboek, pp. 890–964. Bloomington, Ind.: Indiana University Press.

INRA

Table 1.1 Table 2, page 154 in Blum, M. S. (1982). Pheromonal bases of insect sociality: communications, conundrums and caveats. *Les Colloques de l'Institut National de la Recherche Agronomique (INRA)*, **7**, 149–162.

Iowa State University Press

Fig. 9.7 Fig. 1.9, page 12 in Houpt, K. A. (1998). *Domestic animal behavior for veterinarians and animal scientists*, 3rd edn. London: Manson Publishing/ The Veterinary Press. Iowa State University Press, publisher and copyright holder

John Wiley

Fig. 2.3 Fig. 1, page 13 in Matsumura, K., Nagano, M. & Fusetani, N. (1998). Purification of a larval settlement-inducing protein complex (SIPC) of the barnacle, *Balanus amphitrite. Journal of Experimental Zoology*, **281**, 12–20. Copyright © 1998 Wiley-Liss. Reprinted by permission of Wiley-Liss Inc., a subsidiary of John Wiley & Sons, Inc.

Fig. 2.5 Fig. 2.25, page 38 in Pawliszyn, J. (1997). *Solid phase microextraction. Theory and practice.* Copyright © 1997 John Wiley. Reprinted by permission of John Wiley & Sons, Inc.

Fig. 9.1 Fig. 9.1, page 204 in Christensen, T. A. & White, J. (2000). Representation of olfactory information in the brain. In *The neurobiology of taste and smell*, 2nd edn, ed. T. E. Finger, W. L. Silver & D. Restrepo, pp. 201–232. Copyright © 2000 Wiley-Liss. Reprinted by permission of Wiley-Liss Inc., a subsidiary of John Wiley & Sons, Inc.

Fig. 9.3 Fig. 9.9, page 220 in Christensen, T. A. & White, J. (2000). Representation of olfactory information in the brain. In *The neurobiology of taste and smell,* 2nd edn, ed. T. E. Finger, W. L. Silver & D. Restrepo, pp. 201–232. Copyright © 2000 Wiley-Liss. Reprinted by permission of Wiley-Liss Inc., a subsidiary of John Wiley & Sons, Inc.

Fig. 9.8 Fig. 2, page 229 in Vickers, N. J., Christensen, T. A. & Hildebrand, J. G. (1998). Integrating behavior with neurobiology: odor-mediated moth flight and olfactory discrimination by glomerular arrays. *Integrative Biology*, **1**, 224–230. Copyright © 2000 Wiley-Liss. Reprinted by permission of Wiley-Liss Inc., a subsidary of John Wiley & Sons, Inc.

Fig. 9.11 Figs. 1 and 2, page 792 in Thomas, J. H. (1993). Chemosensory regulation of development in *C. elegans*. *Bioessays*, **15**, 791–797. Copyright © 2002 Wiley Periodicals, Inc. Reprinted by permission of Wiley Periodicals, Inc., a Wiley Company.

Karger, Basel

Fig. 1.11 Fig. 1, page 267 in Dulka, J. G. (1993). Sex pheromone systems in goldfish: comparisons to vomeronasal systems in tetrapods. *Brain Behavior and Evolution*, **42**, 265–280.

Fig. 2.2 Fig. 1, page 262 in Mason, R. T. (1993). Chemical ecology of the red-sided garter snake, *Thamnophis sirtalis parietalis*. *Brain Behavior and Evolution*, **41**, 261–268.

Kluwer Academic Publishers

Reprinted with kind permission from Kluwer Academic Publishers:

Fig. 1.3 Fig. 1, page 26 in Sorensen, P. W. & Stacey, N. E. (1999). Evolution and specialization of fish hormonal pheromones. In *Advances in chemical signals in vertebrates*, ed. R. E. Johnston, D. Müller-Schwarze & P. W. Sorensen, pp. 15–48. New York: Kluwer Academic Publishers/Plenum Press.

Fig. 1.10 Fig. 3, page 339 in Sachs, B. D. (1999). Airborne aphrodisiac odor from estrous rats: implication for pheromonal classification. In *Advances in chemical signals in vertebrates*, ed. R. E. Johnston, D. Müller-Schwarze & P. W. Sorensen, pp. 333–342. New York: Kluwer Academic Publishers/Plenum Press.

Fig. 2.1 Fig. 12.2, page 347 in Birch, M. C. (1984). Aggregation in bark beetles. In *Chemical ecology of insects*, ed. W. J. Bell & R. T. Cardé, pp. 331–354. London: Chapman and Hall.

Fig. 2.4 (bottom) Fig. 2, page 429 in Brown, R. E., Roser, B. & Singh, P. B. (1989). Class I and class II regions of the major histocompatibility complex both contribute to individual odors in congenic inbred strains of rats. *Behavior Genetics*, **19**, 659–674.

Fig. 2.6 (bottom) Fig. 2, page 491 in Monnin, T., Malosse, C. & Peeters, C. (1998). Solid-phase microextraction and cuticular hydrocarbon differences related to reproductive activity in queenless ant *Dinoponera quadriceps*. *Journal of Chemical Ecology*, **24**, 473–490.

Fig. 3.14 Fig. 2, page 203 in Löfstedt, C. (1990). Population variation and genetic control of pheromone communication systems in moths. *Entomologia Experimentalis et Applicata* **54**, 199–218.

Box 6.3 Fig. Fig. 11.7, page 312 in Gullan, P. J. & Cranston, P. S. (1994). *The insects: an outline of entomology*. London: Chapman and Hall.

Fig. 7.5 Figs. 1 and 2, page 720, Fig. 3, page 722 and Fig. 4, page 723 in Deneubourg, J. L., Goss, S., Franks, N. & Pasteels, J. M. (1989). The blind leading the blind–modeling chemically mediated army ant raid patterns. *Journal of Insect Behavior*, **2**, 719–725.

Box 7.1 Fig. Fig. 1, page 189 in Fitzgerald, T. D. & Gallagher, E. M. (1976). A chemical trail factor from the silk of the eastern tent caterpillar *Malacosoma americanum* (Lepidoptera: Lasiocamidae). *Journal of Chemical Ecology*, **2**, 564–574.

Box 7.3 Fig. (top) Fig. 3, page 35 in Fitzgerald, T. D. & Edgerly, J. S. (1982). Site of secretion of the trail marker of the eastern tent caterpillar. *Journal of Chemical Ecology*, **8**, 31–39.

Fig. 8.1 Fig. 7, page 162 in Stoddart, D. M. (1980). *The ecology of vertebrate olfaction*. London: Chapman and Hall.

Fig. 10.4 Fig. 1, page 1194 in Ford, N. B. & Low, J. R. (1984). Sex pheromone source location by garter snakes: a mechanism for detection of direction in non-volatile trails. *Journal of Chemical Ecology*, **10**, 1193–1199.

Fig. 10.14 Fig. 4, page 71 in Vickers, N. J. (1999). The effects of chemical and physical features of pheromone plumes upon the behavioral responses of moths. In *Advances in chemical signals in vertebrates*, ed. R. E. Johnston, D. Müller-Schwarze & P. W. Sorensen, pp. 63–76. New York: Kluwer Academic Publishers/Plenum Press.

Fig. 11.1 Fig. 12.4, page 348 in Birch, M. C. (1984). Aggregation in bark beetles. In *Chemical ecology of insects*, ed. W. J. Bell & R. T. Cardé, pp. 331–354. London: Chapman and Hall.

Fig. 12.4 Fig. 35.3, page 409 and Fig. 35.4, page 410 in Trumble, J. T. (1997). Integrating pheromones into vegetable crop production. In *Insect pheromone research: new directions*, ed. R. T. Cardé & A. K. Minks, pp. 397–410. New York: Chapman and Hall.

Fig. 12.5 Fig. 37.4, page 437 in Borden, J. H. (1997). Disruption of semiochemical-mediated aggregation in bark beetles. In *Insect pheromone research: new directions*, ed. R. T. Cardé & A. K. Minks, pp. 421–438. New York: Chapman and Hall.

Box 13.2 Table 2, page 325 in Preti, G. & Wysocki, C. J. (1999). Human pheromones: releasers or primers, fact or myth. In *Advances in chemical signals in vertebrates*, ed. R. E. Johnston, D. Müller-Schwarze & P. W. Sorensen, pp. 315–332. New York: Kluwer Academic Publishers/Plenum Press.

Table 13.3 Table 1, page 71 in Labows, J. N. & Preti, G. (1992). Human semiochemicals. In *Fragrance: the psychology and biology of perfume*, ed. S. Van Toller & G. H. Dodd, pp. 69–90. London: Elsevier Science.

Table 13.4 Table 5, page 82 in Labows, J. N. & Preti, G. (1992). Human semiochemicals. In *Fragrance: the psychology and biology of perfume*, ed. S. Van Toller & G. H. Dodd, pp. 69–90. London: Elsevier Science.

Fig. 13.5 Fig. 1, page 1476 in Zeng, X. N., Leyden, J. J., Lawley, H. J., Sawano, K., Nohara, I. & Preti, G. (1991). Analysis of characteristic odors from human male axillae. *Journal of Chemical Ecology*, **17**, 1469–1492.

Table A1 Table 5B, page 179 in Howse, P. E., Stevens, I. D. R. & Jones, O. T. (1998). *Insect pheromones and their use in pest management*. London: Chapman and Hall.

Appendix Fig. 5.1.47, page 45 in Howse, P. E., Stevens, I. D. R. & Jones, O. T. (1998). *Insect pheromones and their use in pest management*. London: Chapman and Hall.

Marine Biological Laboratory, Woods Hole

Fig. 10.10 Fig. 2, page 130 in Atema, J. (1996). Eddy chemotaxis and odor landscapes-exploration of nature with animal sensors. *Biological Bulletin*, **191**, 129–138.

MIT Press

Fig. 10.12 Fig. 2.2, page 28 in Stein, B. E. & Meredith, M. A. (1993). *The merging of the senses*. Cambridge, Mass.: MIT Press. © 1993 MIT Press.

National Academy of Sciences, USA

Fig. 3.4 Figs. (I) and (II), page 50 and Fig. 1F, page 51 in Eisner, T. & Meinwald, J. (1995). Defense-mechanisms of arthropods. 129. The chemistry of sexual selection. *Proceedings of the National Academy of Sciences of the United States of America*, **92**, 50–55. Copyright 1995 National Academy of Sciences, U.S.A.

Fig. 3.5 Fig. 2, page 9225 in Dussourd, D. E., Harvis, C. A., Meinwald, J. & Eisner, T. (1991). Pheromonal advertisement of a nuptial gift by a male moth (*Utetheisa ornatrix*).

Proceedings of the National Academy of Sciences of the United States of America, **88**, 9224–9227. Copyright the authors, by permission.

National Research Council of Canada

Fig. 3.10 Fig. 2, page 2285 in Atema, J. (1986). Review of sexual selection and chemical communication in the lobster, *Homarus americanus. Canadian Journal of Fisheries and Aquatic Sciences*, **43**, 2283–2390.

Nature Publishing Group

Fig. 3.2 (right) Fig. 1, page 25 in Moore, P. J., Reagan-Wallin, N. L., Haynes, K. F. & Moore, A. J. (1997). Odour conveys status on cockroaches. *Nature*, **389**, 25. © Nature.

Fig. 3.9 Fig. 1, page 596 in Deutsch, J. C. & Nefdt, R. J. C. (1992). Olfactory cues influence female choice in 2 lek-breeding antelopes. *Nature*, **356**, 596–598. © Nature.

Fig. 8.5 (bottom) Fig. 2, page 334 in Ono, M., Igarashi, T., Ohno, E. & Sasaki, M. (1995). Unusual thermal defence by a honeybee against mass attack by hornets. *Nature*, **377**, 334–336. © Nature.

Table 9.1 Table 1, page 795 in Leinders-Zufall, T., Lane, A. P., Puche, A. C., Ma, W. D., Novotny, M. V., Shipley, M. T. & Zufall, F. (2000). Ultrasensitive pheromone detection by mammalian vomeronasal neurons. *Nature*, **405**, 792–796. © Nature.

Fig. 9.3 Fig. 1, page 275 in Lancet, D. (1991). The strong scent of success. *Nature*, **351**, 275–276. © Nature.

Box 9.4 Fig. Fig. 4, page 285 in Sobel, N., Prabhakaran, V., Desmond, J. E., Glover, G. H., Goode, R. L., Sullivan, E. V. & Gabrieli, J. D. E. (1998). Sniffing and smelling: separate subsystems in the human olfactory cortex. *Nature*, **392**, 282–286. © Nature.

Fig. 9.5 Fig. 1, page 287 in Pilpel, Y. & Lancet, D. (1999). Good reception in fruitfly antennae. *Nature*, **398**, 285–287. © Nature.

Fig. 10.1 Fig. 2, page 230 in Bradshaw, J. W. S., Baker, R. & Howse, P. E. (1975). Multicomponent alarm pheromones of the weaver ant. *Nature*, **258**, 230–231. © Nature.

Fig. 11.6 Fig. 1, page 421 in Schiestl, F. P., Ayasse, M., Paulus, H. F., Löfstedt, C., Hansson, B. S., Ibarra, F. & Francke, W. (1999). Orchid pollination by sexual swindle. *Nature*, **399**, 421–422. © Nature.

Fig. 13.2 Fig. 1, page 177 in Stern, K. & McClintock, M. K. (1998). Regulation of ovulation by human pheromones. *Nature*, **392**, 177–179. © Nature.

New York Academy of Sciences

Box 9.5 Fig. Fig. 2, page 364 in Wood, R. I. (1998). Integration of chemosensory and hormonal input in the male Syrian hamster brain. *Annals of the New York Academy of Sciences*, **855**, 362–372.

Ohio State University

Fig. 8.2 Fig. 2, page 17 in Nault, L. R. (1973). Alarm pheromones help aphids escape predators. *Ohio Report*, **58**, 16–17.

Oxford University Press

Fig. 1.4 Fig. 1.1, page 19 in Flood, P. (1985). Sources of significant smells: the skin and other organs. Reprinted from *Social Odours in Mammals*, Volume 1 and 2, edited by Richard E. Brown & David W. Macdonald (1985), by permission of Oxford University Press.

Fig. 3.11 Fig. 3.7c, page 83 in Russell, E. M. (1985). The metatherians: order Marsupialia. Reprinted from *Social Odours in Mammals*, Volume 1 and 2, edited by Richard E. Brown & David W. Macdonald (1985), by permission of Oxford University Press.

Fig. 3.13 Fig. 3, page 432 in Linn, C. E. & Roelofs, W. L. (1989). Response specificity of male moths to multicomponent pheromones. *Chemical Senses*, **14**, 421–437.

Fig. 5.6 Fig. 11.1, page 488 in Macdonald, D. W. (1985). The rodents IV: suborder Hystrico-morpha. Reprinted from *Social Odours in Mammals*, Volume 1 and 2, edited by Richard E. Brown & David W. Macdonald (1985), by permission of Oxford University Press.

Fig. 6.1 (original) Fig. 1, page 104 in Sherman, P. W., Lacey, E. A., Reeve, H. K. & Keller, L. (1995). The eusociality continuum. *Behavioral Ecology*, **6**, 102–108.

Fig. 8.6 (middle, bottom) Fig. 1, page 261 and Fig. 2, page 262 in Mathis, A. & Smith, R. J. F. (1993). Chemical alarm signals increase the survival-time of fathead minnows (*Pimephales promelas*) during encounters with northern pike (*Esox lucius*). *Behavioral Ecology*, **4**, 260–265.

Table 13.1 Table 7.4, page 184. © Alan F. Dixson 1998. Reprinted from *Primate Sexuality. Comparative Studies of the Prosimians, Monkeys, Apes, and Human Beings*, by Alan. F. Dixson (1998), by permission of Oxford University Press.

Fig. 13.4 Fig. 1, page 35 in Gower, D. B. (1990). Quantification of odorous 16-androstene steroids in vertebrates. In *Chemical signals in vertebrates 5*, ed. D. W. Macdonald, D. Müller-Schwarze & S. Natynczuk, pp. 34–47. Oxford: Oxford University Press.

Fig. 13.6 Fig. 3, page 40 in Doty, R. L., Hall, J. W., Flickinger, G. L. & Sondheimer, S. J. (1982). Cyclical changes in olfactory and auditory sensitivity during the menstrual cycle: no attenuation by oral contraceptive medication. © IRL Press 1982. Reprinted from *Olfaction and Endocrine Regulation*, edited by W. Breipohl (1982) by permission of Oxford University Press.

Princeton University Press

Fig. 4.1 Fig. 11.8, page 247 in Vogel, Steven. (1994). *Life in moving fluids: the physical biology of flow*. Copyright © 1981 by Willard Grant Press. First Princeton Pb Print 1983. © 1984, 2nd edn, by P.U.P. Reprinted by permission of Princeton University Press.

R. H. Wright Lectures, Simon Fraser University

Box 9.2 Fig. (c) Fig. 3, in Kaissling, K.-E. (1987). R. H. Wright Lectures on Insect Olfaction.

Royal Society of London

Fig. 2.6 (top, middle) Fig. 1, page 1324 and Fig. 3, page 1325 in Peeters, C., Monnin, T. & Malosse, C. (1999). Cuticular hydrocarbons correlated with reproductive status in a queenless ant. *Proceedings of the Royal Society of London Series B-Biological Sciences*, **266**, 1323–1327.

Fig. 3.6 (b) Fig. 1, page 246 in Wedekind, C., Seebeck, T., Bettens, F. & Paepke, A. J. (1995). MHC-dependent mate preferences in humans. *Proceedings of the Royal Society of London Series B-Biological Sciences*, **260**, 245–249.

Fig. 3.7 Fig. 2, page 1302 in Penn, D. & Potts, W. (1998). MHC-disassortative mating preferences reversed by cross-fostering. *Proceedings of the Royal Society of London Series B-Biological Sciences*, **265**, 1299–1306.

Fig. 7.2 (left) Fig. 1, page 1527 in Kaib, M., Husseneder, C., Epplen, C., Epplen, J. T. & Brandl, R. (1996). Kin-biased foraging in a termite. *Proceedings of the Royal Society of London Series B-Biological Sciences*, **263**, 1527–1532.

Fig. 7.6 Fig. 1, page 1562 in Bonabeau, E., Theraulaz, G., Deneubourg, J. L., Franks, N. R., Rafelsberger, O., Joly, J. L. & Blanco, S. (1998). A model for the emergence of pillars, walls and royal chambers in termite nests. *Philosophical Transactions of the Royal Society of London Series B-Biological Sciences,* **353**, 1561–1576.

Fig. 7.7 Fig. 3a, page 1565 in Bonabeau, E., Theraulaz, G., Deneubourg, J. L., Franks, N. R., Rafelsberger, O., Joly, J. L. & Blanco, S. (1998). A model for the emergence of pillars, walls and royal chambers in termite nests. *Philosophical Transactions of the Royal Society of London Series B-Biological Sciences,* **353**, 1561–1576.

Box 10.3 Fig. Fig. 6, page 718 in Tsuda, A. & Miller, C. B. (1998). Mate-finding behaviour in *Calanus marshallae* Frost. *Philosophical Transactions of the Royal Society of London Series B-Biological Sciences,* **353**, 713–720.

Scientific American

Fig. 7.6 page 141 in Lüscher, M. (1961). Air-conditioned termite nests. *Scientific American* **205**, 138–145.

Fig. 9.9 page 105 in Carter, C. S. & Getz, L. L. (1993). Monogamy and the prairie vole. *Scientific American,* **268**, 100–106.

Sinauer Associates

Table 1.2 Table 3, page 246 in Alcock, J. (1989). *Animal behaviour. An evolutionary approach,* 4th edn.

Table 3.1 Table 4, page 469 in Alcock, J. (1998). *Animal behaviour. An evolutionary approach,* 6th edn.

Fig. 5.2 Fig. 18.9, page 600 in Bradbury, J. W. & Vehrencamp, S. L. (1998). *Principles of animal communication.*

Fig. 8.6 (top) Fig. 9.38, page 331 in Alcock, J. (1998). *Animal behaviour. An evolutionary approach,* 4th edn.

SIR Publishing (The Royal Society of New Zealand)

Fig. 12.3 Fig. 2, page 93, in Suckling, D. M., Shaw, P. W., Khoo, J. G. I. & Cruickshank, V. (1990). Resistance management of light-brown apple moth, *Epiphyas postvittana* (Lepidoptera, Tortricidae) by mating disruption. *New Zealand Journal of Crop and Horticultural Science,* **18**, 89–98.

Society for Reproduction and Fertility

Fig. 1.9 Fig. 1, page 2 in Kikuyama, S. & Toyoda, F. (1999). Sodefrin: a novel sex pheromone in a newt. *Reviews of Reproduction,* **4**, 1–4.

Springer-Verlag

Box 5.1 Figs. Fig. 12, page 37 and Fig. 18, page 54 in Hölldobler, B. & Wilson, E. O. (1978). The multiple recruitment systems of the African weaver ant *Oecophylla longinoda* (Latreille) (Hymenoptera: Formicidae). *Behavioral Ecology and Sociobiology,* **3**, 19–60. Copyright Springer-Verlag.

Fig. 5.4 Fig. 1, page 416 in Gosling, L. M. & McKay, H. V. (1990). Competitor assessment by scent matching – an experimental test. *Behavioral Ecology and Sociobiology,* **26**, 415–420. Copyright Springer-Verlag.

Table 7.2 Table 11, page 57 in Hölldobler, B. & Wilson, E. O. (1978). The multiple recruitment systems of the African weaver ant *Oecophylla longinoda* (Latreille) (Hymenoptera: Formicidae). *Behavioral Ecology and Sociobiology,* **3**, 19–60. Copyright Springer-Verlag.

Box 7.3 Fig. (middle) Fig. 7c, page 29 in Hölldobler, B. & Wilson, E. O. (1978). The multiple recruitment systems of the African weaver ant *Oecophylla longinoda* (Latreille) (Hymenoptera: Formicidae). *Behavioral Ecology and Sociobiology*, **3**, 19–60. Copyright Springer-Verlag.

Fig. 9.2(c) (left) Fig. 5b, page 159 in Steinbrecht, R. A. (1999). Olfactory receptors. In *Atlas of arthropod sensory receptors. Dynamic morphology in relation to function*, ed. E. Eguchi, Y. Tominaga & H. Ogawa, pp. 155–176. Tokyo: Springer Verlag. Copyright Springer-Verlag.

Fig. 9.2(c) (right) Fig. 6a, page 130 in Theisen, B., Zeiske, E., Silver, W. L., Marui, T. & Caprio, J. (1991). Morphological and physiological studies on the olfactory organ of the striped eel catfish, *Plotosus lineatus*. *Marine Biology*, **110**, 127–135. Copyright Springer-Verlag.

Fig. 10.3 Fig. 14, page 126 in Hangartner, W. (1967). Spezifität und Inaktivierung des Spurpheromons von *Lasius fuliginosus* Latr. und Orientierung der Arbeiterinnen in Duftfeld. *Zeitschrift für Vergleichende Physiologie*, **57**, 103–136. Copyright Springer-Verlag.

Fig. 10.13(c) Fig. 1, page 430 in Willis, M. A. & Arbas, E. A. (1991). Odor-modulated upwind flight of the sphinx moth, *Manduca sexta* L. *Journal of Comparative Physiology A-Sensory Neural and Behavioral Physiology*, **169**, 427–440. Copyright Springer-Verlag.

Fig. 11.2(a) Fig. 2, page 567 in Raffa, K. F. & Klepzig, K. D. (1989). Chiral escape of bark beetles from predators responding to a bark beetle pheromone. *Oecologia*, **80**, 566–569. Copyright Springer-Verlag.

Fig. 11.2(b) Fig. 2, page 19 in Raffa, K. F. & Dahlsten, D. L. (1995). Differential responses among natural enemies and prey to bark beetle pheromones. *Oecologia*, **102**, 17–23. Copyright Springer-Verlag.

Sveriges Entomologiska

Box 2.1 Fig. Fig. 2, page 127 in Löfstedt, C. (1986). Sexualferomoner och reproduktiv isolering hos nattfjärilar. *Entomologisk Tidskrift*, **107**, 125–137.

Université Laval

Fig. 3.3 (right) Fig. 3, page 412 in Rasmussen, L. E. L. (1998). Chemical communication: an integral part of functional Asian elephant (*Elephas maximus*) society. *Ecoscience*, **5**, 410–426.

University of Chicago Press

Fig. 1.2 (left) Fig. 1, page S126 in Endler, J. A. (1992). Signals, signal conditions, and the direction of evolution. *American Naturalist Special*, **139**, 125–153. © University of Chicago Press.

Fig. 1.8 Fig. 2, page S69 in Alberts, A. C. (1992). Constraints on the design of chemical communication-systems in terrestrial vertebrates. *American Naturalist*, **139**, 62–89. © University of Chicago Press.

Fig. 8.7 Fig. 1, page 655 in Chivers, D. P., Brown, G. E. & Smith, R. J. F. (1996). The evolution of chemical alarm signals – attracting predators benefits alarm signal senders. *American Naturalist*, **148**, 649–659. © University of Chicago Press.

Table 11.2 Table 4, page 429 in Zuk, M. & Kolluru, G. R. (1998). Exploitation of sexual signals by predators and parasitoids. *Quarterly Review of Biology*, **73**, 415–438. © University of Chicago Press.

University of Kansas Press

Fig. 7.5 (original) Figs. 5 and 6, page 325 in Rettenmeyer, C. W. (1963). Behavioral studies of army ants. *University of Kansas Science Bulletin*, **44**, 281–485.

University of Washington

Fig. 10.13d Fig. 2, page 270 in Johnsen, P. B. (1981). A behavioral control model for homestream selection in migratory salmonids. In *Proceedings of the salmon and trout migratory behavior symposium*, ed. E. L. Brannon & E. O. Salo, pp. 266–273. University of Washington, School of Fisheries.

Urban & Fischer Verlag

Fig. 11.10 (original) Fig. 3, page 583 in Hölldobler, B. (1970). Orientierungsmechanismen des Ameisengastes *Atemeles* (Coleoptera, Staphylinidae) bei der Wirtssuche. In *Verhandlungen Der Zoologischen Gesellschaft (Würzburg, 1969) (Zoologischer Anzeiger Supplement, Vol. 33)*, ed. W. Herre, pp. 580–585. Leipzig: Akademische Verlagsgesellschaft, Geest & Portig K. G.

US Department of Agriculture (USDA)

Box 6.2 Fig. (bottom) from USDA Carl Hayden Bee Research Center web page http://gears.tucson.ars.ag.gov/beebook/queen/info.html downloaded 13 April 2002.

Westview Press

By kind permission of Westview Press, a member of Perseus Books Group:

Fig. 1.6 Fig. 1.1, page 4 in Billen, J. & Morgan, E. D. (1998). Pheromone communication in social insects: sources and secretions. In *Pheromone communication in social insects: ants, wasps, bees, and termites*, ed. R. K. Vander Meer, M. D. Breed, K. E. Espelie & M. L. Winston, pp. 3–33.

Fig. 7.1 Fig. 11.3, page 273 and Fig. 11.4, page 275 in Schmidt, J. O. (1998). Mass action in honey bees: alarm, swarming and role of releaser pheromones. In *Pheromone communication in social insects: ants, wasps, bees, and termites*, ed. R. K. Vander Meer, M. D. Breed, K. E. Espelie & M. L. Winston, pp. 257–292.

Fig. 8.4 Fig. 11.1, page 259 in Schmidt, J. O. (1998). Mass action in honey bees: alarm, swarming and role of releaser pheromones. In *Pheromone communication in social insects: ants, wasps, bees, and termites*, ed. R. K. Vander Meer, M. D. Breed, K. E. Espelie & M. L. Winston, pp. 257–292.

Fig. 9.12 Fig. 12.7, page 309 in Vargo, E. L. (1998). Primer pheromones in ants. In *Pheromone communication in social insects: ants, wasps, bees, and termites*, ed. R. K. Vander Meer, M. D. Breed, K. E. Espelie & M. L. Winston, pp. 293–313.

World Scientific Publishing

Fig. 9.2(b) (left) Fig. 3, page 95 in Kaissling, K. E. (1998). Olfactory transduction in moths: I. Generation of receptor potentials and nerve impulses. In *From structure to information in sensory systems*, ed. C. Taddei-Ferretti & C. Musio, pp. 93–112.

Index

Italic numbers refer to figures, boxes and tables

Organisms are nearly all indexed under taxonomic names in main entries; where common names are used a cross-reference is provided to the taxonomic name (except for groups). Taxonomic and/or common names are used in subentries.

Aardwolf, see *Proteles cristatus*
Absolute configuration, 306–307
Acanthomyops (ant), *5*
Accessory olfactory bulb (AOB), *165, 183, 184,* 201
(5R,6S)-6-Acetoxy-5-hexadecanolide, 75
Acomys cahirinus (spiny mouse)
  kin recognition cues, 110
Acroneuria carolinensis (stonefly), 236
Across-fibre patterning, 171
Actias luna (silk moth)
  antenna as sieve or paddle, 167
Active space, 209
  diffusion in air and water, 210–212
  multi-component pheromones, 212, *214*
  trail, *215*
Acyrthosiphon pisum (pea aphid)
  suicide hypothesis, 152
Adrenal glands
  female mice, Lee–Boot effect, 195
Aeolidia papillosa (nudibranch sea-slug)
  predator marked by prey pheromone, 148
Aeshena umbrosa (dragonfly nymph)
  predator, 157
African mole rats, see *Heterocephalus glaber* (naked mole rat)
Aggregation pheromones, 74
  aposematic insects, 75
  commercial exploitation
    managing bark beetles, *266*
    stopping marine fouling, *266*
  dilution of risk, 75
  ecophysiological benefits, 79
    assembly pheromone and ticks, 80
    overwintering garter snakes, 80
  intraspecific eavesdropping, 236
  males only signal until females come, 82
  Schistocerca gregaria (desert locust), *76*
  in space

oviposition pheromones, 75, *76,* 77
Panulirus interruptus (lobster), 75
  settlement of marine invertebrates, *26,* 77–79
  in time (synchronisation)
    coordinating external fertilisation, 9, *91,* 41–43
    larval release, 77
Agrotis ipsilon (black cutworm moth)
  male response activated by juvenile hormone, 189
Alarm pheromones
  commercial exploitation
    for aphid control, 265
    hatchery trout taught to recognise predator, 254
  costs to responding
    aphid drop, 152
  costs to sender
    eavesdropping by predators, 229, 234
  evolution from pre-existing chemical cues, 9
  evolution in related individuals, 147–157
    family groups, 147, *148,* 149
    kin selection, 148–157
  evolution in unrelated individuals, 157–162
    Diadema antillarum (sea urchin), 158
    fish, 158–162
  predator labelling with host pheromone
    Aeolidia papillosa (sea-slug), 148
    predatory fish, 161
  propaganda by
    plants to repel aphids, 244, *245*
    prey to escape ants, *244*
    robber bees, 243
    slave-making ants, 241, *244*
Alcelaphus buselaphus cokei (Coke's hartebeest)

  self marking and presentation, *92*
Aleochara curtula (rove beetle), 61
Alkylmethoxypyrazine, 75
Allee effects, 75
  feeding, 79
    Dendraster excentricus (sand dollar), *80*
  mating, 79
    external fertilisation, 79
    internal fertilisation, 79
  overcoming tree defences
    bark beetles, 81, *81*
  for wildlife conservation, 254
Allelochemicals, 2
Allomarking, 111
Allomones, 2
  deceit, propaganda, 230
Alloparental care, 123
  egg dumping, 230
Alpheus heterochaelis (snapping shrimp)
  modulation of visual threat signals by pheromones, 18
Alternative mating strategies
  satellite males
    eavesdropping parasitoid pressure, 233
    Nauphoeta cinerea (African cockroach), 60
  she-males
    Aleochara curtula (rove beetle), 61
    Thamnophis sirtalis parietalis (garter snake), 60
Alzheimer's disease, diagnosis, 297
Amblyomma hebraeum (tick)
  pest management, *262*
Amniotic fluid, *199*
Amphibians
  Bufo borealis (western toad)
    tadpoles, alarm pheromone, escape from predator, 157
  Litoria splendida (magnificent tree frog)
    peptide sex pheromone, 21
    tadpoles, kin, and alarm

Amphibians (*cont.*)
  pheromones, 157
  see also Salamanders and newts
Amphid sensilla, 203
Amphiprion (anemone fish), 238, *239*
Amygdala
  interaction between main olfactory
    and vomeronasal organ (VNO)
    outputs, hamster, 185
Anal gland
  Castor canadensis (beaver)
    variation between families, 96
  Meles meles (European badger), 111
  Mustela erminea (stoat)
    repel prey, 265
Anas platyrhynchos (mallard duck), 21
Andrena nigroaenea (solitary bee)
  duped by orchid, 240, *242*
3α-Androstenol, 253, 288, *289*
  steroid sex pheromones in pig, 20
  in truffles, use of sows to find, 253
5α-Androstenone, 191, 288, *289*
  Boar Mate™, 252, 253, 254
Anguilla rostrata (American eel)
  zigzag behaviour, 227
Animal welfare
  understanding olfactory world, 254
Anomala osakana (Osaka beetle), 17
Anonymous signals
  social insects, 102
Anosmias
  induced changes on exposure, 191
  specific, 291
Antagonists, 66, 68, *187*, 227, 260
Antelopes
  Kobus kob (Uganda kob)
    lekking, *58*
  Kobus leche (Kafue lechwe)
    lekking, *58*
  Litocranius walleri (gerenuk), *89*
  Oreotragus oreotragus
    (klipspringer antelope)
    scent marks eavedropped by
    ticks, 233
  Ourebia ourebi (oribi), 87
    border maintenance by
      multimale groups, 96
    territories, 96
  Raphicerus melanotis (grysbok), 13
Antennal lobe, *165*
  macroglomerular complex
    (MGC), *165*
  main, *165*

Anthonomus grandis (boll weevil), 20
  pest management, *263*
Anthopleura elegantissima (sea anemone)
  alarm pheromone (anthopleurine),
    148, *149*
Anthopleurine, 148
  soluble polar pheromone, 15
Anti-aggregation pheromones, 235
Antilocapra americana (pronghorn
  antelope)
  alert signals, *148*
Antorbital glands
  territorial and self marking
    Alcelaphus buselaphus cokei
      (Coke's hartebeest), *92*
  territorial marking by antelope,
    89, 233
Aphaenogaster, see *Novomessor*
Aphrodisiac pheromones, *see* Male
  pheromones
Aphrodisin, *182*
Apis cerana japonica (Japanese
  honeybee), 154–155, *156–157*
Apis mellifera, see Honeybees
Aplysia (sea-slug)
  peptide sex pheromone, 15
Apocephalus paraponerae (phorid fly)
  eavesdrops host pheromone, 233
Apolipoprotein D (apoD), 290
Aposematism
  insects, aggregation, 75
  toad tadpoles, 157
Applications of pheromones
  animal husbandry
    primer pheromones, 252
    signalling pheromones, 253
  aquaculture
    hatchery trout taught to
      recognise predator odour, 254
    induce sperm production in
      carp, 254
  beneficial insects
    honeybees, 255
  captive breeding rare species, 254
    kin recognition, mate choice, 254
  commercialisation, 267–268
    economics, 268
    public policy, 268
  pest management
    alarm pheromones, aphids, 265
    deterrent odours, 264–265
    lure and kill or mass trap,
      260, 261–264

  marine fouling organisms, 266
  mating disruption, 257–260,
    261, 267
  monitoring, 256, *256*, 257
  pest resistance to pheromones, 267
  primer pheromones, 265–266
  push-pull or stimulo-deterrent
    diversionary strategies,
    265, *266*
  self-protecting plants, 266
pets
  cats, spray marking, 254
  minipigs, 254
Tuber melanosporum (truffles),
  finding, 253
Aquatic pheromones
  alarm pheromones
    Diadema antillarum (sea
      urchin), 158
    fish, 158
    toad tadpoles, 157
  coordination of external fertilisation
    Carassius auratus (goldfish), 9, *19*
    Nereis succinea (polychaete worm), *42*
  Dendraster excentricus (sand dollar)
    larval settlement, *80*
  Diadema antillarum (sea urchin)
    alarm responses, 158
  eicosanoids (PUFAs)
    barnacle hatching pheromone, 77
  Homarus americanus (lobster)
    precopulatory mate guarding,
    61, *62*
  peptides
    Aplysia (sea-slug), 15
    Nereis succinea (polychaete
      worm), 42
    sodefrin (newt), 15, *16*
  polar pheromone
    anthopleurine, sea anemone,
    15, 148, *149*
  prostaglandins
    Carassius auratus (goldfish), *19*,
    27–28
  Temora longicornis (copepod), *213*
  see also Pheromones, aquatic habitats;
    Signals, design, aquatic habitats
Arachnids
  mites
    Varroa, 116, 262
  phalangids (harvestmen)
    use ant alarm pheromone,
    defence, 244

spiders
  bolas spiders, 240, *243*
  Habronestes bradleyi, attracted to
    fighting ants, 229
  Linyphia litogiosa, 59
ticks
  guanine, assembly pheromone, 80
  Ixodes neitzi, 233
Argyortaenia velutinana (moth)
  male pheromone response wider
    than range females produce, 69
Argyropelecus hemigymnus
  (deep-sea hatchet fish), 207
Arms race
  bark beetles and their
    predators, 234
Arthropodin, *see* Settlement-inducing
  protein complex (SIPC)
Artificial insemination (AI), 253
Assembly pheromone
  ticks, 80
Assortative mating, 65
Asymmetric tracking, 68
Atemeles pubicollis (beetle social para-
  site of *Myrmica* ants), 246
Atta (leaf-cutter ants), 114, 263
  Atta texana
    sensitivity of workers to trail
     pheromone, *136*

'Badges' of status
  Nauphoeta cinerea (African
    cockroach), 44
Bark beetles
  push-pull or stimulo-deterrent
    diversionary strategies, *266*
  see also Aggregation pheromones;
    Allee effects; Arms race;
    Dendroctonus; Exploitation of
    pheromones by organisms; *Ips*;
    Mice; Pheromones
Barnacles
  disrupting settlement, 266
  egg hatching pheromone, 77
  larval settlement, *26*, *78*
Beetle aggregation pheromones, 81
Behavioural manipulation by semen
  components
  Drosophila melanogaster, 63
  humans? 64
  Thamnophis sirtalis parietalis
    (garter snake), 63
Benzaldehyde, 5, 243

Bioassays, *see* Pheromone identification, bioassays
Birds
  Anas platyrhynchos (mallard duck)
    possible sex pheromone, 21
  eavesdropping vole scent marks by
    ultraviolet cues
  Buteo lagopus (rough-legged
    buzzard), 232
  Falco tinnunculus (kestrel), 232
  glands
    anal, *21*
    sebaceous, *21*
    uropygidial, *21*
  nest recognition, *21*
  pheromones, *21*
  procellariiform seabirds
    olfaction, *21*
  zigzag flight upwind, *225*
cis-(Z)-α-Bisbolene epoxide, 233
Boar Mate™, 252, 253, 254
Boiga irregularis (brown tree snake)
  pest management, *262*
Bombykol
  identification of, 25
Bombus, *see* Bumblebees
Bombyx mori (silk moth), 23
  antennae, design, 166
Bos taurus (bulls and cows), 253
Boundary layer
  effects on olfactory sensors, 174
  turbulence, 219
Brain, *see* Olfaction
endo-Brevicomin, 235
exo-Brevicomin, 5
Bruce effect
  memory and pregnancy block in
    mice, 201
  pregnancy block by male
    pheromones, 195
Bufo borealis (western toad)
  tadpoles, alarm pheromone, 157
Bumblebees (*Bombus*)
  marking visited flowers, 85
Butenandt, Adolf, 23
Buteo lagopus (rough-legged
  buzzard), 232
2-*sec*-Butyl-dihydrothiazole
  (thiazole or SBT), 193, 194

C_{20} polyunsaturated fatty acids
  (PUFAs) (eicosanoids), 77
Caenorhabditis elegans (nematode), 203

Callinectes sapidus, (blue crab)
  rheotaxis, 224
Callithrix jacchus (common marmoset)
  subordinate female ovulation
    suppressed, 197
Camponotus socius (ant)
  group recruitment, *134*
Canis
  Canis latrans (coyote)
    urine repells rodents, 265
  Canis lupus (grey wolf)
    border marks in territory, 97
    suppression of subordinate
     reproduction, 124
  Canis lupus familiaris, see Dogs
  Canis sinensis (Ethiopian wolf)
    border marks in territory, 97
Carassius auratus (goldfish)
  primer and releaser
    pheromones, *19*
  sex pheromones evolved from
    hormones, 9, 27–28
Castes
  differences
    pheromone secretions, ants, *114*
    responses to pheromones, social
     aphids, 153
    social parasites copy odour badges
     of termite host castes, 248
  see also Honeybees
Castor canadensis (beaver)
  anal glands, 96
  avoid marks at low population
    densities, 91
  castoreum, 96
  kin recognition, 111
  pest management, 265
  suppression of subordinate
    reproduction, 124
Castoreum, 96
Central-place foragers
  naked mole rat 133
  social insects, 133
Ceratitis capitata (Mediterranean
    fruit fly)
  leks, 58, 59
  eavesdropped by predator, 232
Ceratovacuna lanigera (social aphid)
  soldiers respond, non-soldiers
    flee, 153
  soldiers attack predator, *153*
Cerocebus albigena (mangabey,
    primate), 271

c-*fos*, 182
Chemical camouflage, 247–248
Chemical mimicry, 248–249
Chemical structure, 302
Chemoreception, *164*
Chirality, 305
5β-Cholestane-3,24-dione
    trail pheromone, *Malacosoma*, 130
Chrysopa slossonae (lacewing larvae)
    escapes detection by chemical
        camouflage, 247
cis (in chemical name), 305
Citral, 243
Clean-up enzymes, 177
Clethrionomys glareolus (bank vole)
    females prefer odours of dominant
        male, 46
Cloacal gland, *16*
Coccinella septempunctata (ladybird
    beetle)
    aposomatic pheromones, 75
Colony defence
    coordinated response to
        danger, 155
Colony-level selection
    honeybees, 107
Colophina monstrifica (social
    aphid), *153*
Combinatorial coding
    glomeruli, 170
Communal nesting
    enabled by synchronous
        breeding, 196
    mouse preference for kin, *54*, 55
Communication, 2, 3
    privacy in, 16
    sex communication channel, 66
    see also Composite signals;
        Pheromones; Signals
Composite (combined) signals
    chirp and odours, naked mole rat
        foragers, 136
    parallel sensory channels, 17
    pheromone and stridulation
        recruitment
        Aphaenogaster (*Novomessor*)
            (ant), 140
Conflict between the sexes
    Linyphia litogiosa (spider), 59
    Nicrophorus (burying beetle), 59
    semen components
        Drosophila melanogaster, 63
        humans, 64

Thamnophis sirtalis parietalis
    (garter snake), 63–64
Conscious and unconscious responses
    to odours, 185
    human responses, 185
        electroencephalogram (EEG), 185
        functional magnetic resonance
            imaging (fMRI), 185
Contact chemoreception
    ants detecting colony identity, 4
Contests, 44–46
Coolidge effect, 57, 202
    Gryllus bimaculatus (cricket), 57
    male rodents, 57
    Vipera berus (adder), 57
Cooperative breeders, 123–127
    plural breeders, 192
        subordinate female ovulation
            suppressed, 124, 126, 127,
            197–198
    singular breeders, 124
Cooperative signal, 122
Coptotermes (termite), *130*
Coremata, *10*, 11
    honest signal, *3*
    sexual selection, 47, *49*
Cornicle secretions
    aphids, 150
Corpora allata, 190
    termites, 119
Corynebacterium
    bacteria in human armpits, 291
Costs (of signalling)
    eavesdropping by parasites
        Ixodes neitzi (ticks), 233
        parasitoid insects, 232
    eavesdropping by predators, 231
        ant-decapitating flies, 234
        lekking *Ceratitis capitata* (Mediter-
            ranean fruit fly), 232
        spiders attracted to fighting
            ants, 229
    low metabolic cost of pheromones
        Anthonomus grandis
            (boll weevil), 20
        Litoria splendida (magnificent tree
            frog), 21
        long-distance communication,
            207
    risk
        investigating marks, beavers, 89
    time
        marking, antelope, 87

Countermarking, *see* Over-marking
Courtship
    Cerocebus albigena (mangabey,
        primate), *271*
    Creatonotus gangis (arctiid moth), *10*
    Drosophila melanogaster (fruit fly),
        18, 70
    Grapholitha molesta (oriental fruit
        moth), 7
    Homarus americanus (lobster), 61, *62*
    humans, 277
    Macaca mulatta (rhesus monkey), 281
    newts, *16*
    salamanders , pheromone delivery
        patterns, 4
    species recognition, 64, 70, 72
    Crataegus mollis (hawthorn), 85
    Creatonotus gangis (arctiid moth)
        coremata, *3*, *10*
    Cremogaster dohrni artifex (attendant
        ant of butterfly), 238
    Crocuta crocuta (hyena)
        expose scent glands on approach, 91
Cross-fostering experiments
    Acomys cahirinus (spiny mouse)
        diet cues, 110
    Odocoileus hemionus hemionus (black-
        tailed deer), 199
    honeybees, 106
    MHC preferences of mice, 55
    Spermophilus beldingi (Belding's
        ground squirrel), 110
Crotalaria
    plant source of pyrrolizidine alka-
        loids (PAs), 47
Crustaceans
    Alpheus heterochaelis (snapping
        shrimp)
        modulation of visual threat by
            pheromones, 18
    barnacles
        larval settlement, *26*
        Semibalanus balanoides (barnacle),
            77
        settlement behaviour, *78*
    copepods
        Lepeophtheirus salmonis (salmon ec-
            toparasite), control, 262
    crabs
        Callinectes sapidus (blue crab), 224
        Rhitropanopeus harrisii (mud crab),
            77
    Homarus americanus (lobster)

contest, 44
precopulatory mate guarding, 61, *62*
tropotaxis, 222
Panulirus interruptus (California spiny lobster)
aggregation for defence, 75
plankton
Temora longicornis (copepod), *213*
Culex (mosquito), 263
oviposition pheromone, 75
Cuticular hydrocarbons
ant trail pheromones, 140
colony recognion
honeybees, 106
ants, 108
deception, 247–249
Polistes fuscatus (paper wasp), 107
deception, 61
dipteran sex pheromones, 70
dipteran speciation, 71
Gryllus bimaculatus (cricket)
mate choice, 51, 57
solid phase microextraction (SPME) from living ants, *33*
Cydia pomonella (codling moth)
pest management, 258
Cynictis penicillata (yellow mongoose)
scent marking, 90
Cynops pyrrhogaster (red-bellied newt)
courtship, *16*
sodefrin, 15, *16*
Cypris larva (barnacle), 78
L-Cysteine-glutathione disulfide, 42

Danaidone, 4
Danaus gilippus (queen butterfly)
transfer pheromone direct to female, 4
Darwin, Charles
sexual selection on pheromones, 37
Dauer larva
Caenorhabditis elegans (nematode), 203
Dear-enemy phenomenon (territorial defence), 99, 112
Deer
Odocoileus hemionus columbianus (black-tailed deer)
alert signals, 28, 147–148
composite (combined) alarm signal, 17
scent marking, 91

2,3-Dehydro-*exo*-brevicomin (brevicomin or DHB), 5, 193, *194*
2,5-Demethylpyrazine, 194
Dendraster excentricus (sand dollar), 80
Dendroctonus (bark beetles), 81
Dendroctonus brevicomis
eavesdropped by predator, 234
Dendroctonus frontalis, 82
male inhibition by male pheromones, 235
Dendroctonus micans, 81, *81*
Dendroctonus montanus, 81
Dendroctonus pseudotsugae, 82
Dendroctonus valens, 81
Dermacentor variabilis (tick)
pest management, *262*
Desmognathus ochrophaeus (salamander)
male injects pheromone into female, 4
Development
age
hamster, 189
honeybee worker responses, 190
honeybee, queen mandibular pheromone (QMP), *121*, *122*
Agrotis ipsilon (black cutworm moth), 189
caste differences in social insects
pheromone production and perception, 190
learning
male hormone surge to vaginal odours (hamster), 182
maternal behaviour, 191, 200
Rhagolitis pomonella (apple maggot fly), 190–191
social insects, 107, 248
Schistocerca gregaria (desert locust), *76*
sensitive period
ewe learning odour of her lamb, 199, *200*
kin recognition, 110
newly adult ant, 248
sex differences in brain
mammals, 190
moths, 190
switch
dauer pheromone, 203
social insects, 119, 204, *204*
Diadema antillarum (sea urchin)
alarm responses, 158

Diastereoisomers, 307
Diet
effect on odour
Acomys cahirinus (spiny mouse), 110
humans, 274
quality and mate choice
Microtus pennsylvanicus (meadow vole), 46
Utetheisa ornatrix (tiger moth)
hydroxydanaidal (HD), 47
2,5-Dimethylpyrazine
female mice, Lee-Boot effect, 195
N,N-Dimethyluracil, *130*
Dinoponera quadriceps (ant)
cuticular hydrocarbons, non-destructive study by solid phase microextraction (SPME), *33–34*
Diomedea (albatross)
navigation, food location, *21*, 225
Dipsosaurus dorsalis (iguana)
composite (combined) signal with ultraviolet cue, 14
Disassortative mating
MHC genotype, 55
(*Z,Z,E*)-Dodecatrienol, *130*, 140
(*Z*)-7-Dodecen-1-yl acetate, 1
Dogs (*Canis lupus familiaris*)
ability to track odour trail, and detect its direction, 218
as detector of cow oestrus, 253
distinguishing humans by smell, 274
forensic use of smell, 296
olfactory sensory sensitivity, 167
scent marking landmarks, 99
Dolichotis patagonum (mara)
males directly mark females, *100*
Dominance
Homarus americanus (lobster), 44, 61
mice, 39, 93, 193, *194*
Microcebus murinus (lesser mouse lemur), 196
Microtus pennsylvanicus (meadow vole), 46
Nauphoeta cinerea (African cockroach), 40, 44, *45*
queen, social insects, 118
in queenless ants, *34–34*
social mammals, 124
Drakaea glyptodon (orchid) dupes wasp, *241*

Drosophila, 36, 71, 188, 209
  Drosophila adiastola subgroup, 72
  Drosophila mauritiana, 72
  Drosophila melanogaster, 63, 72
    conflict between the sexes, 63
    courtship, pheromones, and
      speciation, 18, 70, *71*
    genetic engineering, 35–36, 70
  Drosophila sechellia, 72
  Drosophila simulans, 71
Dufour's gland, *11, 130*
  ant trail, *139*
  Lasioglossum zephyrum (sweatbee), 108
  parasitoid wasps, 84, *84*
  queen bee egg mark, 122
  slave-making ants, 241, *244*

Eavesdropping, *see* Exploitation of
    pheromones by organisms
Echinoderms
  Dendraster excentricus (sand dollar)
    larval settlement, feeding
      benefit, *80*
  Diadema antillarum (sea urchin)
    alarm responses, 158
Eciton (army ants), foraging, 141–143
Egg dumping, 230
  fish, 231
  insects, 230
Egg hatching pheromone
  Semibalanus balanoides (barnacle), 77
Eicosanoids (PUFAs) (prostaglandins
    in vertebrates)
  barnacle hatching pheromone, 77
Electroantennogram (EAG), 29, *30*
Electroencephalogram (EEG)
  response to odours, 185
Electroolfactogram (EOG), 189
Electrovomerogram (EVG), *194*
Elephants
  African (*Loxodonta africana*), 47
    mate guarding, 62
  Asian (*Elephas maximus*), 1, 5, 47
    musth, testosterone elevation, *48*
    pheromone shared with moths, *2*
  musth
    female response, 47
    male aggressiveness, 47
    temporal gland secretion, 47, *48*
    testosterone elevation, 47, *48*
    urine, 47
  young males smell different from
    adults, 47

Elephas maximus, see Elephants, Asian
Enantiomers, 17, 305
Enemy specification, 155
Enoclerus lecontei
  eavesdropping, 234, *236*
Entgegen (E), opposite, 308
Environmental odour cues and
    recognition, 103, 110
Epiphyas postvittana (light brown apple
    moth)
  pest management, 257, *258*, 259
Ephestia
  Ephestia elutella (tobacco moth), 47
  Ephestia kuehniella (flour moth)
    pest management, *263*
Equus (horse), flehmen to oestrous
    urine, *179*
Esox lucius (pike), 254
  labelled with prey alarm
    pheromone, 159–161
  naïve prey learn odour, 159
  predator attraction, interference
    hypothesis, 162, *162*
Estrus, *see* Hormones; Oestrus
Estrogen, *see* Hormones; Oestrogen
4-Ethyl-2-methoxyphenol, 44
Ethyl-*trans*-cinnamate, *7*
Euglossine bees
  courtship needs plant perfume
    oils, 12
Eupoecilia ambiguella (grape moth)
  pest management, *258*
Eusociality, 116
  conflict over reproduction in
    societies, 118
  continuum (reproductive skew),
    117, *117*
  social insects and social mammals,
    parallels, 127
  see also Social insects; Social
    mammals
Evolution, *see* Pheromones,
    evolution
Experience, *see* Development;
    Learning
Experimental methods, *see*
    Pheromone identification
Exploitation of pheromones by
    organisms
  arms race, 230
  code breaking, 229
  deception
    bolas spiders, 240, *243*

cost of excluding deceivers, 249
    pollination by sexual deception,
      240, *241*, *242*
    predators, guests and parasites of
      social insects, 244–249
    propaganda, 241–244, *245*
  deception by mimic chemicals
    enabled by shared biochemistry
      in all life, 230
  eavesdropping, 229
    aggregation pheromones, 81, 234,
      235, *236*
    alarm pheromones, 229, 234
    egg dumping, 230–232
    enabled by characteristics of
      olfaction, 230
    intraspecific, sex pheromones as
      aggregation pheromones, 236
    prey responses to predator
      pheromones, 154–155, *156–157*,
      236, 243, 264–265
    sex pheromones, 231–232
    territorial marking
      pheromones, 233
  intraspecific
    hormones evolving into
      pheromones, 8
  mutualism, 237
    'communication' between leaf-
      cutter ants and fungus, 238
    ants and lycaenid butterfly
      caterpillars, 237
    aphids tended by ants, 237
    sea anemones and anemone fish,
      238–240
External fertilisation
  coordination by pheromones, 41
  Carassius auratus (goldfish), 9, *19*
  Nereis succinea (polychaete
    worm), *42*

Fabré, Jean-Henri, 23
Faeces, territorial marking
  antelopes (oribi), 96
  Meles meles (European badger), 98
  salamanders, 95
Falco tinnunculus (kestrel), 232
Farnesene, 130, 193, *194*
  produced by wild potato, 244, *245*
Feliway™ (synthetic cheek gland
    pheromone), 254
Female choice, *see* Mate choice; Sexual
    selection

Female pheromones
  effects on females
    biases sex ratio in pregnancy, 197
    delayed puberty, *194, 195*
    oestrus synchronisation, 196, 282
    social insects, *115–116, 119, 121, 123*
    subordinate female ovulation
      suppressed, 197
    suppressed oestrus cycles, *194, 195*
  effects on males
    Microcebus murinus (lesser mouse
      lemur), 196
Femoral gland
  Iguana iguana (lizard)
    testosterone influence, 46
Fish
  alarm pheromones, 158–162
  Amphiprion (anemone fish
      symbiotic with sea anemones),
      238, *239*
  Anguilla rostrata (American eel)
    zigzag behaviour, 227
  Argyropelecus hemigymnus (deep-sea
      hatchet fish)
    biggest nasal organ for body size,
      vertebrate, 207
  Carassius auratus (goldfish)
    pheromones from hormones, 9,
      19, 27–28, 43, 189
  Esox lucius (pike)
    labelled with prey alarm
      pheromone, 159–161
    naïve prey learn odour, 159
    odour to teach hatchery rainbow
      trout, 254
    predator attraction, interference
      hypothesis, 162, *162*
  Gadus morhua (cod)
    zigzag orientation in plume, *225*
  Gasterosteus aculeatus (stickleback)
    MHC and mate choice, 55
  Lepomis cyanellus (green sunfish)
    alloparental carer, 230–231
  Notropis umbratili (redfin shiner)
    interspecific egg dumper, 231
  Oncorhynchus mykiss (rainbow trout)
    taught to recognise predator
      odour, 254
  Petromyzon marinus (marine lamprey)
    control, *262*
  Phoxinus phoxinus (European
      minnow)
    alarm pheromone, 158

Plotsus lineatus (catfish)
  scanning electron microscopy of
    olfactory epithelium, *169*
Premnas (anemone fish symbiotic
    with sea anemones), 238
Primiphales promelas (fathead min-
    now), 158
Puntius (androgen effects on elec-
    troolfactogram (EOG)), 189
salmon
  zigzag orientation in plume, *225*
Flame ionisation detector (FID), *30*
Flehmen
  Bos taurus (bulls and cows), 253
  Elephas maximus (Asian elephant), 48
  Equus (horse), to oestrous urine, *179*
Flicks, antennal
  behaviour of smelling, 174
  Homarus americanus (lobster), 174
Floaters, 99
Fluctuating asymmetry, 49
  humans, 280
  Panorpa japonica (scorpion fly), 49
Formica subintegra (slave-making ant),
    241, 244
Formicoxenus (shampoo ant)
  social parasite of ants (*Myrmica*),
    247
Frequency dependant responses
  parasitoid wasp superparasitism,
    188
Fright response, 158–159
Functional groups, 302
Functional magnetic resonance
    imaging (fMRI)
  humans, *174, 175*, 185

Gadus morhua (cod)
  zigzag orientation in plume, *225*
Galago (bushbaby)
  nocturnal, use of pheromones, 13
Galaxolide, 185
Gargaphia solani (aubergine lace bug)
  intraspecific egg dumping, 230
  subsocial care, alarm pheromone,
    147
Garter snake, see *Thamnophis sirtalis
    parietalis*
Gas liquid chromatography (GC), *see*
    Pheromone identification,
    separating chemicals
Gasterosteus aculeatus (stickleback), 55
Genetics

Caenorhabditis elegans (nematode)
  using mutants, 203
Drosophila, 35–36, 70
humans, perception of odours,
    specific anosmias, 291–292
moths, pheromone signal, 69
olfactory receptor proteins (OR), 169
Geometrical isomers
  Entgegen (E), *Zusammen* (Z), 308
Geraniol
  alarm pheromone
    Gargaphia solani (aubergine lace
      bug), 147
Gestalt model
  colony odour, 108
Glands, 9, *10, 11*
  adrenal
    female mice, Lee–Boot effect, 195
  anal
    birds, *21*
    Castor canadensis (beaver), 96
    Meles meles (European badger),
      111
    Mustela erminea (stoat), 265
  antorbital
    territorial marking by antelope
      (gerenuk) 89, *233*
  cloacal, *16*
  closable, due to eavesdropping
    predators, 233
  Dufour's
    ant trail, *139*
    Lasioglossum zephyrum (sweatbee),
      108
    parasitoid wasps, 84, *84*
    queen bee egg mark, 122
    slave-making ants, 241
  femoral
    Iguana iguana (lizard), 46
  labial
    termites, 138
  mandibular
    Trigona (stingless bee), 136
  musk, 148
  Nasonov
    honeybees, *131*
  poison
    ant, 131
  postcloacal
    salamander, 95
  postpharyngeal
    ants, 108, 248
  preorbital, *see* antorbital

Glands, (cont.)
  preputial
    mice, 193
  pygidial
    ant, 134
  sebaceous
    birds, 21
    humans, 284
    signal life, 14
    some vary more within a
      species, 113
  sternal
    ant, 130
    Malacosoma americanum (tent
      caterpillar), 130
    Polistes (wasp), 107
    termite, 130, 137
  subcaudal
    Meles meles (European badger), 111
  sweat
    apocrine, 284
    eccrine, 284
  temporal
    elephants, 46
  uropygidial
    birds, 21
Glomeruli
  combinatorial coding, 170–174
  functional units of processing, 172
L-Glutamic acid, 42
L-Glutamine, 42
Gossyplure, 303
G-proteins, 166
Grandlure, 263
Grapholitha molesta (oriental fruit moth)
  courtship display of hair pencils, 7
  pest management, 258
  response to pheromone blend, 67
'Green beard' phenomenon, 104, 105
Gryllus bimaculatus (cricket)
  Coolidge effect, 57
  mate choice for optimal
    outbreeding, 51
Guanine, 80

Habronestes bradleyi (spider), attracted
    to fighting ants, 229
Hair pencils
  moth, 7
Hamilton's 'selfish-herd' theory
  fish shoals and alarm
    pheromone, 159
Hamilton's rule, kin selection, 116

Hawaiian Drosophila fruit flies, 72
  Drosophila adiastola subgroup, 72
  speciation, 72
Helicoverpa zea (corn ear worm
    moth), 187
Helogale undulata (dwarf mongoose)
  individual recognition, 112
(Z)-7-Heneicosene, 61
Heneicosane, 84
2-Heptanone, 194
Herpestes auropunctatus (Indian
    mongoose)
  individual recognition, 112
Heterocephalus glaber (naked mole
    rat), 125
  colony odour, parallels with social
    insects, 112
  queen suppression of worker
    reproduction, 124
  recruitment of foragers, 136
Heterodera glycines (soybean cyst
    nematode)
  pest management, 258, 260
E-10,Z-12-Hexadecandien-1-ol, 25
Z-9-Hexadecenal, 130
Hexanoic acid, 130
Hinterland marking, 97–98
Homarus americanus (lobster), 44
  antennal flicks, 174
  precopulatory mate guarding, 61, 62
  tropotaxis, 222
Homo sapiens, see Humans
Honest signal, 2, 44, 45
  queen pheromone
    honeybees, 121, 121
  reflecting biological state of
    marker
    territorial scent marks, 89
  Utetheisa ornatrix (tiger moth), 47,
    49, 49, 50
Honeybees (Apis mellifera)
  Africanised bees, monitoring
    spread of, 256, 256
  cell capping, odour cues, 116
  colony odour sources, 106
  colony-level signature, 106
  comb wax odour cues, 106
  manipulating behaviour for bee-
    keeping
    delaying swarming, 255
    directing pollination, 255
    luring swarms, 255
  marking visited flowers, 85

  multiple paternity in nest
    blending of template colony
      odour, 105
  Nasonov pheromone
    marking new nesting cavity, 131
    swarming, 131
  pheromones, summary, 115
  queen mandibular pheromone
    (QMP), 121
    as honest signal, 121
    as sex pheromone, 19
    rapid colony response to loss, 121
    transmission, 121
  queen-rearing nepotism, 107
  queen retinue, 116
  sting pheromone attracts other
    bees to sting, 154
  swarming, 121
  Varroa mite, 116, 262
Hormones
  cortisol and recognition of human
    babies, 276
  humans
    birth, 201
    cortisol and recognition of
      human babies, 276
  juvenile hormone (JH)
    Agrotis ipsilon (black cutworm
      moth), 189–190
    social insects, 204, 204
  oestrogen, 198
  oxytocin in ewes, 199, 200, 201
  pheromone influence on hormones
    female stimulates testosterone
      secretion in lesser mouse
      lemur, 196
    male hormone surge to vaginal
      odours in hamster, 182
    puberty acceleration in rodents,
      ungulates and non-human
      primates, 193
    responses induced via olfactory
      system, 185
    sexual activity in Petaurus
      breviceps (sugar glider), 196
  testosterone
    androgen effects on electroolfac-
      togram, fish, 189
    femoral gland, Iguana iguana
      (lizard), 46
    graded response of scent glands
      in Microtus pennsylvanicus
      (meadow vole), 46

hamster male brain needs to respond, 189

male sexual response in mammals, 183

musth and temporal gland secretion in elephants, 47, *48*

see also Primer pheromones; Female pheromones; Male pheromones

Host-discrimination, 83

Host-marking pheromones (HMP), 74, 83–85

evolution of, 83

host-discrimination, 83

individual recognition, 85

Human leucocyte antigen (HLA), *see* Humans, mate choice; Major histocompatibility complex; Mammals, mate choice

Humans

advertisement of oestrus, 281

'copulins', 281

unresolved in humans, 282

age, changes with

perception, 271

secretions, 275

sensitivity, 297

anosmias

induced changes on exposure, 191

specific, 291

antiperspirants and deodorants, 288

axilla (armpit) hair and odour creating bacteria, *288*

behavioural manipulation by semen hormones? 64

candidate compounds, 285

axillae (armpits), 287–291

glands, 284

sites over body, *286, 287*

cleanliness not necessarily a virtue, 276

cultural and social aspects of odours, 273

osmologies, 273

Napoleon, 273

sexual attraction, 273

importance of odours in human behaviour and biology, 274

mate choice, 277

5α-androstenone, 277

body symmetry and, 280

kibbutz or Westermarck effect, 280

human leucocyte antigen (HLA), 52, 278

major histocompatibility complex (MHC), 52–56, 278–280

sexually selected traits, 277

mothers and newborn babies, 275

correlation of mother cortisol and baby recognition, 276

importance of natural odours, 275

speed of learning, 275

odour preferences

learning, 271

neonates different from older children and adults, 271

odours and memory

experimental tests, 276

Proust, 276

olfaction

mapping brain activity, 174, *175,* 185

olfactory receptors

genes and pseudogenes, *170*

perception of odours, 291

changes during menstrual cycle, 292, *293*

changes with puberty, 292

specific anosmias, 291, 292

variation between individuals, 291

perfumes, 273, 279

phenotype matching, 276

primer pheromones

age of female puberty, 285

menstrual synchrony, 282–285

steroids and mood, 295

recognition, 274

children, sibs, parents, 275

security blankets, 276

self, partners, 275

T-shirt sniff tests, 274

semen components, 64

truffles, 253, 288

using human odours, 295

'human pheromones', 298

forensics, 296

medical diagnosis, 296–297

mood changers, 298

vomeronasal organ (VNO), putative, 294

genetic evidence, 295

location, *294*

neuroanatomical and histological evidence, 295

see also Primates

Hutterites

MHC and partner choice, 280

Hydrobates pelagicus (storm petrel)

nest recognition, 21

Hydrocarbons, *see* Cuticular hydrocarbons

Hydroides (polychaete worm)

plankton settlement, 79

3-Hydroxy-2-butanone, 44

6-Hydroxy-6-methyl-3-heptanone, 193, *194*

Hydroxydanaidal (HD), 50

derived from plant pyrrolizidine alkaloids in diet, 47

Hypothalamus, 184

Microtus ochrogaster (prairie vole), *198*

oxytocin

maternal behaviour, 199, *200*

Hypoxanthine 3-N-oxide, *158*

Iguana iguana (lizard)

forest habitat, volatile pheromone, 14–15

femoral gland

testosterone influence, 46

Imprinting

mate recognition, 198

odours, recognising kin and colony members, 199

Individual recognition

generalisation across gland odours, 113

Helogale undulata (dwarf mongoose), 112

Herpestes auropunctatus (Indian mongoose), 112

Homarus americanus (lobster), 46

humans, 274–275

mechanisms, within a species some glands vary more than others, 113

Meles meles (European badgers), 112

mice, 112

pair bonding

Niveoscincus microlepidotus (lizard), 113

Proteles cristatus (aardwolf)

'dear-enemy' recognition, 113

Information centre
  Malacosoma americanum (tent caterpillar), *133*
Inosine, *42*
Integrated pest management (IPM), 255
  pheromones in control of tomato pests, 259
Ips (bark beetles)
  Ips paraconfusus, 235
    discovery of synergy, 35
  Ips pini
    bioassays, *24*
  Ips typographus
    pest management, 264
(R)-(−)-ipsdienol, 234
(S)-(+)-ipsdienol, 234
Iridomyrmex (ants), 130
Isomers, 304
  constitutional
    functional group, 304
    positional, 304
  naming of, 305
  stereoisomers
    chirality and enantiomers (optical isomers), 305–307
Ixodes neitzi (tick)
  responds to scent marks of host antelope, 233

(R)- and (S)-Japonilure enantiomers
  use by sympatric beetles, 17
Juvenile hormone (JH)
  Agrotis ipsilon (black cutworm moth)
    male response activated by, 189
  social insects, 204
    ants, 204
    termites, *119*

Kairomones, 2
  eavesdropping, 230
Kangaroos
  mate guarding by male, *62*
Keifera lycopersicella (tomato pin worm)
  pest management, *258*, *259*, *261*
Kibbutz or Westermarck effect
  human mate choice, 280
Kin recognition 103–105
  mate choice
    conservation projects, 254
    mechanisms, *103*
    low metabolic costs, *103*
Kin selection, Hamilton's rule, 116

Kobus
  Kobus kob (Uganda kob), 58
  Kobus leche(Kafue lechwe), 58

Labelled lines, 187
Labial gland
  termites, 138
Lasioderma serricorne (cigarette beetle)
  aggregation pheromone, 83, *84*
Lasioglossum zephyrum (sweatbee)
  guard bees discriminate relatedness, 108
Lasius (ants)
  trail pheromone, 130
  Lasius alienus
    alarm pheromone, causes panic and escape, 155
  Lasius fuliginosus
    trail following, *216*
  Lasius niger
  foraging effort, food value, 139
Lateral olfactory tract, *183*
Learning
  fish, conditioned alarm response to predator odours, 159, 191, 254
  mammals
    induced peripheral changes in response to odours after contact, 191
    maternal behaviour, 191
    needed to distinguish oestrous from dioestrous female odours, 191
  maternal
    ewes bond quicker on subsequent births, 200
  odour stimulation and noradrenaline release in olfactory bulbs, 199
  Rhagolitis pomonella (apple maggot fly), 190–191
  see also Development
Lee–Boot effect, 194, 195
Leks, 57
  Ceratitis capitata (Mediterranean fruit fly), 58
  cost, eavesdropped by predators, 231–232
  Kobus kob (Uganda kob), *58*
  Kobus leche (Kafue lechwe), 58
  Lutzomyia longipalpis (sandfly), 58
  Xylocopa fimbriata (carpenter bee), 57

Lemur catta (ring-tailed lemur)
  group territorial 'stink fights', *90*
Leontopithecus rosalia (lion tamarin)
  suppression of subordinate reproduction, 127
Lepeophtheirus salmonis (salmon ectoparasite)
  pest management, *262*
Lepomis cyanellus (green sunfish)
  alloparental carer, 230–231
Leptothorax acervorum (ant)
  tandem running recruitment, 134
Leptotyphlops dulcis (blind snake), 247
Lestrimelitta limao (robber bee), 243
Linoleic acid, *84*
Linyphia litigiosa (spider)
  conflict between the sexes, 59
Liriomyza sativae (leafminer fly), 259
Litocranius walleri (gerenuk), 89
Litoria splendida (magnificent tree frog)
  peptide sex pheromone, 21
Lobesia botrana (grape vine moth)
  pest management, *258*
Lordosis, 20, 184, 253
Loxodonta africana, see Elephants, African
Luteinising hormone (LH), 183
  Microtus ochrogaster (prairie vole), 198
Luteinising hormone-releasing hormone (LHRH), *198*
Lutzomyia longipalpis (sandfly), *263*
  lekking, 58

Macaca mulatta (rhesus monkey), 281
Macroglomerular complex (MGC), 165
Macrotermes (termite)
  nest building, 143–144
Maculinea rebeli (lycaenid butterfly)
  manipulates ant host, 249
Main olfactory bulb (MOB), 165
Main olfactory epithelium (MOE), 165, 178, 179, *182*, *183*
Main olfactory system
  detection of pheromones, 181
    Oryctolagus cuniculus (European rabbit), 181
    Sus scrofa (pig), 184
  sheep maternal behaviour, 201
  see also Olfaction
Major histocompatibility complex (MHC)

bacterial flora and odours, *53*
Bruce effect, 195
communal nesting, mouse
   preference for kin, *54, 55*
familial imprinting
   odour preferences for difference
      in mates, 55
human leucocyte antigen (HLA),
   52, 278
kin recognition in other vertebrates
   (fish, birds, reptiles), 56
mate choice, 52
   MHC diversity, *Gasterosteus
      aculeatus* (stickleback), 55
   mice and humans, *54*
   mutual distinguishing MHC of
      odours inrats, mice and
      humans, 52
   not universal, 56
   odours as marker for degree of
      kinship, 52
   semi-natural experiments, 52
preferences
   learning, 55
   women taking oral
      contraceptives, *54*
selection pressures maintaining
   diversity of MHC alleles, *55, 56*
disassortative mating, 55
higher implantation rates if dissimilar, 55
reduced chance of inbreeding if
   dissimilar, 55
source of the odours, *52–54*
Major urinary proteins (MUPs),
   194, 195
longevity of signal, slow release,
   193
elephants, 195
mice, 195
Malacosoma (tent caterpillars)
   trail pheromone, *130*
   Malacosoma americanum, *133*
      laying trail, *135*
Male pheromones
   diverse release structures, *7, 10, 70*
   effects on females
      leks, 57–59
      mice, Bruce effect (pregnancy
         block), 195
      oestrus induction and puberty
         acceleration (Whitten and
         Vandenbergh effects), 192

oestrus synchronisation, mice,
   194
effects on males, 196
   sexual activity of subordinates
      suppressed, 196
selection by female to avoid hybrid
   matings, 70
Mammals
   group, kin, family, and individual
      recognition, 109
   individual recognition, 112
      Helogale undulata (dwarf
         mongoose), 112
      Herpestes auropunctatus (Indian
         mongoose), 112
      Meles meles (European badger), 112
      mice, 112
   kin recognition
      Acomys cahirinus (spiny mouse),
         110
   mate choice
      major histocompatibility
         complex (MHC), 52–56
   mother–infant recognition, 109
   scent marking, 87
   scent sources, summary, *10*
   secretory glands, 11
   see also Humans; Primates; Social
      mammals
Man, *see* Humans
Mandibular gland
   Trigona (stingless bee), 136
Manduca sexta (tobacco horn moth)
   zigzag orientation upwind, *225*
Marine invertebrates, alarm
   responses, 158
Marking enemies for further attack,
   155
Marsupials
   kangaroos
      mate guarding by male, 62
   Monodelphis (opossum)
      vomeronasal organ (VNO) needed
         for oestrus induction, 181
   Petaurus breviceps (sugar glider)
      scent marks affect sexual
         activity, 196
Mastophora
   Mastophora bisaccata (bolas spider),
      243
   Mastophora hutchinsoni (bolas
      spider), 240
Mate choice

avoiding lethal alleles, 't-complex'
   in mice, 56
diet quality
   Microtus pennsylvanicus (meadow
      vole), 46
fluctuating asymmetry
   humans, 280
   Panorpa japonica (scorpion fly), 49
MHC diversity, *see* Major
   histocompatibility complex,
   mate choice
optimal outbreeding, 51
   Gryllus bimaculatus (cricket), 51
   Lasioglossum zephyrum (sweat-
      bees), 51
parasites, 51
   mice, preference for uninfected
      mates, 51
paternal investment and female
   choice, 47
   Utetheisa ornatrix (tiger moth), 47
pheromones
   Ephestia elutella (tobacco moth), 47
Mate guarding, *see* Sperm competition
   and mate guarding
Maternal behaviour and recognition
   humans, 275
   sheep, 199
Mating disruption, as pest control,
   258
Medial amygdaloid nucleus, *183*
Medial preoptic area (MPOA), *182, 183*
Megaponera (ant), *130*
Meles meles (European badger)
   anal glands, 111
   border and hinterland latrines,
      98, *98*
   individual recognition, 112
   subcaudal glands, 111
Memory and pregnancy block (Bruce
   effect) in mice, 201
   accessory olfactory bulb (AOB), 201
Mesocricetus auratus (hamster) , *183*
   aphrodisin, 182
   over-marking, 100
   self-matching, *105*
Messor barbarus (harvester ant)
   matching trail patterns to food
      distribution, 139
4-Methyl-3-heptanone, 234
6-Methyl-5-hepten-2-one, 243
(*E*)-3-Methyl-2-hexenoic acid ((*E*)-3M2H
   or TMHA), 289

Methyl-4-methylpyrrole-2-carboxylate, *130*, *136*

2-Methyliazolidine, 44

MHC, *see* Major histocompatibility complex

Mice (*Mus musculus domesticus*)
Bruce effect
memory and pregnancy block, 201
individual recognition, 112
Lee–Boot effect, 195
male pheromones
oestrus induction and puberty acceleration in females (Whitten and Vandenbergh effects), 192
marking rates, dominants and subordinates, 94, *94*
mate choice
preference for uninfected mates, 51
mate choice to avoid lethal alleles, 't-complex', 56
pheromone marks
importance for animal welfare, 254
scent marking by dominant male, 93
signal sites, urine posts, 94
similar pheromone to bark beetle, 5
't-complex', 56
Vandenbergh effect, 192
Whitten effect, 192

Microcebus murinus (lesser mouse lemur)
female urine stimulates spermatogenesis and testosterone secretion, 196
pheromones and sex ratios, 197
sexual activity of subordinates suppressed, 196

Microdon mutabilis (syrphid fly)
matches its ant host, 248

Microtus
Microtus ochrogaster (prairie vole), 198
unfamiliar male odours, stimulation of oestrus, *198*
Microtus pennsylvanicus (meadow vole)
suppression of subordinate reproduction, 124, 126, 197–198

graded response of scent glands to testosterone implants, 46

Microtus pinetorum (pine vole)
suppression of subordinate reproduction, 127

Miriamide, *84*

Mitral/tufted cells, *171*, *172*
Bruce effect (memory of mate), 201
memory for lamb, 200

Molluscs
Aplysia (sea-slug), peptide sex pheromone, 15
snails, direction of a trail, 217

Monodelphis (opossum)
vomeronasal organ (VNO) needed for induction of oestrus, 181

Monogamous partner recognition, 202

Monozygotic twins
human, dogs have difficulty distinguishing, 274
sheep, lamb recognition experiment, 109

Mother–infant recognition
humans, 275
sheep, 109

Mouse, *see* Mice

Multi-component pheromones, 187
alarm pheromones in social insects, 154
ant, alarm pheromone signal, 212, *214*
asymmetric tracking, 68
major and minor components, 66
male response wider than range females produce, *69*
response is to whole blend, moths, 67
specificity, 17, 66
synergy
insects, 35
mammals, 195

Musca domestica (house fly)
pest management, *263*

Mus musculus domesticus, see Mice

Mushroom body, 165

Musk, 185

Musk gland, 148

Mustela
Mustela erminea (stoat), anal glands, 265
Mustela putorius furo (ferret), 190

Musth, *see* Elephants

Myrmecaphodius excavaticollis (beetle)
parasite of *Solenopsis* (fire ant), 247

Myrmecophilous invertebrates, 245

Myrmica (ants), 247
Myrmica sabuleti
matching foraging effort to food value, *139*

Myzus persicae (aphid)
repelled by alarm pheromone from plant, 244

Naming pheromones
functional groups
prefixes and suffixes, 303

Napoleon, 273

Nasonov pheromone
honeybees, *131*, 255

Nasutitermes (termite), *120*, 155
marking enemies for further attack, 155
recruitment pheromones, 140

Nauphoeta cinerea (African cockroach)
'badges' of status, 44
honest signal, 44, 45
male contest and female choice, 40, 44
satellite males, 60

Nematodes
Caenorhabditis elegans (model system), 203
Heterodera glycines (soybean cyst nematode)
mating disruption with sex pheromone, *258*, *260*
number of olfactory receptor types, *172*

Nemeritis (parasitoid wasp)
host marking pheromone, 84

(*E,E,E*)-Neocembrene-A (*E*-6-cembrene A), 140

Nereis succinea (polychaete worm)
pheromones coordinate external fertilisation, 42

Newts, *see* Salamanders and newts

Nezara viridula (southern green stink bug)
sex pheromone blend change due to parasitoid, 233

Nicrophorus (burying beetle), 59

Niveoscincus microlepidotus (lizard)
individual recognition
pair bonding, 113

Noradrenaline, *198*

role in Bruce effect, 201
Norepinephrine (NE), *see* Noradrenaline
Notropis umbratili (redfin shiner fish)
　interspecific egg dumper, 231
Novomessor (ant)
　recruitment, 140
　signals, 18
Nuptial dance
　polychaete worms, *42*

Odocoileus hemionus columbianus (black-tailed deer), *see* Deer
Odorant-binding proteins (OBP), *169*, 177, 178
Odour sieve, 167
Oecophylla
　Oecophylla longinoda (African weaver ant)
　　alarm pheromone signal, 212, *214*
　　laying trail, *135*
　　territories, 88–89
　Oecophylla smaragdina
　　territories, 89
Oestrogen, *see* Hormones, oestrogen
Oestrus
　Microtus ochrogaster (prairie vole), *198*
　rats, 181, 191, 193, 196
　synchronisation
　　humans, 282–283, 285
　　mice, *194*
　　rats, 196
Olfaction
　across-fibre patterning, 171
　amygdala
　　interaction between main olfactory and vomeronasal organ (VNO) outputs, hamster, 184
　behaviour of smelling (sniffs and flicks), *174*
　brain
　　macroglomerular complex (MGC) not needed for response to all pheromones, 188
　　mapping activity, 172, *173*, 175–176, 185
　　moths and sex pheromones, *165*, 186–188
　　outputs modulated, 178, 182
　　similarities in vertebrates and insects, *165*
　combinatorial coding, 170
　glomeruli, 170–174

integration of olfactory and visual inputs, 178
main olfactory system
　functional overlap with vomeronasal olfactory system (VNO), 180–86
mapping brain activity,
　fluorescing dyes, 172, *173*
　functional magnetic resonance imaging (fMRI)
　humans, *174*, *175*, 185
memories for odours
　renewal of olfactory sensory neurons (OSNs), 173
odotypes
　hypothesis for coding features of odour molecules, 170, *171*
olfactory receptors
　types (and organisation)
　　nematodes, insects, and mammals, *172*
　sensitivity
　　predator olfactory sensory neurons (OSNs) as sensitive as prey's, 231
smell (olfaction) vs. taste (gustation), *164*
specificity
　clean-up enzymes, 177
　odorant-binding proteins (OBPs), 177
synaptic organization, *172*
　insect antennal lobe, *172*
　mammalian olfactory bulb, *172*
temporal coding in brains, 174
　fast oscillations of neural activity, 174
　humans, *175*
　fast oscillations, synchronisation, 174
Olfactory bulb
　ewe learns individual odour of lamb, 199, *200*
Olfactory cues
　recognition
　　mates, species, kin, 198
Olfactory organs
　amphid sensillae (nematode), 203–204
　Argyropelecus hemigymnus (deep-sea hatchet fish)
　　biggest nasal organ for body size in vertebrates, 207

functional design, *166–168*, 174
lobsters
　antennule sensory hairs, *174*, 221
sensitivity
　dogs, 167
　moth antennae, *166*
Olfactory receptor proteins (OR), 166
　evolution from G-protein-coupled receptors, 176
　genes, 169–170
　humans, *170*
Olfactory sensory neurons (OSNs), *165*, 166
　renewal, 173
　similarity across animal kingdom, *169*
　stem cells, 169
Olfactory space, 169
Oncorhynchus mykiss (rainbow trout)
　taught to recognise predator odour, 254
Ondata zibethicus (muskrats)
　monitoring, 257
Ophrys sphegodes (orchid)
　dupes solitary bee, 240, *242*
Optical isomers, 305
Optimal outbreeding
　mate choice, 51
Oreotragus oreotragus (klipspringer antelope)
　scent marks eavesdropped by ticks, 233
Orientation behaviour, 206
　active space, 209
　diffusion, 210–212
　Argyropelecus hemigymnus (deep-sea hatchet fish)
　　mate location, 207
　arresting, 207
　chemical plumes, 206
　combining information from different senses, 223
　directly guided, 207
　idiothetic, 207, *208*
　indirectly guided, 207, *208*
　kinesis, 207, *208*
　　klinokinesis, *208*
　　orthokinesis, *208*
　odour 'landscape', 206
　odour concentration gradient, 207
　odour stimulus
　　plumes, 210–227
　　short range diffusion, 210–213, *214*

Orientation behaviour, (*cont.*)
  trails, 210, *213*, 214–218
  precise stimuli, 207
  ranging, 209
    strategies, 209
  sampling
    sequential, 209
    simultaneous, 209
  scale
    communication strategies, 212, 221
  self-steered, 207
    counter-turning, 225
  taxis, 207, *208*
    anemotaxis, 222
    klinotaxis, *208*
    rheotaxis, 222
    teleotaxis, *208*
    tropotaxis, *208, 213*, 222
  teleology, 207
  threshold (K), 210
  turbulence
    transmission in currents, 158
  virtual-reality chemical goggles, 223
Oryctolagus cuniculus (European rabbit)
  nipple pheromone, 181, 183
Ostariophysi, 158
Ostrinia nubilalis (European corn borer moth)
  pheromone signal genetics, 69
Ourebia ourebi (oribi), territories, 96
Over-marking, 99–100
  Castor canadensis (beaver), 96
  message centre, 99
  scent blending, 99
  scent masking, 99
  used in mate choice in house mice? 94
Oviposition-marking pheromones, *see* Host-marking pheromones
Oviposition pheromones
  Simulidae (blackfly), 75
  Culex (mosquito), 75, 77
  Schistocerca gregaria (desert locust), 76
Ovis aries, see Sheep
Oxytocin
  peptide hormone, 199, *200*

Pachycondyla (ants)
  Pachycondyla laevigata
    mass recruitment, 134
  Pachycondyla obscuricornis
    individual hunter, 134

Panorpa japonica (scorpion fly)
  symmetrical males more attractive, 49
Panulirus interruptus (California spiny lobster)
  aggregation for defence, 75
Paraleptophlebia adoptiva (mayfly), 236
Paraponera clavata (ant)
  eavesdropped by parasitoid fly, 233
Parasitoid insects
  flies
    Apocephalus paraponerae (phorid), 233
    Trichopoda pennipes (tachinid), 233
  host suicide hypothesis, 152
  wasps
    host marking, 84
    Telenomus euproctidis, 232
    Trichogramma pretiosum, 85
Parkinson's disease, 297
Paternal investment
  Utetheisa ornatrix (tiger moth), 47, 49, 50
Pectinophora gossypiella (pink bollworm moth)
  males, wider pheromone response than range females produce, 69
  pest management, 258, 267
Peptide pheromones, 15
  aphrodisin, *182*
  Litoria splendida (magnificent tree frog), 21
  sodefrin, 15, *16*
  specificity, 16
Perfumes, 273, 279
Periglomerular cells, 172
Periplaneta americana (American cockroach)
  sex pheromone, 16
Periplanone-B
  sex pheromone, *Periplaneta americana* (American cockroach), 16
Petaurus breviceps (sugar glider)
  scent marks affect sexual activity, 196
Petromyzon marinus (marine lamprey)
  pest management, *262*
Pharmacophagy, 12
Pheidole (ants)
  Pheidole dentata
    enemy specification, 155
  Pheidole pallidula

  matching foraging effort to food value, 139
Pheromone binding proteins (PBPs), 177
Pheromone identification, 25
  bioassays, 25–29
    appropriate concentrations, 28, 188
    examples, *24*, 25–29
    subtractive approach, 35
    Y-maze, *133*
  by comparative approach
    ants, 32
    mice, 32
  challenges, 23, 36
  collection of chemical signals, 29
    aeration and cold trap, *24*
    antelope scent marks, 29
    entrainment, Poropak-Q™, Tenax™, cold trap, 31
    solid phase microextraction (SPME), 31, *32, 33–34*
    solvent washes or whole glands, 29, 31
  composite (combined) signals, 29
  electrophysiology
    electroantennogram (EAG), 29, 31
    electroolfactogram (EOG), 29, 189
    electrovomerogram (EVG), *194*
    single cell recording (SCR), 29
  humans
    functional magnetic resonance imaging (fMRI), *174*, 185
    organoleptic tests, 289
    'scratch & sniff' odorants, 297
    T-shirt sniff tests, 274
    twins studies, 292
  ionised air
    tracking plumes, *221*
  model systems, 36, 203
  neurotransmitter dopamine as plume tracer, *221*
  new techniques
    brain imaging, 36
    genetic analysis, 36
    genetic engineering, 35–36, 70
    genomics, 35
  separating chemicals
    fractionation, 25, 31
    gas liquid chromatography (GC or GLC), 30, 32
    gas liquid chromatography–mass spectroscopy (GCMS), 32

high performance liquid
chromatography (HPLC), *26*
signal redundancy
deer alarm signal, *28*
Pheromones, 1
advertising
Homarus americanus (lobster), 61
aquatic habitats, 15
assembly
guanine, tick, 80
calling pattern, by mating system
and habitat
stored-product insects, 82
caste differences
mandibular gland secretions,
114, 115
changes with age
elephant males, 47
humans, 275
queen mandibular pheromone
(QPM) blend, 122
convergent, 4–6
cooperative breeding, 113–127
inhibition or suppression of
subordinate reproduction, 126
coordinating external
fertilisation, 41
Nereis succinea (polychaete
worm), *42*
Carassius auratus (goldfish), 9, *19*
costs of signalling, *see* Costs
cuticular hydrocarbons
dipteran sex pheromones, 70
social insects, colony odours,
106, 108
direct methods of transfer, 4
into bloodstream, 4
to female, 4
dominance status, 44
eavesdropped by visual cues, 232
egg hatching pheromone (barnacle)
eicosanoids (PUFAs), 77
evolution
alarm pheromones, 9
clues from geographical
variation, 69, 232–234, 248
eavesdropping, 8, 234
effect of habitat, 82, 97–98
enabled by olfactory receptor
proteins and combinatorial
brain circuits, 176
from existing chemical cues, 9,
19, *42*, *43*, 189

honeybees, sociality, 114
sensory drive, 6, *7*
glands
mammals, summary, *10*
wasps, bees, ants, and termites,
summary, *11*
glycoproteins
barnacle settlement, *78*
salamander, 4
graded response
vole scent glands to
testosterone, 46
hormone-based, 9
host marking, 83–85
insect
molecular weight related to
function, *14*
larval aggregation pheromones
bark beetles, 81
long distance
produced by females, 41
produced by males, 41
longevity
major urinary proteins (MUPs), 193
major urinary proteins (MUPs)
elephants, 195
mice, 195
mammals
aphrodisin, *182*
detection by main olfactory
system, 181
molecular weight related to
habitat, 13, *15*
marking of patches
bees, 85
ladybirds, 85
parasitoid wasps, 85
metabolic cost, low
Anthonomus grandis (boll weevil), 20
Litoria splendida (magnificent tree
frog), 21
molecular weight related to habitat
aquatic, 15
terrestrial, 13, *15*
multi-component
alarm pheromones in social
insects, 154
moths, 17, 66
provide specificity in insects and
vertebrates, 17
recruitment, 141
multiple messages in odours
mammals, 102

social insects, 102
nipple
Oryctolagus cuniculus (European
rabbit), 181, 183
oviposition, 75, *76*, 77
peptides
aphrodisin, 182
larval settlement pheromone,
barnacle, 78
Litoria splendida (magnificent tree
frog), 21
marine invertebrates, 15
pumping pheromone, crab, 77
sodefrin, 15, *16*
pheromone binding proteins (PBPs)
polar molecules
anthopleurine, 15
prostaglandins, 9
pumping, 77
queen egg marking, honeybee, 122
queen mandibular pheromone
(QMP), 121, 204
queen pheromones in social
insects, 118–123
release as pulses rare, 221
reproduction in social groups, 113
sex ratio effects
Microcebus murinus (lesser mouse
lemur), 197
sources
plants, 11, 12, 234
symbiotic bacteria, 11, 111
waste products, 80
stereoisomers
importance in invertebrates and
vertebrates, 17
steroids, 9
synergy in multi-component
pheromones
insects, 35
mammals, 35, 195
terrestrial habitats, 13–15
transmission in currents
fish alarm pheromone, 158
see also Aggregation pheromones;
Alarm pheromones;
Antagonists; Glands;
Honeybees; Host-marking
pheromones; Pheromone identification; Primer pheromones;
Recruitment pheromones;
Releaser pheromones; Sex
pheromones; Signals; Territories

Philopatry, *125*

Phoxinus phoxinus (European minnow)
  alarm pheromone, 158

Pieris (cabbage butterfly)
  host-marking pheromone, *84*

Pigs
  minipig pets, 254
  see also *Sus scrofa*

Piriform cortex, *165*

Plankton,
  commercial applications, disrupting settlement, 266
  settlement
    barnacle, 26, 78
    Dendraster excentricus (sand dollar), 80
    Hydroides (polychaete worm), 79
  3D trails
    Temora longicornis (copepod), 213

Plants
  deception by wild potatoes, 244
  bee pheromones from orchids, 12
  pyrrolizidine alkaloids (PAs), 12, 47
  tree defences against bark beetles, 81
  see also Pheromones, sources, plants

Platysoma cylindrica, eavesdropping, 234, 236

Plethodon
  Plethodon cinereus (red-backed salamander)
    contest, 46
    feeding and breeding territories, 46
    scent matching hypothesis and territory, 95
  Plethodon jordani (salamander)
    pheromone transfer by male, 4
    postcloacal gland, 95

Plotsus lineatus (catfish), *169*

Plumes
  turbulence, visualisation, *219*
  tracking plumes, 221
    ionised air, *221*
    neurotransmitter dopamine, 221

Podisus maculiventris (spined soldier bug)
  eavesdropped by parasitoid, 233

Pogonomyrmex badius (harvester ant)
  alarm pheromone signal, 212

Poison gland, ant, *130*

Polistes fuscatus (paper wasp)
  kin recognition mechanisms, 107

Popilla japonica (Japanese beetle), 17

Postcloacal gland
  salamander, 95

Postpharyngeal gland
  ants, 108
  host source of hydrocarbons for parasite, 248

Precopulatory mate guarding, *see* Sperm competition and mate guarding

Predators and parasitoids, 232

4-Pregnen-17α-,20β-diol-3-one (17,20β-P), 9

Pregnancy block, male pheromones (Bruce effect), 195

Premnas (anemone fish), 238

Preorbital gland, *see* Antorbital gland

Preputial gland
  mice, 193

Primates
  Callithrix jacchus (common marmoset)
    subordinate female ovulation suppressed, 197
  Cerocebus albigena (mangabey), *271*
  Leontopithecus rosalia (lion tamarin)
    cooperative breeding not odour controlled, 127
  Microcebus murinus (lesser mouse lemur)
    bias of offfspring sex ratios, 197
    female urine stimulates spermatogenesis and testosterone secretion, 196
    sexual activity of subordinates suppressed, 196
  micro-osmic, 271
  olfactory receptors, 169–170
    genes, 170
  Saguinus fuscicollis (saddle-back tamarin)
    multiple messages in odours, 102
    subordinate female ovulation suppressed, 197
  scent glands and scent marking, *272*
  see also Humans

Primer pheromones, *18*
  can be fast acting, 20
  Carassius auratus (goldfish), *19*
  gregarising factor in egg foam
    Schistocerca gregaria (desert locust), *76*

juvenile hormone (JH)
  social insects, 204
  ants, *204*

maturation synchrony
  Schistocerca gregaria (desert locust), 76

pest management, 266

reproduction, 192
  daylength and pheromones in sheep, 192
  similiarity of action in mammals and insects, 20

social insects
  queen effects in ants, *204*

termites
  caste control, 119

Primiphales promelas (fathead minnow)
  alarm pheromone, 158

3-Propyl-1,2-dithiolane, 265

Prostaglandins, 9
  15-keto-prostaglandin-F-2α (15-keto-PGF2α), 9
  fish peripheral response, 189
  prostaglandin F-2α (PGF2α), 9

Prostephanus truncatus (larger grain borer beetle), 256
  males only signal until females come, 82
  monitoring, 256

Proteles cristatus (aardwolf)
  'dear-enemy' individual recognition, 113

Protocerebrum, *165*

Proust, memory, 276

Pseudaletia (moth), *5*

Pseudogates
  termites, 119

Pseudogenes, 170

Puberty acceleration, *see* Hormones

PUFAs (C_{20} polyunsaturated fatty acids), 77

Puntius (cyprinid fish), 189

Pygidial gland, ant, 134

Pyrrolizidine alkaloids (PAs), 12, 47

Queen mandibular pheromone (QMP), *see* Pheromones

Queen pheromones in social insects, *see* Pheromones

Racemate or racemic mixture, 306

Ranging behaviour, 209

Raphicerus melanotis (grysbok), 13

Rattus norvegicus (Norway rat)
  follows trails from good food
    sources, 137
Recognition mechanisms, 103
  direct familiarisation, 104
  indirect familiarisation (phenotypic
    matching), *104*
  Microtus ochrogaster (prairie vole),
    197–198
  phenotypic matching hypothesis
    field tests, 111
  recognition allele ('green beard'
    phenomenon), *104*, *105*
  fire ants, *105*
  sensitive period for learning, *105*
  social insects
    learning needed before
      discrimination, 103–109
  usually by learning, *103–104*
Recruitment pheromones
  alarm
    Apis cerana japonica (Japanese
      honeybee), 154, *156–157*
    marking enemies for further
      attack, *153*, *154*, 155
  ants
    competition strategy, 140
    group recruitment, *134*
    laying trail, *135*
    mass communication, 129, *134*
    scout ants, 134
    tandem running, *134*
    trail specificity, *134*
    worker sensitivity to trail
      pheromone, *136*
  aphids, social
    soldiers attack enemy, 153, *153*
  competition strategy, 140
    ants, 140
    Trigona (stingless bee), 140
  convergence in marking
    behaviour, *135*
  coordinated attack
    Vespa mandarinia japonica (giant
      hornet), 154–155, *157*
  expand colony diet, 136
  food distribution, recruitment
    type, 136
  longevity related to food
    supply, 137
  Malacosoma americanum (tent
    caterpillar)
    laying trail, *135*

matching foraging effort to food
    value, 138
  Rattus norvegicus (Norway rat)
    follow trails from good food
      sources, 137
  scout ants, 134
  termites, 137
    laying trail, *135*
    trail specificity, *135*
  trail specificity, 140
  see also Alarm pheromones; Self-
    organising systems
Reinforcement, 65
Releaser pheromones, *18*
  Carassius auratus (goldfish), *19*
  mammal, 20
  nipple
    Oryctolagus cuniculus (European rab-
      bit), 181
Reproductive character
    displacement, 65
Reproductive skew, 116, *117*
Response to pheromones, factors
    affecting, 188, 191
  context
    aphids tended by ants, 237
    honeybees, 188
    parasitoid wasps, 188
  hormones
    central nervous system
      effects, 189
    peripheral effects, 189
  receiver's characteristics, 188
  see also Learning; Development
Reticulitermes flavipes (termite)
  recruitment pheromones, 140
Reynolds numbers (Re), 167
  copepod pheromone trails, *213*
Rhagolitis
  Rhagolitis cerasi (cherry fruit fly)
    pest management, 264
  Rhagolitis pomonella (apple
    maggot fly)
    Crataegus mollis (hawthorn),
      original host, 85
    evolution of host-marking
      behaviour, 85
    experience of marking
      pheromone needed for
      response, 190–191
    host marking pheromone, 84
Rhitropanopeus harrisii (mud crab)
  pumping pheromone, 77

Ritualised signals, *3*
Ropalidia marginata (paper wasp)
  non-kin may join colony, 108

Saguinus fuscicollis (saddle-back
    tamarin)
  multiple messages in odours, 102
  subordinate female ovulation
    suppressed, 197
Salamanders and newts
  courtship pheromone delivery
    patterns, 4
  Cynops pyrrhogaster (red-bellied
    newt)
    courtship, 16
    sodefrin, 16
  Desmognathus ochrophaeus
    male injects pheromone into
      female, 4
  Plethodon cinereus (red-backed
    salamander)
    postcloacal gland, 95
    scent matching, 95
Sanguinis fuscicollis (saddle-back
    tamarin)
  multiple messages in odours, 102
  subordinate female ovulation
    suppressed, 197
Satellite males, *see* Alternative mating
    stategies
Saturnia pavonia (emperor moth)
  early demonstration of
    pheromone, 23
Scale
  boundary layer
    effects on olfactory sensors, *174*
  chemical plumes, 206
  communication strategies, 212
  design of olfactory organs, *166–168*
  diffusable pheromones in water, *213*
  Reynolds numbers (Re), 167, *213*
Scent marking
  border maintenance hypothesis, 90,
    96–97
  Castor canadensis (beaver), 96
  ecological factors, patterns between
    and within related species, 91
  males directly mark females
    Dolichotis patagonum (mara), *100*
  marking behaviour of house mice, *94*
  non-territorial mammals, 100
  response by mice varies by
    competitive ability, 94

Scent marking (cont.)
  salamanders, nose-tapping, 95
  scent fence hypothesis, 91
  scent matching hypothesis, 90,
    91–96
    mice, 93–94
    reducing cost of territorial
      defence, 93
    salamanders, 95
    self-marking and presentation, 92
    testing the hypothesis, mice, 95
  using vertebrate scent marks to
    census, 257
Schedorhinotermes (termites), 5
Schedorhinotermes lamianus
  foraging and soldiers/workers,
    138
  pheromones guide food
    collection, 138
Schistocerca gregaria (desert locust)
  gregarising factor in egg foam, 76
  maturation synchrony, 76
  oviposition pheromone, 76
  pheromones in life cycle, 76
Scramble competition, 43–44
  Carassius auratus (goldfish), 43
  moth males responding to female
    pheromones, 43
  selection for male moth sensitivity,
    43, 166
  Thamnophis sirtalis parietalis
    (red-sided garter snake), 43
Sebaceous gland
  birds, 21
  humans, 284
  signal life, 14
Sebum, 13
Self-organising systems, 141
  ant foraging, 141
    Eciton (army ants), 142
    matching resources to effort, 143
    modelling food distribution, 142
    stochastic effects, 143
  termite nests, 143–145
  stigmergy, 143
'Selfish herd' theory, 159
Semen
  behavioural manipulation by
    Drosophila melanogaster, 63
    humans? 64
    Thamnophis sirtalis parietalis
      (red-sided garter snake), 63–64
Semibalanus balanoides (barnacle)

egg hatching pheromone
  eicosanoids (PUFAs), 77
Semiochemicals, 1
Sensitive periods, 201
  ewe learning odour of her lamb,
    199, 200
  hormonal state and learning, 199
  sex ratio effects
    Microcebus murinus (lesser mouse
      lemur), 197
  termite caste development, 205
Sensory drive, 6, 7
Serricorone, 84
Settlement-inducing protein complex
  (SIPC), 78
Settlement of marine vertebrates, 77–79
Sex differences
  in brain
    macroglomerular complex (MGC)
      in male moths, 190
  response to pheromones
    Carassius auratus (goldfish), 189
  sensitivity to 3α-androstenol in
    sows, 253
  see also Development
Sex pheromones
  aphrodisin, 182
  Aplysia (sea-slug)
    peptide sex pheromone, 15
  arachnids
    Linyphia litogiosa (spider), 59
  crustaceans
    Homarus americanus (lobster), 61, 62
    Temora longicornis (copepod), 213
  exploitation
    commercial, 253, 254
    natural, 231, 232–233
  insects
    Ceratitis capitata (Mediterranean
      fruit fly), 58, 59
    Lutzomyia longipalpis (sandfly), 58
    Nicrophorus (burying beetle), 59
    Periplaneta americana (American
      cockroach), 16
    Xylocopa fimbriata (carpenter
      bee), 57
  mammals, role of experience, 182
  sodefrin, Cynops pyrrhogaster (newt),
    15, 16
  speciation, 64
  see also Development; Female
    pheromones; Male
    pheromones

Sex ratio
  Microcebus murinus (lesser mouse
    lemur)
    bias of sex ratio in pregnancy,
      197
Sexual selection, 37
  contests, same-sex, 38
    Homarus americanus (lobster), 44
    Nauphoeta cinerea (African cock-
      roach), 40, 44
  female choice
    elephants, 47
    Ephestia elutella (tobacco moth), 47
    Nauphoeta cinerea (African
      cockroach), 40, 44
  male characters
    coremata, 3, 10, 11, 47, 49
  mate choice, 37–38
  mate quality and courtship, 46
  mechanisms
    direct benefit, 38, 39
    imprinting, 40
    indicator, 40
    runaway ('Fisherian', 'sexy sons'),
      38, 39, 59
  scramble competition, 43–44
  sexual competition, 38
  which sex should call? 40
'Sexy faeces' hypothesis
  Plethodon cinereus (red-backed
    salamander), 46
She-males, see Alternative mating
  strategies
Sheep (Ovis aries)
  mate choice not universal, 56
  maternal behaviour, 199–201
  mother–infant recognition, 109
  reproduction, 192
  synchronised oestrus, 252
Signals, 1–2
  anonymous
    social insects, 102
  butterflies
    visual cues and pheromones, 14
  composite, 17–18
    parallel sensory channels, 17
    with ultraviolet cues, 14
  cooperative, 122
  design
    aquatic habitats, 15
    contrasting modalities, 12
    habitat and daily activity, 13
    function and environment, 13

lizard signals in contrasting
habitats, 14
longevity, selection on, 13
molecular weight and function,
13
molecular weight and habitat,
13–15
terrestrial habitats, 13
Dipsosaurus dorsalis (iguana)
composite signal with ultraviolet
cue, 14
dishonest, 121
distinguishing chemical signals
from noise, 206
honest, 3, 22, 44, *45*
queen pheromones
honeybees, 121, *121*
territorial scent marks, 89
Utetheisa ornatrix (tiger moth),
47, 49, *49*, *50*
low metabolic cost of pheromones
Anthonomus grandis (boll
weevil), 20
Litoria splendida (magnificent tree
frog), 21
mechanisms for specificity
peptides (amino acid sequence), 16
unique molecules, 16
multi-component, 17
stereoisomerism, 17
modulation
intensity, 17
by sound, 18
of visual threat by
pheromones, 18
ON–OFF
clean-up enzymes, 177
recruitment, 129
ants, 134
redundancy, 17
deer alert signal, *28*, 148
ritualised, *3*
sexually selected, 37
synergy, 35
see also Alarm pheromones;
Pheromones, evolution
Simulidae (blackfly)
oviposition pheromone, 75, 263
Singular breeders, 124
Sitophilus (beetle), 83
Smell, *see* Olfaction
Snakes
Boiga irregularis (brown tree snake)

control, *262*
detecting trail direction, 217, *217*
Leptotyphlops dulcis (blind snake)
follows ant pheromone trails, 247
tongue-flicking
trail following, 215
Thamnophis sirtalis parietalis
(red-sided garter snake)
overwintering aggregations, 80
Vipera berus (adder)
Coolidge effect, 57
Sniffing
behaviour of smelling, *174*
Social insects
alarm pheromones, 152–157
ants
Atta texana, *136*
Oecophylla longinoda (African
weaver ant), 129, 212, *214*
Solenopsis invicta (fire ant), 123
aphids
alarm pheromones, 152–153
Ceratovacuna lanigera, *153*
Colophina monstrifica, *153*
caste differences
pheromone secretions, 114
reproductives and workers, 118
response, 153
caterpillars
Malacosoma americanum (tent
caterpillar), 133
colony odours
variation, 248
guard bees
defense of colonies, 105
discrimination of relatedness, 108
recognition mechanisms, 105
Polistes fuscatus (paper wasps)
kin recognition mechanisms, 107
Ropalidia marginata
non-kin may join colony, 108
parallels with social mammals, 127
pheromones, 114
primer pheromones
development of ant castes, *204*
queen mandibular pheromone
(QMP), 121
action via juvenile hormone (JH)
levels, 204
queen pheromones, 118–123
control or cooperative signal, 118
evolution, 122
fire ants, 123

reproductive division of labour, 116
termites
colony recognition, 109
territories
Oecophylla longinoda (African
weaver ant), *88–89*
Oecophylla smaragdina (ant), *89*
Trigona (stingless bees)
odour beacons rather than trails,
136
worker policing, 122
see also Honeybees; Recruitment
pheromones; Self-organising
systems
Social mammals
clan recognition, 111
Meles meles (European badger) , 111
cooperative breeding, 124–127
evolution, 125
inhibition or suppression of
subordinate reproduction, 126
control by pheromones, 124, 126
control not by pheromones, 124,
125, 127
kin recognition
Castor canadensis (beaver), 111
Spermophilus beldingi (Belding's
ground squirrel), 110
parallels with social insects, 112,
127
primer pheromones and
reproduction, 123–124
recruitment, 136
Heterocephalus glaber (naked mole
rats), 134
reproductive division of labour, 116
shared colony odour, parallels
with social insects, 112
Sodefrin, peptide pheromone, 15, *16*
Solanum berthaultii (wild potato),
255, *245*
Solenopsis (fire ants), *130*, 137
Solenopsis invicta
enemy specification, 155
mass communication
recruitment, 134
queen mutual inhibition, 123
Solid phase microextraction (SPME),
31, *32*
Speciation, 64–72
allopatric, 64, 65
Drosophila melanogaster
cuticular hydrocarbons, 70

Speciation, (cont.)
  reproductive character
    displacement, 65
  male pheromones, selection by
    female to avoid hybrid
    matings, 70
  signal shift, 68
    asymmetric tracking, 68
  sympatric, 64, 65, 66
Specificity of male moth responses to
    pheromones, 68
Specific-Mate Recognition System, 65
Sperm competition and mate
    guarding, 61–64
  cryptic female choice, 63
    Tribolium castaneum (flour beetle),
      63
    Utethiesa ornatrix (tiger moth), 63
  precopulatory mate guarding,
    61–62
  consorting by male kangaroos, 62
  elephants, 62
  Homarus americanus (lobster), 61, *62*
Spermophilus beldingi (Belding's
    ground squirrel)
  kin recognition cues, 110
Spindasis lohita (lycaenid butterfly),
    238
Squalene, 43
'Standing behaviour', *see* Lordosis
Stem cell, olfactory, *168–169*
Stereoisomers, 305
Sternal gland
  ant, 130
  Malacosoma americanum (tent
    caterpillar), 130
  Polistes (paper wasp), 107
  termites, 130, 137
Steroids, 9
  5β-cholestane 3,24-dione
    Malacosoma trail pheromone, 130
Stigmergy, *see* Self-organising systems
Stimulo-deterrent diversionary
    strategy (SDDS), 265
Stochastic effects, 143
Stored product insects
  calling pattern, by mating system
    and habitat, 82
  Lasioderma serricorne (cigarette
    beetle), 83
  Prostephanus truncatus (larger grain
    borer beetle) , 82, 256
  Sitophilus (beetle), 83

Trogoderma granarium (Khapra
    beetle), 83
Stria terminalis, *183*
Subcaudal gland
  Meles meles (European badger), 111
Superparasitism, 83
Sus scrofa (pig)
  female response
    3α-androstenol, 20, 184, 253
    5α-androstenone, 184
  lordosis by oestrous sow to male
    pheromones, 20, 184, 253
Swarming, *see* Honeybees
Sweat glands
  apocrine, 284
  eccrine, 284
Synergy, 35
Synomones, 2
  mutualism, 230

Tadpoles, kin, and alarm
    pheromones, 157
Talpa europaea (European mole), 91
Taste
  smell (olfaction) vs. taste
    (gustation), *164*
  taste buds, *164*
Taxis, 207, *208*
't-complex' in mice, 56
Telenomus euproctidis (parasitoid wasp)
  eavesdrops host, 232
Teleology, 207
Temnochila chlorodia (beetle), 234
Temora longicornis (copepod), 3D trails,
    213
Temporal gland
  elephants, 46
Termites
  beetle mimics its termite host, 248
  caste-change pheromones, 119
  colony recognition odours, 109
  developmental pathways, *120*
  pest management, *263*
  primer pheromones, 119
  soldiers, control of numbers, *119*
Termitophilous invertebrates, 245
Territories, 87
  costs (of signalling)
    Oreotragus oreotragus (klipspringer
      antelope), 233
  dear-enemy phenomenon, 99
  economics of scent marking, 97
  patterns, 97

variation by habitat in
    Hyaenidae, 98
feeding and breeding
  Plethodon cinereus (red-backed
    salamander), 46
group territorial defence
  Lemur catta (ring-tailed lemur)
    'stink fights', *90*
'owner advantage'
  Oecophylla longinoda (African
    weaver ant), *88*
scent marking
  border maintenance hypothesis,
    90, 96–97
  composite signals with
    ultraviolet cue, 99
  lamp-post effect, 99
social insects
  Oecophylla longinoda (African
    weaver ant), *88–89*
Testosterone, *see* Hormones,
    testosterone
Tetramorium caespitum (ant), foraging,
    143
Thamnophis sirtalis parietalis (garter
    snake)
  conflict between the sexes
    semen components, 63–64
  detecting trail direction, 217, *217*
  male pheromone, squalene, 43
  overwintering aggregations, 80
  she-males, 60
Thanasimus dubius
  eavesdropping, 234
Thiazole, *see* Butyl dihydrothiazole
Ticks
  Amblyomma hebraeum
    pest management, *262*
  Dermacentor variabilis
    pest management, *262*
  guanine, assembly pheromone, 80
  Ixodes neitzi responds to scent marks
    of host antelope, 233
Tools to study pheromones, *see*
    Pheromone identification
Tongue-flicking, snakes, 215
Trail pheromones, *see* Recruitment
    pheromones
trans (in chemical name), 305
Tree defences, resins and toxic
    chemicals, *81*
Tribolium castaneum (flour beetle), 82
  cryptic female choice, 63

Trichogramma pretiosum (parasitoid wasp), 85
Trichoplusia ni (cabbage looper moth)
  both sexes call, 41
  mutation in pheromone blend, 68
  male response requires hormone, 189
Trichopoda pennipes (tachinid)
  selection presure on host, 233
Trichopsenius frosti (beetle)
  matches its termite host, 248
(Z)-7-Tricosene, 61
(E)-4-Tridecenyl acetate, 259
Trigona (stingless bees), 5, 136, 140
  Trigona angustula
    resists pheromone propaganda by robber bee, 243
  Trigona subterranea
    victim of pheromone propaganda by robber bee, 243
Trogoderma granarium (Khapra beetle), 83
Trophallaxis
  oral, *119*
  proctodeal, *119*
Truffles (*Tuber melanosporum*)
  3α-androstenol, 253
  use of sows to find, 253
  human taste for, 288
Tuber melanosporum, see Truffles
Tufted cells, *172*

Ultraviolet (UV) cues
  birds eavesdrop vole scent marks by UV, 232
  Dipsosaurus dorsalis (iguana), composite signal with UV, 14
Undercrowding
  Allee effects, 75
Urine
  active only on contact
    Microtus ochrogaster (prairie vole), 197–198
  dominant male effective, 192
  Homarus americanus (lobster) contest, 44
  major urinary proteins (MUPs)
    elephants, 195
    mice, 195
  reflects hormonal and dominance states in rodents, 46
Uropygidial gland, 21
Utetheisa ornatrix (tiger moth)
  female choice, *3*, 63

honest signal, 47, 49, *49*, *50*
hydroxydanaidal (HD) from diet, 47
sexual selection, *39*, 70

Vaginocervical stimulation during birth
  hormonal releasers, 200
Vandenbergh effect (puberty acceleration), 192
Vanillic acid, 260
Varroa mite
  pest management, *262*
  response to honeybee larval odours, 116
Ventral amygdalofugal pathway, *183*
Verbenone, 235
Veromessor (ant)
  benzaldehyde as defensive compound, *5*
Vertebrates
  androgens
    gland secretion rates, 46
  dual olfactory system, 178
  pheromones to attract mates, 41
  see also Amphibians; Antelopes; Deer; Dogs; Fish; Mammals; Mice; Pheromones; Primates; Salamanders and newts; Social mammals; Wolves; *other taxonomic entries*
Vespa mandarinia japonica (giant hornet), 154, *156–157*
Vespula (wasps)
  Vespula germanica, 155
    eavesdropping prey leks, 232, 231
  Vespula vulgaris, 155
Vipera berus (adder), Coolidge effect, 57
Vomeronasal organ (VNO), *165*, 178–185
  female VNO, urine contact, 198
  induction of oestrus
    VNO needed, 181
    VNO not needed, 181
Vomeronasal amygdala, *165*
Vomeronasal olfactory system, 165
  functional overlap with main olfactory system, 180–186
  in humans? 294
  memory and pregnancy block (Bruce effect) in mice, 201
  molecular architecture and differences from main olfactory system (MOE),

179–180
  mapping to glomeruli, 180
  receptor proteins, 179
  receptor specificity, 179
  response to proteins, 180
  some pheromones detected by main olfactory system (MOE), 181
  trail following by snakes, 215–217

Westermarck or kibbutz effect
  human mate choice, 280
Whitten effect (oestrus induction), 192
Wolves
  Canis simensis (Ethiopian)
    border marks in territory, 97
  Canis lupus (grey)
    border marks in territory, 97
    control of reproduction, 124
Worker policing, 122

Xylocopa fimbriata (carpenter bee), leks, 57

Yponomeuta (small ermine moths)
  calling time, 66
  host plants, 66, *67*
  pheromone blend, 66, *67*

Zaspilothynnus trilobatus (thynnine wasp)
  duped by orchid, *241*
Zigzags
  swimming or flying up plumes, 224
Zootermopsis nevadensis (termite)
  nymph laying trail, *135*
Zusammen, (Z) together, 308